Protecting Visibility in National Parks and Wilderness Areas

Committee on Haze in National Parks and Wilderness Areas
Board on Environmental Studies and Toxicology
Commission on Geosciences, Environment, and Resources

National Research Council

NATIONAL ACADEMY PRESS
Washington, D.C. 1993

NATIONAL ACADEMY PRESS   **2101 Constitution Ave., NW**   **Washington, DC 20418**

NOTICE: The project that is the subject of this report was approved by the Governing Board of the National Research Council, whose members are drawn from the councils of the National Academy of Sciences, the National Academy of Engineering, and the Institute of Medicine. The members of the committee responsible for the report were chosen for their special competencies and with regard for appropriate balance.

This report has been reviewed by a group other than the authors according to procedures approved by a Report Review Committee consisting of members of the National Academy of Sciences, the National Academy of Engineering, and the Institute of Medicine.

The project was supported by the Department of Energy under grant number DE-FG01-90FE62072, the Department of the Interior under contract number 14-01-0001-89-C-39, the State of Arizona Salt River Project Agricultural Improvement and Power District under grant number VN08016CAS, the Environmental Protection Agency, Chevron, and the United States Department of Agriculture under grant number 91-G-018.

Library of Congress Catalog No. 93-83079
International Standard Book No. 0-309-0:4844-3

B-065

Front cover: Computer-simulated picture of natural and current average conditions for Shenandoah National Park, Virginia. Relative humidity is the same for the simulations (Trijonis, J., W. Malm, M. Pitchford, and W.H. White, 1990).

Printed in the United States of America

Committee on Haze in National Parks and Wilderness Areas

ROBERT A. DUCE *(Chair)*, Texas A & M University, College Station, TX
JACK CALVERT *(Vice Chair)*, National Center for Atmospheric Research, Boulder, CO
GLEN R. CASS, California Institute of Technology, Pasadena, CA
JOHN E. CORE, WESTAR Council, Portland, OR
H. WILLIAM ELDER, Retired, Florence, AL
PETER H. MCMURRY, University of Minnesota, Minneapolis, MN
PAULETTE B. MIDDLETON, State University of New York at Albany (University Cooperation for Atmospheric Research, Boulder), Albany, NY
CRAIG N. OREN, Rutgers University School of Law, Camden, NJ
JOSEPH M. PROSPERO, University of Miami, Miami, FL
PERRY J. SAMSON, University of Michigan, Ann Arbor, MI
IAN M. TORRENS, Electric Power Research Institute, Palo Alto, CA
JOHN TRIJONIS, Santa Fe Research Corporation, Bloomington, MN
WARREN H. WHITE, Washington University, St. Louis, MO

Project Staff

RAYMOND A. WASSEL, Program Director
KATHLEEN J. DANIEL, Project Director (until February 1991)
ROBERT B. SMYTHE, Program Director (until August 1991)
LEE R. PAULSON, Editor
RUTH R. CROSSGROVE, Staff Editor
ANNE M. SPRAGUE, Information Specialist
WILLIAM H. LIPSCOMB, Research Assistant (until June 1992)
FELITA S. BUCKNER, Senior Program Assistant
BOYCE AGNEW, Project Assistant (until July 1991)

Board on Environmental Studies and Toxicology

iv

Commission on Geosciences, Environment, and Resources

Preface

National parks and wilderness areas are among our nation's greatest treasures. Ranging from inviting coastal beaches and beautiful shorelines to colorful deserts and dramatic canyons to towering mountains and spectacular glaciers, these regions inspire us as individuals and as a nation. A vital part of the enjoyment of these natural wonders is the ability to see them clearly. However, in many of these areas haze has diminished the visibility and significantly affected our enjoyment of nature. Much of this haze is derived from human activities. Congress recognized the importance of visibility in national parks and wilderness areas in the 1977 Clean Air Act and its 1990 amendments. The Clean Air Act specifically established a goal of correcting and preventing anthropogenic visibility impairment in these regions, but relatively little progress has been made toward attaining that goal.

Early in 1990 the National Research Council established the Committee on Haze in National Parks and Wilderness Areas to address a number of questions related to visibility and its degradation in these pristine areas, including methods for determining anthropogenic source contributions to haze and for considering alternative source control measures. As part of its charge, the committee completed an interim report in late 1990, *Haze in the Grand Canyon*, that evaluated the National Park Service's Winter Haze Intensive Tracer Experiment (WHITEX) report of the causes of wintertime haze in the region between the Grand Canyon and Canyonlands National Park.

In this final report, the committee addresses the broader-scale issues of regional haze. This report discusses visibility conditions in the United States, the legal and institutional context of visibility protection programs, the scientific aspects of haze formation and visibility impairment, methods for identifying and apportioning the components of haze, and the relationship of emission controls to visibility.

The task undertaken by this committee was not an easy one. The subject is controversial, not in the least because significant economic implications can result from the determination of the sources responsible for haze in these regions. Nevertheless, the committee members addressed their charge with enthusiasm and commitment—working long hours during meetings and at their home locations, including numerous discussions over the telephone—to produce what I believe is a fair and accurate assessment of the current understanding of this issue. I have never worked with a more dedicated group of individuals.

The committee held seven meetings, including sessions at Grand Canyon and Yosemite National Parks. During our meetings, we were provided with information by many organizations, including the National Park Service, the Bureau of Reclamation, the Department of Interior's Office of Environmental Quality, the Environmental Protection Agency, the Department of Energy, the Forest Service, the Arizona Salt River Project and their consultants, and Chevron Corporation.

F. Sherwood Rowland of the University of California at Irvine provided the committee with information on the use of tracers in the environment. Michael Walsh, an independent consultant, provided information on relationships between mobile source emissions and visibility impairment. Robert Charlson provided the committee with valuable input during the early part of the study.

The federal liaison group, representing the sponsoring agencies, and the Arizona Salt River Project were helpful in providing information and data to the committee whenever it was needed. The committee received very useful information and perspectives from many individuals, including the following:

John Bachmann, Environmental Protection Agency
Robert Bauman, Environmental Protection Agency
C. Shephard Burton, consultant to Salt River Project
James Byrne, Forest Service

Stanley Coloff, Bureau of Land Management
Jonathan Deason, Department of Interior
Carol Ellis, Southern California Edison Company
Robert Farber, Southern California Edison Company
Mark Green, Desert Research Institute
Peter Hayes, Salt River Project
Eugene Hester, National Park Service
Donna Lamb, Forest Service
William Malm, National Park Service
Gregory McRae, Carnegie Mellon University
Peter Mueller, Electric Power Research Institute
John O'Gara, Department of Defense
Janice Peterson, Forest Service
Roger Pielke, Colorado State University
William Pierson, Desert Research Institute
Marc Pitchford, Environmental Protection Agency
Richard Poirot, Vermont Agency of Natural Resources
Benjamin Radecki, Bureau of Reclamation
David Sandberg, Forest Service
Jerry Shapiro, consultant to Salt River Project
David Stonefield, Environmental Protection Agency
Denise Swink, Department of Energy
Ivar Tombach, consultant to Salt River Project
Edward Trexler, Department of Energy
Jan van Wagtendonk, National Park Service
Patrick Zimmerman, National Center for Atmospheric Research

Of particular importance was the dedicated and effective support provided to the committee by the National Research Council staff. James Reisa, the director of the Board on Environmental Studies and Toxicology, and program director Robert Smythe (until August 1991) provided us with valuable advice and oversight. Project directors Kathleen Daniel (until February 1991) and Raymond Wassel worked closely and effectively with the committee through the report's preparation. They were each very sensitive to the controversial nature of the issues discussed by the committee and played a vital role in working with the committee to develop a consensus on major issues. They deserve a significant amount of the credit for this report. Lee Paulson did an excellent job as editor. Other staff who contributed greatly to the effort were research assistant William Lipscomb, who helped in the final stages; Ruth Crossgrove, who provided editorial assistance; Tania Williams, who prepared the document for publication; Felita Buckner, Boyce Agnew, and Sandi Fitzpatrick as project assistants; information specialist Anne Sprague; and other dedicated staff of BEST's Technical Information Center.

We hope that this report will provide useful recommendations and guidance as the United States works toward a national visibility policy and develops a framework for protecting and preserving the natural visibility in our national parks and wilderness areas—both for present and future generations of Americans.

Robert A. Duce
Chairman
January 7, 1993

Other Recent Reports of the Board on Environmental Studies and Toxicology

Contents

Protecting Visibility
in National Parks
and Wilderness Areas

Protecting Visibility in National Parks and Wilderness Areas

Executive Summary

Many visitors to America's national parks and wilderness areas are unable to enjoy some of the beautiful and dramatic views that would prevail in the absence of air pollution. Scenic vistas in most U.S. parklands are often diminished by haze that reduces contrast, washes out colors, and renders distant landscape features indistinct or invisible.[1] The National Park Service (NPS) has reported that visibility impairment caused by air pollution occurs in varying degrees at many park monitoring stations virtually all the time. Today, the average visual range in most of the western United States, including national parks and wilderness areas, is 100-150 km (about 60-100 miles), or about one-half to two-thirds of the natural visual range that would exist in the absence of air pollution.[2] In most of the East, including parklands, the average visual range is less than 30 km (about 20 miles), or about one-fifth of the natural visual range.[3]

Visibility degradation in parklands is a consequence of broader regional-scale visibility impairment. The causes of this impairment are well understood. Most impairment is caused by fine particles that ab-

---

[1] Haze degrades visibility primarily through the scattering or absorption of light by fine atmospheric particles. Visibility is the degree to which the atmosphere is transparent to visible light.

[2] Visual range is defined as the greatest distance at which a large black object can be discerned against the horizon sky.

[3] The natural visual range in the East is less than that in the arid West.

sorb or scatter light. Some of these particles (primary particles) are emitted directly to the atmosphere; others (secondary particles) are formed in the atmosphere from gaseous precursors. Visibility-reducing particles and their precursors can remain in the atmosphere for several days and can be carried tens, hundreds, or thousands of kilometers downwind from their sources to remote locations, such as national parks and wilderness areas. During transport, the emissions from many sources mix together to form a uniform, widespread haze known as regional haze.

Most visibility impairment is caused by five particulate substances (and associated particulate water): sulfates, organic matter, elemental carbon (soot), nitrates, and soil dust. The major cause of reduced visibility in the East is sulfate particles, formed principally from sulfur dioxide (SO_2) emitted by coal combustion in electric utility boilers. In the West, the other four particle types play a relatively greater role than in the East. The causes and severity of visibility impairment vary over time and from one place to another, depending on meteorological conditions, sunlight, and the size and proximity of emission sources.

Congress in 1977 established a national goal of correcting and preventing pollution-related visibility impairment affecting large national parks and wilderness areas, termed "mandatory Class I areas."[4] However, the federal government and the states have been extremely slow in developing an effective visibility protection program. The present program lacks sufficient resources, and it targets few of the major types of sources of visibility impairment in Class I areas. As a result, little progress has been made toward the national visibility goal established by Congress 15 years ago.

The Clean Air Act includes two emissions control programs specifically concerned with visibility in national parks and wilderness areas. One of these, the Prevention of Significant Deterioration (PSD) program, is directed mainly at new sources; the other, a visibility protection program, largely is aimed at existing sources.

---

[4]These are national wilderness areas and national memorial parks larger than 5,000 acres, national parks over 6,000 acres, and international parks. Any such area must have been in existence on August 7, 1977, the date the Clean Air Act Amendments of 1977 were signed into law, to be considered a mandatory Class I area.

The PSD program requires that each new or expanded "major emitting facility" locating in a "clean air area" install the "best available control technology", and it establishes increments (allowable increases) that limit the cumulative increases in pollution levels in clean air areas. Many large national parks and wilderness areas are designated as Class I areas and therefore are subject to the most stringent increments. The PSD program has protected visibility to some extent by reducing the *growth* of emissions of pollutants that contribute to regional haze. The program's requirement that major new sources locating in clean air areas install the best available control technology has been particularly important.

But the limits on growth in air pollutant concentrations established by the PSD program have been only partially effective. First, the restrictive Class I increments apply only to large parks created before enactment of the Clean Air Act Amendments of 1977; many other scenic areas receive no special protection. Second, it is not even clear that the Class I increments ensure effective protection against new sources that might cause visibility impairment. The increments do not distinguish between particles in the 0.1-1.0 μm range—which have the greatest potential to degrade visibility—and larger particles. Moreover, increments focus on the concentration of pollution at a given time and place; but visibility impairment depends on the total magnitude of fine particulate matter between an object and an observer.

A 1990 report by the U.S. General Accounting Office (GAO) discussed other flaws in the PSD program. The GAO found that federal land managers had not fully met their responsibilities to review PSD permit applications, due to lack of time, staff, and data and also due to the failure of the U.S. Environmental Protection Agency (EPA) to forward permit applications. Moreover, many sources of visibility impairment in national parks and wilderness areas are exempt from PSD requirements, because they are considered minor sources, or because they existed before the PSD program took effect in the 1970s.

The other visibility protection program under the Clean Air Act requires states to establish measures to achieve "reasonable progress" towards the national visibility goal and to require the installation of the "best available retrofit technology" on large sources contributing to visibility impairment in mandatory Class I areas. In 1980, EPA issued rules aimed primarily at controlling "plume blight" (impairment due to visible plumes from nearby individual sources). At that time, EPA also expressed its intention to regulate regional haze at some future date "when improvement in monitoring techniques provides more data on source-specific levels of visibility impairment, regional-scale models become more refined, and scientific knowledge about the relationships between air pollutants and visibility impairment improves." More than a decade later, despite major advances in monitoring techniques, regional-scale models, and scientific knowledge of visibility impairment, EPA has yet to issue rules for regulating regional haze. Instead, EPA's rules require only the regulation of impairment that is attributable to individual sources through the use of simple techniques. This has greatly weakened the visibility program's effectiveness. Fourteen years passed until the first pollution source was required to control its emissions under this program.

Emission-control measures already adopted or planned will not solve the nation's visibility problems. The acid rain control program established by the 1990 Clean Air Act Amendments has been predicted to reduce SO_2 emissions in the East by about 36% by 2010. That reduction probably will improve visibility in much of the East but will eliminate only a fraction of the anthropogenic visibility impairment. In the West, where most Class I areas are located, projections done for EPA indicate that the acid rain control program will halve, but not entirely prevent, expected *growth* in SO_2 emissions between now and 2010.

THE CHARGE TO THE COMMITTEE

This report was prepared by the National Research Council's Committee on Haze in National Parks and Wilderness Areas. The committee was convened by the Council's Board on Environmental Studies and Toxicology in collaboration with the Board on Atmospheric Sciences and Climate of the Commission on Geosciences, Environment, and Resources. The committee's members have expertise in meteorology, atmospheric chemistry, air-pollution monitoring and modeling, statistics, control technology, and environmental law and public policy. The committee's work was sponsored by the U.S. Department of the Interior (National Park Service, Bureau of Reclamation, and Office of Environmental Quality), U.S. Department of Energy, U.S. Environmental Protection Agency, U.S. Department of Agriculture (Forest Service), the Arizona Salt River Project, and Chevron Corporation.

The committee was charged to develop working principles for assessing the relative importance of anthropogenic emission sources that contribute to haze in Class I areas and for considering various alternative source control measures. It also was charged to recommend strategies for filling critical scientific and technical gaps in the information and data bases on (1) methods for determining individual source contributions, (2) regional and seasonal factors that affect haze, (3) strategies for improving air-quality models, (4) the interactive role of photochemical oxidants, and (5) scientific and technological considerations in choosing emission control measures.

In 1990, the committee published an interim report, *Haze in the Grand Canyon*, which evaluated the National Park Service's Winter Haze Intensive Tracer Experiment (WHITEX) report on the causes of wintertime haze in the region between the Grand Canyon and Canyonlands National Park. The WHITEX report by NPS had asserted that the Navajo Generating Station (NGS), a large coal-fired power plant in Page, Arizona, is a principal contributor to visibility impairment in Grand Canyon National Park (GCNP). Our committee's interim report concluded that, at some times during the study period, NGS contributed significantly to haze in GCNP, but that WHITEX failed to quantitatively determine the fraction of sulfate particles and resulting haze attributable to NGS emissions. The committee identified flaws in the models used to estimate NGS's contribution, in the interpretation of those models, and in the data base. The committee found that sources other than NGS appeared to account for a significant fraction of haze observed in GCNP during the study period. Thus, if NGS emissions were to be controlled, visibility impairment in GCNP would be reduced but not eliminated.

THE COMMITTEE'S APPROACH TO ITS CHARGE

In this final report, the committee examines patterns of visibility degradation and haze-forming pollutant concentrations in various parts of the United States resulting from natural and anthropogenic sources of gases and particles (Chapter 2). It considers the regulatory and institutional frameworks for efforts to improve and protect visibility, including the Clean Air Act (Chapter 3). This report also reviews the scientific understanding of haze formation and visibility impairment, including the meteorological and chemical processes responsible for the transport and transformation of gases and particles in the atmosphere, as well as chemical and physical measurement techniques (Chapter 4). The approach of first relating source emissions to aerosol composition, and then relating aerosol composition to visibility, is fundamental to most of the committee's analyses.

In evaluating methods for source identification and apportionment, this report considers the technical adequacy (including degree of uncertainty), flexibility, and difficulty of implementation of the various approaches (Chapter 5). In discussing control techniques, the report describes the emissions reduction potential of various control measures and illustrates the translation of control measures into a rough prediction of effects on visibility (Chapter 6). The report also considers policy implications of scientific knowledge about visibility and recommends approaches to remedy scientific and technical gaps that limit present understanding of source effects on visibility and the ability to evaluate control measures (Chapter 7).

GENERAL CONCLUSIONS AND RECOMMENDATIONS

The complete design of a program for protecting and improving visibility in Class I areas must involve many policy issues outside the bounds of science and the committee's expertise. However, present scientific knowledge about visibility impairment in Class I areas has several important implications for policy makers.

Progress toward the national goal of remedying and preventing man-made visibility impairment in Class I areas (Clean Air Act, Section 169A(a)) will require regional programs that operate over large geographic areas and limit emissions of pollutants that can cause regional haze.

Most visibility impairment in national parks and wilderness areas (Class I areas) results from the transport by winds of emissions and secondary airborne particles over great distances (typically hundreds of miles). Consequently, visibility impairment is usually a regional problem, not a local one. Regional haze is caused by the combined effects of emissions from many sources distributed over a large area, rather than of individual plumes caused by a few sources at specific sites. As a result, a strategy that relies only on influencing the location of new sources, although perhaps useful in some situations, would not be effective in general. And of course, such a strategy would not remedy the visibility impairment caused by existing sources until those sources are replaced.

A program that focuses solely on determining the contribution of individual emission sources to visibility impairment is doomed to failure. Instead, strategies should be adopted that consider many sources simultaneously on a regional basis, although assessment of the effect of individual sources will remain important in some situations.

Because haze is caused by the combined effects of the emissions of many sources, it would be an extremely time-consuming and expensive undertaking to try to determine, one source at a time, the percent contribution of each source to haze. For instance, the efforts to trace the contribution of the Navajo Generating Station to haze in the Grand Canyon National Park took several years and cost millions of dollars without leading to quantitatively definitive answers. Moreover, there are (and will probably continue to be) considerable uncertainties in ascertaining a precise relationship between individual sources and the spatial pattern of regional haze.

Assessment of the contribution of individual sources to haze will remain useful in some situations. For instance, a regional emissions management approach to haze could be combined with a strategy to assess whether locating a new source at a particular location would have especially deleterious effects on visibility. In Chapter 5, the committee has set out working principles for attributing visibility impairment to individual sources.

Visibility impairment can be attributed to emission sources on a regional scale through the use of several kinds of models. In general, the best approach for evaluating emission sources is a nested progression from simpler and more direct models to more complex and detailed methods. The simpler models are available today and could be used as the basis for designing

regional visibility programs; the more complex models could be used to refine those programs over time.

After identifying which pollutants are impairing visibility for a given region, it is useful to apportion visibility impairment among contributing sources to the extent possible so that the relative effectiveness of alternative control measures can be evaluated. Source apportionment models of varying degrees of accuracy and complexity can be used to analyze regional haze problems, although no single source-apportionment method is necessarily best for all visibility problems. Simpler methods are most effective in the early stages of source apportionment, with the more complex methods being applied, if necessary, to resolve difficult technical issues.

For regional haze problems, the committee recommends use of the following models, in order of increasing sophistication:

* *Speciated rollback models.* These are simple, spatially averaged models that assume changes in pollutant concentrations to be directly proportional to changes in regional emissions of these pollutants or their precursors.

* *Hybrid combinations of chemical mass balance receptor models with a different source-oriented secondary particulate mass formation model, and used with empirical data for pollutant scattering and absorption efficiencies.* Receptor models are models that infer source contributions by characterizing atmospheric aerosol samples, often using chemical elements or compounds in those samples to identify emissions from particular source types. Hybrid models are formed by combining two or more separate modeling techniques.

* *Hybrid combinations of mechanistic models for transport and secondary particulate mass formation with measured particle size-distribution data to facilitate light scattering calculations.* Mechanistic models are 3-dimensional, computer-based models that simulate the atmospheric transport, dispersion, chemical conversion, and deposition of pollutants as faithfully as possible.

Speciated rollback models are available now; in Chapter 6 the committee uses such a model to illustrate apportionment of regional haze. The recommended hybrid combinations could be assembled from available components.

To assess the contribution of an existing single source to visibility impairment, photographic and other source identification methods could be used in simple cases. More complex situations require the use of hybrid combinations of chemical mass-balance or tracer techniques with secondary particle models that include explicit transport calculations and an adequate treatment of background pollutants. For complex applications that require the greatest sophistication, the most advanced reactive plume models available should be used with measured data on particle properties in such plumes and should be accompanied by an adequate treatment of background pollutants.

To assess new single sources, the most advanced reactive plume models available should be used with measured data on particle properties in the plumes of similar sources and accompanied by an adequate treatment of background pollutants.

To analyze a single source at the regional scale, a description of the source in question should be inserted into an appropriately chosen multiple-source description of the regional haze problem.

The next step in designing a visibility protection strategy is to determine whether methods for controlling visibility-impairing emissions exist or can be developed and to assess the effects of alternative sets of controls. The committee's analysis of one control scenario indicates that application of commercially available emission controls would reduce but not eliminate anthropogenic visibility impairment; the greatest improvement would be in the East. (This analysis should not be construed as endorsing a technology-based or any other specific control strategy.)

Visibility policy and control strategies might need to be different in the West than in the East.

Haze in the East and in the West differ in important ways. Haze in the East is six times more intense than in the West because of the much higher levels of pollution in the East. Were all anthropogenic pollution to disappear, visibility would still be greater (by about 50 percent) in the West. In relatively clean areas, small increases in pollutant concentrations can markedly degrade visibility; increases of the same magnitude are less noticeable in more polluted areas. Hence, visibility in Class I areas in the West is particularly vulnerable to increased levels of pollution. Moreover, the West contains most of the nation's large national parks and wilderness areas, which can be fully appreciated only when visibility is excellent. The East, however, contains a large population to enjoy the benefits of any improvement in visibility in that region.

In the East, sulfates derived from SO_2 emissions from coal-fired power plants account for about one-half of all anthropogenic light extinction. Reductions in these emissions are expected to occur in the next two decades as a result of the 1990 Clean Air Act Amendments' acid rain control program. In the West, no single source category dominates; therefore, an effective control strategy would have to cover many source types, such as electric utilities, gasoline- and diesel-fueled vehicles, petroleum and chemical industrial sources, forest-management burning, and fugitive dust.

Efforts to improve visibility in Class I areas also would benefit visibility outside these areas.

Because most visibility impairment is regional in scale, the same haze that degrades visibility within or looking out from a national park also degrades visibility outside it. Class I areas cannot be regarded as potential islands of clean air in a polluted sea.

Reducing emissions for visibility improvement could help alleviate other air-quality problems, just as other types of air-quality improvements could help visibility.

The substances that contribute to regional haze also contribute to a variety of other undesirable effects on human health and the environment. For example, SO_2 is a precursor of sulfuric acid in acid rain, oxides of nitrogen (NO_x) and volatile organic compounds (VOCs) are precursors of lower-atmosphere ozone, and fine atmospheric particles are a respiratory hazard. Such particles can influence climate by interacting with incoming solar radiation and by modifying cloud formation. Policy makers should consider the linkages between visibility and other air-quality problems when designing and assessing control strategies.

Achieving the national visibility goal will require a substantial, long-term program.

The national visibility goal is unlikely to be achieved in a short time.

Policy makers might develop a comprehensive national visibility improvement strategy as the basis for further regulatory action, and establish milestones against which progress toward the visibility goal could be measured.

Current scientific knowledge is adequate and control technologies are available for taking regulatory action to improve and protect visibility. However, continued national progress toward this goal will require a greater commitment toward atmospheric research, monitoring, and emissions control research and development.

The slowness of progress to date is due largely to a lack of commitment to an adequate government effort to protect and improve visibility and to sponsor the research and monitoring needed to better characterize the nature and origin of haze in various areas. The federal government has accorded the national visibility goal less priority than other clean-air objectives. Even to the extent that Congress has acted, EPA, the Department of Interior, and the Department of Agriculture have been slow to carry out their regulatory responsibilities or to seek resources for research.

RECOMMENDED RESEARCH

The committee addressed the need to alleviate scientific and technical gaps in the areas of visibility and aerosol monitoring and measurement, source apportionment, and emissions control technology. The committee considered what measures might be taken to understand better the sources of haze, possible means of reducing emissions from those sources, and alternative ways of preventing future visibility impairment in Class I areas.

The committee emphasizes that the need for additional research does not imply that further regulatory action, if otherwise warranted, to improve visibility in Class I areas would be premature. The authority of regulatory agencies to act without complete scientific knowledge is clearly implied in the Clean Air Act. Moreover, visibility impairment is probably better understood and more easily measured than any other air-pollution effect. The remaining gaps in knowledge of visibility are primarily a symptom of the lack of a strong national commitment to enforcing the visibility protection provisions of the Clean Air Act.

Resources for research are limited; therefore, precautions should be taken to ensure that the visibility protection activities of the federal land management agencies, EPA, the Department of Energy, and state and local air agencies are of the highest possible quality. In addition, a greater effort is needed for formal publication of scientific work in independent, peer-reviewed literature.

The committee recommends establishing an independent science advisory panel with EPA sponsorship to help guide the research elements of the national visibility program. This panel could address the need for wider participation by the scientific community in addressing visibility problems.

EPA should build upon and expand its efforts to track the success of the PSD program. In particular, information is needed about the potential of new sources to reduce visibility in Class I areas and about the effects on such areas of the new emissions trading programs of the 1990 Clean Air Act Amendments. EPA's current visibility-screening model needs to be revised to consider the contribution of an individual source to regional haze.

Research on relating human judgments of visibility to objective measures, such as light extinction, should continue. The results should be used to inform decision makers and the public about the perceptibility of predicted visibility changes.

Areas in which research is needed include atmospheric transport and transformations of visibility-impairing pollutants, the development of models that can better apportion haze among sources, and improved instrumentation for routine monitoring and for obtaining data that can be used to evaluate models. Monitoring and research must be closely coordinated. Better models, however, are not enough. Any model, even the simplest or most refined, depends on good empirical data on the airborne particles that cause haze and on their sources. Greater resources are needed to develop these data.

Monitoring Strategies

If national visibility monitoring networks are to achieve their goals, a long-term commitment to establishing and financially supporting these networks is essential. Monitoring programs should be able to relate visibility impairment to its sources on a scale commensurate with regional haze events and the distribution of major emissions sources. Monitoring networks in the East need to be expanded to track visibility improvements associated with reductions in SO_2 emissions. Wind observations should be evaluated to ensure that atmospheric transport is represented accurately.

A consensus should be developed on the specific instrumentation to be used for monitoring light extinction. Standards should be established for the performance characteristics of the instrumentation. Future measurement programs should devote increased attention to quality assurance and control. Strengthening the quality assurance and control program of the Interagency Monitoring of Protected Visual Environments (IMPROVE) network should be a high priority.

Greater attention should be given to the implications that planned changes in airport visibility monitoring hold for research on visibility impairment. Airports should be equipped with integrating nephelometers sensitive enough to measure the range of haze levels encountered in the atmosphere.

Measurement Methods

Current measurement methods permit reasonable estimates of the average contributions of major aerosol constituents to atmospheric visibility impairment. However, several aerosol measurement methods need to be developed or improved for the following:

- Accurate measurement of organic and elemental carbon particles, especially at low concentrations;
- Routine measurement of the water content of airborne particles;
- Measuring particle size distributions;
- Continuous measurement of sulfates, organics, elemental carbon, nitrates, and elemental composition;
- Solar- and battery-powered measurement for use in remote areas.

The committee recommends using high-sensitivity integrating nephelometry for routine visibility monitoring. This technique, which measures the scattering of light from an air sample drawn through an enclosed cell, can provide accurate data at reasonable cost. Nephelometer data can be compared with measured particle concentrations at the same point to determine the contributions of different pollutants to visibility impairment. A readily available, easily serviced, and electronically up-to-date instrument with adequate sensitivity for good and poor visibility is needed. Nephelometer measurements of light scattering should be supplemented with independent measurements of light absorption. Instrumentation for continuous measurements of particle absorption coefficients should be developed.

Source-Apportionment Modeling

Source-apportionment models require better input data on source emissions, along with unified procedures for testing individual sources. Emissions data need to be integrated more accurately into overall emissions inventories. The inventories requiring the most improvement are those for primary organic and elemental carbon particles and gaseous VOCs.

Models should be validated using existing data sets from comprehensive field studies. Mechanistic models and hybrid receptor models should be included in validation studies.

Receptor models require substantial source testing and ambient emissions measurements to improve emissions profiles for sources of haze. Standard protocols for the release and sampling of tracers should be developed, along with field studies to verify these protocols. Inexpensive and relatively short-lived tracers are needed to distinguish the emissions of similar sources.

Research should also continue toward the development of advanced mechanistic models. Two kinds of mechanistic models are especially needed: (1) an advanced reactive model for analysis of visibility-impairing plumes from single sources; and (2) a grid-based, multiple-source regional model for analysis of regional haze problems. The development of such models will require significant refinement in the understanding of processes that affect particle size distributions. Critical processes include atmospheric emissions of particles and gases that play a role in the production of secondary particles, and gas-to-particle conversion. Measurement programs that are intended to acquire such information should be designed in collaboration with modelers to ensure that the results are suitable for model development and validation.

Emission Controls

Continued research and development support by government and industry is needed to improve the cost-effectiveness of existing emissions control technologies and to develop new technologies. Desirable areas for further research and development in the near term include

• Combined NO_x/SO_2 control technologies for power plants and industrial boilers;
• Low-cost, low-temperature selective catalytic reduction for NO_x control;
• Better particulate control for diesel vehicle engines;
• More efficient batteries for electric-powered vehicles, which could provide greater range before recharging.

Long-term emissions reduction research efforts should focus on low-emission, high-efficiency technologies for replacing or repowering fossil-based electricity generation, more efficient energy use technologies, industrial process modification to minimize emissions, and development of low-emission transportation systems.

Better economic modeling techniques are needed to identify the most cost-effective control strategies. Although the costs and effectiveness of many control techniques are known, it is difficult to translate costs for a specific technology into costs for overall emission reductions in an urban area or region.

FUTURE DIRECTIONS FOR PROTECTING AND IMPROVING VISIBILITY

Present scientific knowledge has important implications for the design of programs to protect and improve visibility. What is needed, overall, is the recognition that any effective visibility protection program must be aimed at preventing and reducing regional haze. An effective program must, therefore, control a broad array of sources over a large geographic area. Such a program would mark a considerable break from the present approach of focusing on visible plumes from nearby sources and of attempting to determine the effects of individual sources on visibility impairment.

Although visibility impairment is as well understood as any other air pollution effect, gaps in knowledge remain. Filling these gaps will require an increased national commitment to visibility protection research. We believe that the time has come for Congress, EPA, and the states to decide whether to make that commitment.

Protecting Visibility
in National Parks
and Wilderness Areas

1

Introduction

Many U.S. national parks and wilderness areas—the Grand Canyon, Yosemite, Shenandoah, and many others—are famous for their beautiful and dramatic scenery. Millions of people visit these areas each year to observe and appreciate nature firsthand. Visibility lies at the heart of this experience—the ability to look out over great vistas to see shapes and colors with crystalline clarity. In parts of the Southwest, the views can be spectacular. But such superb visibility is possible only when the air is extremely clean and particle concentrations are low. Even small increases in particle concentrations can substantially degrade visibility.

Fine particles in the atmosphere absorb and scatter light, thus limiting visual range, shifting colors, and obscuring the details of distant objects. These particles are the main constituent of the pervasive haze that impairs visibility over much of the United States.[1] Most of these particles are anthropogenic pollutants, either emitted as particles or formed in the atmosphere by gas-to-particle conversion.

The geographic areas affected by haze have increased with population growth and the spread of industrialization (Clark and Munn, 1986). Haze has become a common landscape feature in many densely populated parts of the United States, especially in the East and in California. But episodes of decreased visibility are also common over large areas

---

[1]Haze degrades visibility primarily through the scattering or absorption of light by fine atmospheric particles. Visibility is the degree to which the atmosphere is transparent to visible light.

19

remote from major centers of population and industrial development. The affected areas include many of the nation's most beautiful national parks and wilderness areas.

Studies have shown that varying degrees of visibility impairment occur at many park monitoring stations virtually all the time and that air pollution is responsible for the impairment (NPS, 1988). In recognition of the deteriorated visibility, the 1977 Amendments to the Clean Air Act (Section 169A) establish a national goal of preventing and remedying visibility impairment due to anthropogenic pollution in mandatory Class I areas, which include most large national parks and wilderness areas in the United States.[2]

This goal will be difficult to achieve because many types of pollutant sources degrade visibility. Many different types of sources produce the same types of chemicals, including those that are most important in visibility degradation. This makes it difficult to assign responsibility unambiguously to a source (for example, a specific coal-burning power plant) or even to a specific class of sources (for example, electric power plants or motor vehicles). Many techniques have been devised to try to resolve these problems. However, the extent to which these techniques can be used in attributing visibility impairment is uncertain, as is their usefulness in estimating the effect that different control strategies might have on visibility.

This report presents working principles for determining the relative importance of anthropogenic contributions to haze in mandatory Class I areas and for assessing various source control measures. It also provides guidelines for alleviating gaps in present knowledge about the sources and formation of haze, air-quality modeling, and emission-control techniques.

This report was prepared by the Committee on Haze in National Parks and Wilderness Areas. The committee was convened by the National Research Council's Board on Environmental Studies and Toxi-

---

[2]Class I areas are areas subject under the Clean Air Act to relatively stringent restrictions on increases in air pollutants over baseline concentrations. Most of the United States is classified as Class II and is subject to less stringent restrictions. Mandatory Class I areas are national parks and wilderness areas not subject to reclassification.

cology in collaboration with the Board on Atmospheric Sciences and Climate. The committee comprised members with expertise in meteorology, atmospheric chemistry, air-pollution monitoring and modeling, statistics, control technology, and environmental law and public policy. The committee's work was sponsored by the U.S. Department of the Interior (National Park Service, Bureau of Reclamation, and Office of Environmental Quality), U.S. Department of Energy, U.S. Environmental Protection Agency, U.S. Department of Agriculture (Forest Service), the Arizona Salt River Project (a political subdivision of Arizona), and Chevron Corporation.

The committee also issued an interim report (NRC, 1990) evaluating the Winter Haze Intensive Tracer Experiment (WHITEX), the 1987 National Park Service (NPS) report on the causes of wintertime haze in the region between the Grand Canyon and Canyonlands National park. The WHITEX report asserted that the Navajo Generating Station (NGS), a large coal-fired power plant in Page, Arizona, is a principal contributor to visibility impairment in Grand Canyon National Park (GCNP). The committee concluded that, at some times during the study period, NGS contributed significantly to haze in the GCNP; however, WHITEX did not quantitatively determine the fraction of sulfate (SO_4^{2-}) particles and resulting haze attributable to NGS emissions.

HAZE FORMATION AND VISIBILITY IMPAIRMENT

Visibility impairment is primarily due to the interaction of light with particles in the atmosphere; gaseous pollutants usually play a small role. Particles interact with light by two important mechanisms: They can absorb light, and they can scatter light in a direction different from that of the incident light. The magnitude of these effects depends on several factors, the most important of which are the size and composition of the particles and the wavelength of the incident light. Despite the complexity of these phenomena, a variety of commonly available techniques makes it possible to characterize the optical properties of the atmosphere and to identify and quantify the pollutants that affect visibility.

Visibility refers to the degree to which the atmosphere is transparent to visible light. Traditional meteorological usage equates visibility with *visual range*, the greatest distance at which a large black object can be distinguished against the horizon sky. One of the principal indices of visibility is the *extinction coefficient*,[3] defined as the fraction of light that is attenuated by scattering or absorption as the light beam traverses a unit of the atmosphere. The extinction coefficient is a measure of the rate at which energy is lost or redirected through interactions with gases and suspended particles in the atmosphere. In a uniform atmosphere, light extinction can be shown to be inversely proportional to visual range. Light extinction is directly related to perceptual cues of overall human judgements of *visual air quality*. The 1977 Clean Air Act Amendments and EPA's implementation of these amendments extend the term visibility to cover freedom from discoloration, reduced contrast, and other visible departures from the natural atmosphere.

This report discusses visibility primarily in terms of physical parameters such as visual range and extinction. However, the ultimate goal of visibility protection programs is to improve visibility as judged by human observers. Therefore it is important to relate measurable physical parameters to human judgments of visibility also known as visual air quality. These judgments are complex and depend on many factors, including sun angle, air color, and scene composition.

Most visibility impairment can be traced to five particulate substances: sulfates, nitrates, organics, elemental carbon, and soil dust. Water, which can combine with sulfates, nitrates, and some organics, is also a major cause of visibility impairment. Elemental carbon, soil dust, and some organics are emitted directly to the atmosphere; these are called primary particles. Sulfates, nitrates, and other organics are formed in the atmosphere from gaseous precursors; these are known as secondary particles. Improving and protecting visibility thus requires removing both gases and particles from source emissions; this is more complex and costly than removing primary particles alone.

---

[3] The extinction coefficient is usually given in units of Mm^{-1} (10^{-6} meters^{-1}) or km^{-1} (10^{-3} meters^{-1}). If the extinction coefficient is 10^{-2} km^{-1}, the intensity of incident light is reduced by approximately 63% over a distance of 100 km (assuming extinction is uniform along the path of transmission). If the extinction coefficient is 1 km^{-1}, then the intensity of incident light is reduced by 63% over a distance of 1 km. The former value is characteristic of a particle-free, unpolluted area and the latter of a highly polluted area.

The complex chemical and physical processes that result in the formation of visibility-impairing particles are not fully understood. However, these processes are known to enhance the effect of human activities on visibility. The particles that scatter light most efficiently per unit mass are those of approximately the same size as wavelengths of visible light (0.4-0.7 μm). As a result of atmospheric processes, a large fraction of anthropogenic airborne particles accumulates in the 0.1 to 1.0 μm diameter range. These particles can survive in the atmosphere for several days and be transported hundreds of kilometers. During transport, the emissions from many different sources can become mixed, making it difficult to assess the effects of individual sources on visibility. This mixing leads to the formation of regional haze, which is now a major threat to visibility in U.S. parklands.

Visibility degradation is closely connected with other effects of air pollution. The human activities that reduce visibility also can impair human health and cause other environmental effects. For example, some of the major effluent gases emitted by fossil-fuel combustion react to form oxidants and acids that are health and environmental hazards as well as visibility-degrading airborne particles. The particles that degrade visibility can also influence climate by interacting with incoming solar radiation and by modifying cloud formation (Trijonis et al., 1990; Charlson et al., 1991). Because of these connections, visibility degradation has become recognized as an indicator of multiple human-health effects and environmental effects resulting from air-pollution all over the world (e.g., Gilpin, 1978; Holgate et al., 1982; Clark and Munn, 1986). Understanding these multiple connections could elucidate the potential environmental benefits of air quality improvement strategies.

DIFFICULTIES IN DEVELOPING
EFFECTIVE PROGRAMS

Scientific Difficulties

Formulating and implementing effective programs to protect and enhance visibility can be impeded by several political, institutional, and scientific barriers. To control visibility impairment, the air-pollution sources must be identified and, to the extent possible, their relative importance must be determined. This information is difficult to obtain because of the multiple sources of haze in national parks and wilderness areas, including industrial and utility operations, cars and other mobile sources, as well as burning for management of forest and crop lands.

The controversy over the Navajo Generating Station, affords an example of the difficulties in estimating source contributions to visibility impairment. NGS emits about 148 Mg (163 tons) of SO_2 per day, making it one of the largest single SO_2 sources in the United States west of the 100th meridian. The plant is located 25 km from the Grand Canyon National Park and 110 km from the Grand Canyon Village tourist area on the Canyon's south rim. NPS, the managing agency for the GCNP, believes that NGS is an important source of SO_4^{2-} particles that cause wintertime haze in the park. In 1987, a large-scale experiment known as the Winter Haze Intensive Tracer Experiment was carried out to investigate the causes of wintertime haze in the region of GCNP and Canyonlands National Park (NPS, 1989). The experiment called for the injection of a tracer, deuterated methane (CD_4), into one NGS stack.

During some haze episodes, significant concentrations of CD_4 were detected at a sampling station on the south rim of the Grand Canyon. NPS analyzed data from WHITEX and issued a final report concluding that NGS was responsible for approximately 70% of the anthropogenic particulate sulfate and approximately 40% of the anthropogenic aerosol-related light extinction during selected wintertime periods of haze at the sampling station. This committee evaluated the NPS WHITEX report and concluded that, at some times during the study period, NGS contributed significantly to haze in GCNP. This assessment was based on evaluation of meteorological, photographic, chemical, and other physical evidence. However, the committee also concluded that the data presented in the WHITEX report were insufficient to determine quantitatively the fraction of SO_4^{2-} aerosol particles and resultant visibility impairment in GCNP that are attributable to NGS emissions. The committee identified flaws in the models used to estimate NGS's contribution, in the interpretation of those models, and in the data base. The committee found that sources other than NGS appear to account for a significant fraction of haze observed in GCNP during the study period. The committee also noted that control of NGS emissions likely would reduce, but not eliminate, wintertime haze at GCNP. Precise quantification of the effect of NGS emissions on surrounding air quality remains to be determined.

Regulatory Difficulties

Efforts to meet the national visibility goal have been hindered by a lack of commitment to meeting the goal. Visibility impairment is at least as well understood as any other effect of air pollution. Yet the nation has not given the same priority to meeting the national visibility goal as it has to addressing other air pollution problems. For instance, Section 169A(f) of the Clean Air Act makes it clear that the EPA is not required to achieve the visibility goal by any particular date. Rather, states are obliged only to make "reasonable progress" towards the goal and the federal government has devoted only modest resources to visibility regulation and research. In contrast, the act requires that the health-based primary air-quality standards be attained within a specified time.

This ambiguity in the act reflects the nation's uncertainty about the visibility goal and the costs of achieving it. Such conflict can be seen even within agencies that have visibility protection as part of their missions. For instance, the Forest Service and NPS routinely use prescribed burning as a forest-management practice, yet such burning conflicts with their responsibility to protect visibility in Class I areas. Within the Department of the Interior (DOI), the NPS mission of preserving natural areas conflicts with DOI's traditional interest in resource development. A striking example of this conflict arose during discussions of whether pollution control equipment should be installed on NGS. This issue placed NPS in conflict with the Bureau of Reclamation, which holds part ownership in the power plant.

A fundamental difficulty linked to those described above has been the general approach taken thus far to visibility impairment. EPA's current regulations require retrofitting of controls only on those sources whose contributions to visibility impairment in mandatory Class I areas can be shown through "visual observation or other techniques the State deems appropriate." A source's contribution to regional haze, however, usually cannot be detected through visual observation. At most, observation and similar techniques (such as photography) can detect visible plumes from individual sources. Visible plumes, however, appear to be minor contributors to visibility impairment in Class I areas.

Efforts to decide whether a particular source is contributing to regional haze have thus far encountered grave obstacles. Studies designed to estimate the effect of a particular source on surrounding visibility are expensive, and the results can be uncertain and controversial. For instance, WHITEX cost more than $5 million, and Congress has appropriated $2.5 million initially for a similar study at the Mohave power station.

SCOPE OF THE REPORT

Recently proposed strategies suggest that visibility regulation should be based on controlling a large number of sources across a broad geographic area, and the Clean Air Act Amendments of 1990 take steps in this direction. In addition to examining methods aimed at assessing the contribution of individual sources to visibility impairment, the committee considered techniques that apportion contributions to regional haze among categories of sources or among geographic regions. The result is a set of working principles that can be used to address either single-source or regional-scale problems.

The committee kept in mind that visibility protection can take many forms and that the choice of form influences the types of analytical tools that can be used to study visibility problems. Estimates of source contributions to visibility impairment should account not only for existing sources, but also for proposed new sources that seek permits under the prevention of significant deterioration program of the Clean Air Act or other visibility-protection programs. The committee stresses that such sources must be examined as part of an overall analysis of the regional haze problem within which the effects of specific sources are embedded. In addition, the committee identified gaps in present understanding of haze formation and set priorities for information-gathering and research in these areas. The committee also established working principles to improve knowledge of these processes and to aid in the development of more effective control strategies.

In a visibility protection program, decisions must be made as to the best strategy to reduce emissions. This report presents working principles for decision makers to use in assessing alternative strategies. These principles are intended to assist in the design of emission-management plans specifically for visibility protection. However, they will also help in the assessment of visibility effects of actions taken primarily for other purposes—for example, the effect on visibility that will result from attainment of the National Ambient Air Quality Standards for inhalable particles.

Chapter 2 describes visibility degradation in different parts of the country resulting from specific natural and anthropogenic sources of gases and particles. Chapters 3 and 4 provide the legal and technical foundation for the report. Chapter 3 presents the regulatory and institutional framework for visibility protection, including the Clean Air Act. That chapter also sets forth broad regulatory paradigms that were noted by the committee as possible frameworks within which future technical analyses might have to be conducted. Chapter 4 describes scientific understanding of haze formation and visibility impairment, including the meteorological and chemical processes responsible for the transport and transformation of gases and particles in the atmosphere. Chapter 4 also reviews atmospheric chemical and physical measurement techniques.

Chapter 5 delineates in some detail various source identification and apportionment techniques. That chapter also discusses the relative difficulty of implementing these approaches and suggests detailed evaluation criteria. Technological and administrative feasibility, economic efficiency, flexibility, balance, and error analyses and biases are considered. Chapter 6 describes the emission reduction potential of various source control measures and briefly illustrates the translation of control measures into a rough prediction of effects on visibility. It provides guidance for decision makers as to the best means to develop control strategies.

Chapter 7 sets forth the committee's recommendations for the best approaches to remedy current scientific and technical gaps that limit present understanding of source effects on visibility and ability to evaluate control measures. Chapter 7 also summarizes the policy implications of the current scientific understanding of visibility impairment. The committee urges that the present program be reoriented to address regional haze and the apportionment of haze among groups of sources rather than the present single-source-oriented approach.

2

Visibility Conditions
in the United States

This chapter briefly characterizes overall visibility conditions in the United States and the emission sources that affect visibility. By comparing current spatial, temporal, and statistical visibility patterns with patterns in airborne particle concentrations, the committee presents information about the causes of visibility degradation. Studies that have related historical visibility trends to historical emission trends are summarized, and major emission source types (natural and anthropogenic), that affect visibility are discussed.

Visibility impairment episodes can range in scale from local plumes to widespread regional haze. The sources of locally visible plumes are easy to identify, for example, the smoke from a power plant stack or from a burning field. However, when plumes are carried by winds, they become more diffuse, and the sources are identified less readily. In regions with many sources, the plumes can merge and become mixed with the emissions from many small sources, such as motor vehicles. The result is a widespread haze in which individual contributions from the various sources are virtually indistinguishable. This latter condition—regional haze—is the main focus of this chapter.

The most intense regional haze in the United States occurs in the East, where haze often is linked to high concentrations of ambient sulfate (SO_4^{2-}). SO_4^{2-} concentrations are highest in the summertime during meteorological conditions that are usually associated with the western half of slow-moving high-pressure systems. Under such stagnant conditions, pollutants from many different sources can accumulate, causing severe and widespread visibility degradation.

It is important to recognize, however, that regional hazes are not necessarily caused by local emissions, nor do they depend on stagnant meteorological conditions. In the absence of precipitation, airborne particles (and their gaseous precursors) can exist in the atmosphere for many days and can be carried great distances by winds. Studies have shown that regional hazes are often associated with transport from distant sources. Haze episodes at eastern sites are well correlated with previous occurrences (more than 36 hours earlier) of low wind speeds (periods of stagnation) in upwind source regions (Samson, 1978). High SO_4^{2-} concentrations at rural areas of upstate New York (including the Adirondack Mountains) most often are associated with winds coming from the south and southwest (Galvin et al., 1978). High SO_4^{2-} concentrations at Shenandoah National Park, Virginia, usually are associated with moderate wind flows from the west to northwest (Wolff et al., 1982).

CURRENT VISIBILITY CONDITIONS

Geographical Patterns

Figure 2-1 shows isopleths of median visual range at rural U.S. sites (Trijonis et al., 1990). The spatial patterns in this map are based on airport observations of visual range which differ from measurements of standard visual range.[1] Airport visual range measurements are based on the identification by human observers of targets at known distances from the observation point, whereas standard visual range is calculated from light extinction measurements. However, airport observations have been calibrated against the results of instrumental studies for standard visual range, and the airport data presented in these figures have been adjusted accordingly.

Figure 2-1 shows that the mountainous Southwest has the best visibility in the country. Median standard visual range exceeds 150 km in the region comprising Utah, Colorado, Nevada, northern Arizona, north-

---

[1] Standard visual range is defined as the greatest distance at which a standard observer can discern a large black object against the horizon sky under uniform lighting conditions.

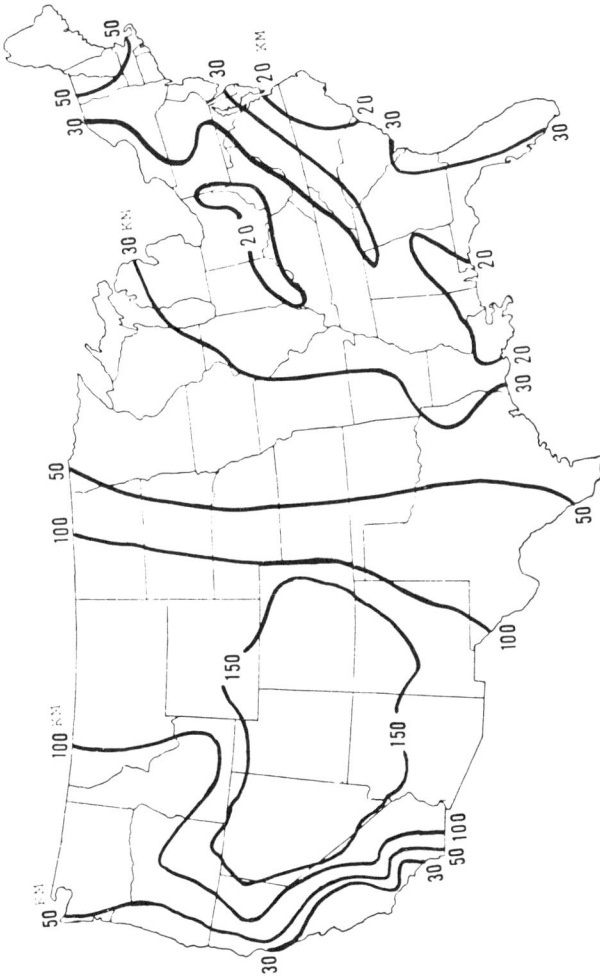

FIGURE 2-1 Estimated median standard visual range (km) for rural (suburban and nonurban) areas of the United States. Values are based on airport median visual ranges multiplied by 1.3 to account for differences in detection thresholds in estimating standard visual range. Data included for all days (all weather conditions). Data are for 1974-1976, but recent studies indicate that current conditions are approximately the same as shown here. Median values are reported instead of averages because a significant portion of the data are less than the lowest extinction threshold (above the farthest visibility marker). Averages, presented in other figures, are highly correlated with median values. Source: Trijonis et al., 1990.

western New Mexico, and southwestern Wyoming. In the adjoining regions to the north and south, median standard visual range is also quite good, exceeding 100 km. However, visual range decreases sharply to the east and west of this area. Median visual range falls to less than 50 km in a narrow band along the northern Pacific coast, less than 30 km in the central valley of California, and to less than 15 km in the Los Angeles basin (Trijonis, 1982a). Although some parts of the East (e.g., New England) have moderately good visibility levels (about 40-60 km), median visual range is generally less than 30 km in the large area east of the Mississippi and south of the Great Lakes.

Observations show a distinct relationship between visual range and altitude. On average, visual range is somewhat greater at higher altitudes than in the surrounding areas (Trijonis, 1982a; Air Resources Specialists, 1988). Many national parks are located at higher elevations than the sites from which the data were obtained for Figure 2-1, and the visual range in some national parks could be as much as 50% higher than indicated in the figure (Trijonis et al., 1990).

The National Park Service (NPS) routinely measures particle concentrations and composition in many national parks and wilderness areas. Most of those parks and wilderness areas are located in the West; consequently, few data are collected for the eastern portion of the country. The geographical patterns in the annual average data are summarized in Figures 2-2 through 2-7, which show fine-particle (less than 2.5 μm diameter) mass concentration (Figure 2-2), fine particulate sulfur (Figure 2-3), fine soil-derived materials (Figure 2-4), and absorption coefficient (Figure 2-5). (The absorption coefficient is directly related to the concentration of elemental carbon.) Figure 2-6 shows the distribution of the remaining fine-particle mass (the total fine-particle mass minus the concentrations of fine sulfate, elemental carbon, and soil particles). Figure 2-7 presents data for estimated nonsulfate hydrogen (the total hydrogen concentration less the hydrogen that is associated with sulfates). The remaining mass and the nonsulfate hydrogen are believed to be qualitatively related to the spatial distribution of organic aerosols.

These figures, as well as rural data sets reported in the National Acid Precipitation Assessment Program (NAPAP) Visibility State of Science and Technology Report (Trijonis et al., 1990), indicate the following differences between the air quality of the rural West (particularly the arid, mountainous Southwest) and that of the rural East (particularly the area south of the Great Lakes and east of the Mississippi):

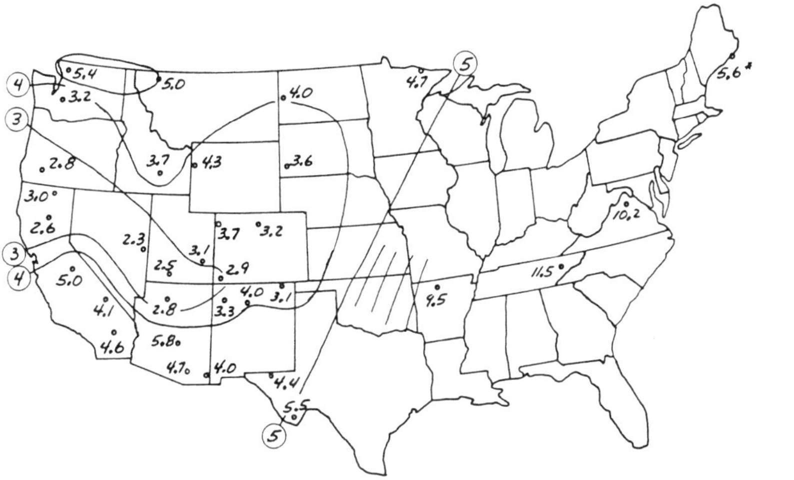

FIGURE 2-2 Average concentrations of fine-particle mass ($\mu g/m^3$) from the National Park Service network, 1983-1986. Source: Eldred et al., 1987 (From Trijonis et al., 1990).

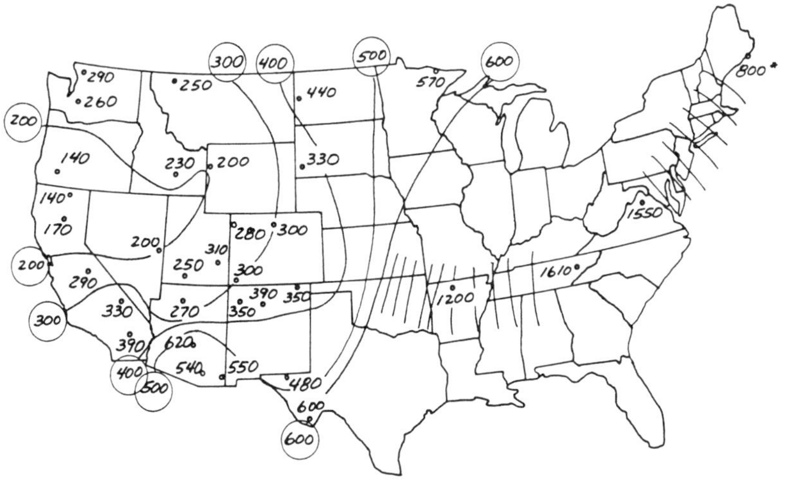

FIGURE 2-3 Average concentrations of fine-particle sulfur (ng/m^3) from the National Park Service network, 1983-1986. Source: Eldred et al., 1987 (From Trijonis et al., 1990).

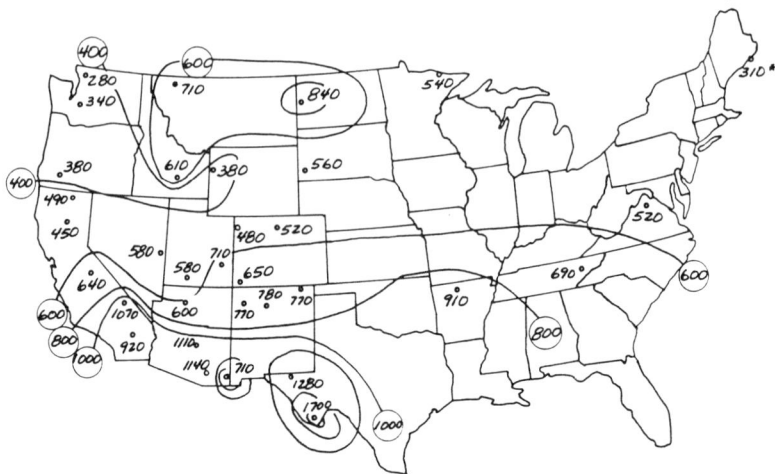

FIGURE 2-4 Average concentrations of fine-particle soil materials (ng/m³) from the National Park Service network, 1983-1986. Source: Eldred et al., 1987 (From Trijonis et al., 1990).

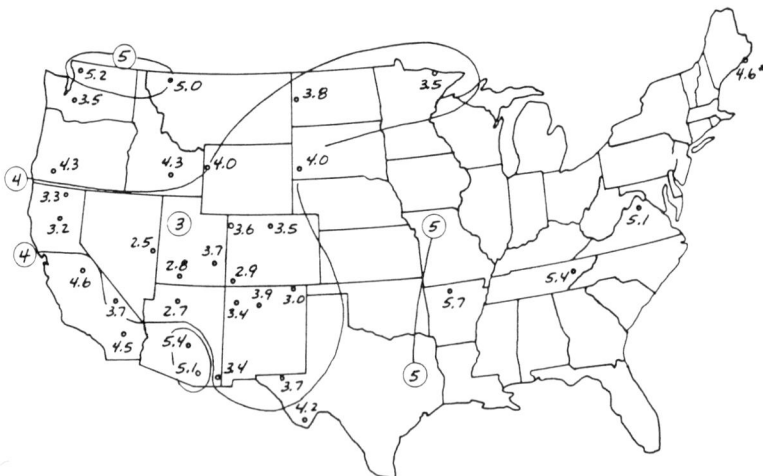

FIGURE 2-5 Average fine-particle absorption coefficient (Mm⁻¹) from the National Park Service network, 1983-1986. Fine elemental carbon concentrations in $\mu g/m^3$ can be estimated by dividing the fine-particle absorption coefficient (in Mm⁻¹) by 10. Source: Eldred et al., 1987 (From Trijonis et al., 1990).

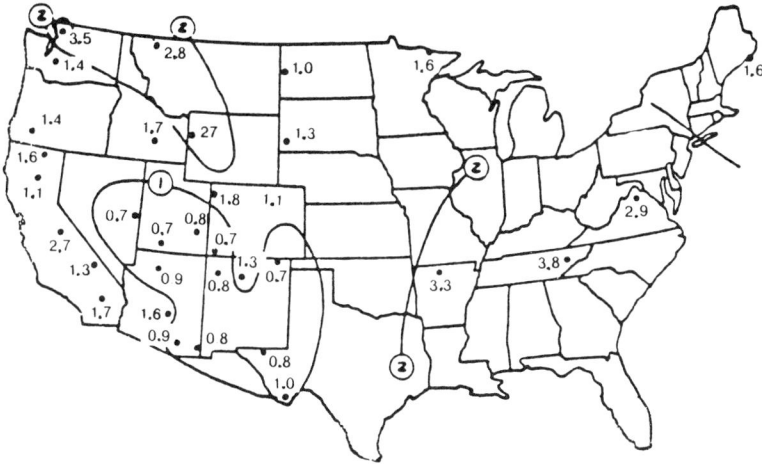

FIGURE 2-6 Average concentrations of remaining fine-particle mass (μg/m^3) from the National Park Service network, 1983-1986. The fine particle remaining mass is defined as the total fine-particle mass minus the concentration of fine sulfate, elemental carbon, and soil particles. Organic carbon is a major component of the remaining fine-particle mass. Source: Eldred et al., 1987 (From Trijonis et al., 1990).

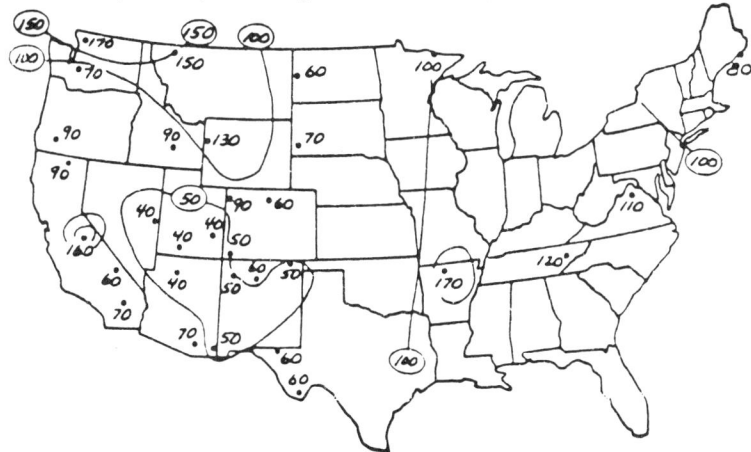

FIGURE 2-7 Average concentrations of fine-particle nonsulfate hydrogen mass (ng/m^3) from the National Park Service network, 1983-1986. The nonsulfate hydrogen is calculated from the total hydrogen concentration by subtracting the hydrogen that is associated with sulfates. Source: Eldred et al., 1987 (From Trijonis et al., 1990).

• SO_4^{2-} concentrations are about six times greater in the rural East than in the rural West.

• Fine soil concentrations are about the same in the rural East and West.

• Elemental carbon and organic particle concentrations (as reflected by the remaining fine particle mass and the fine-particle nonsulfate hydrogen) are about two times greater in the rural East. (This finding agrees with the data for 19 rural sites reported by Shah et al., 1986.)

• Fine nitrate (NO_3^-) concentrations are about the same in both regions, about 0.5 to 1.0 $\mu g/m^3$. (See denuder nitrate data in Trijonis et al., 1990.)

Seasonal Patterns

Two data sets show the seasonal visibility patterns for North America. The first data set is quarterly median light extinction data obtained at NPS automated camera sites for 1986-1988 (see Table 2-1 and Appendix B). The second involves analyses of data at U.S. and Canadian airports as shown in Figures 2-8a and 2-8b.

Both data sets reveal an extremely strong seasonal feature that occurs on a large geographical scale—visibility is lowest in the summer in the region south of the Great Lakes and east of the Mississippi. The seasonality in light extinction measured at the two NPS sites in this region (Great Smokies and Shenandoah) is clearly evident in Figure 2-9; here, light extinction is more than twice as great in the summer as during the other seasons.

The visibility minimum observed during the summer in the East is, in large part, due to maximal SO_4^{2-} concentrations during the summer. Aerosol data show that SO_4^{2-} concentrations in the East are nearly twice as high in the summer as during the rest of the year (Trijonis et al., 1990). Figure 2-10 shows seasonal patterns in the East for extinction, fine-particle mass, and SO_4^{2-} (Trijonis, 1982a); light extinction correlates with total fine-particle mass and fine sulfate, and SO_4^{2-} constitutes about half of the total fine-particle mass.

Neither of the two available data sets are adequate for a definitive analysis of the more subtle seasonal visibility patterns in New England and the western two-thirds of the United States. The NPS data are biased due to treatment of data on days when targets were snow-cov-

TABLE 2-1 Quarterly Median Visual Ranges for National Park Service
Automated Camera Sites, 1986-1988

Site	Seasonal Median Visual Range (km)			
	Dec.- Feb.	March- May	June- Aug.	Sept.- Nov.
Acadia Park, ME	85	64	69	64
Arches Park, UT	140	158	189	166
Bandelier Monument, NM	171	154	156	178
Big Bend Park, TX	168	128	145	185
Black Canyon Monument, CO	134	146	159	140
Bryce Canyon Park, UT	243	182	184	189
Bridger Wilderness, WY	52	60	165	84
Buffalo River, AR	61	43	46	64
Capitol Reef Park, UT	102	144	172	158
Capulin Volcano, NM	156	125	136	110
Carlsbad Caverns Park, NM	156	177	110	137
Chaco Culture NHP, NM	155	168	166	162
Chiricahua Monument, AZ	170	162	134	166
Colorado Monument, CO	119	151	171	156
Craig (BLM), CO	75	123	148	107
Crater Lake Park, OR	105	83	148	66
Craters of the Moon, ID	66	129	140	128
Death Valley Monument, CA	203	133	92	117
Dinosaur Monument, CO	131	133	162	178
Glacier Park, MT	35	58	152	82
Glen Canyon Area, AZ	151	152	149	146
Grand Teton Park, WY	18	119	127	94
Great Basin Park, NV	225	174	195	138
Great Sand Dunes, CO	158	123	114	140
Great Smoky Mountains, TN	49	54	20	50
Green River Area, WY	131	67	176	164
Guadalupe Mountains, TX	150	120	106	125
Isle Royale Park, MI	24	49	66	58
Jarbidge Wilderness, NV	64	24	136	92
Joshua Tree Monument, CA	244	140	115	139
Lake Mead Area, NV	234	143	156	149
Lassen Volcanic Park, CA	107	94	156	164
Lava Beds Monument, CA	154	146	158	112

TABLE 2-1 continued

Site	Seasonal Median Visual Range (km)			
	Dec.- Feb.	March- May	June- Aug.	Sept.- Nov.
Mesa Verde Park, CO	166	152	156	164
Mount Rainier Park, WA	81	82	102	98
Olympic Park, WA	10	59	94	66
Pinnacles Monument, CA	162	114	132	114
Point Reyes, CA	65	50	36	28
Redwood Park, CA	91	69	42	50
Rocky Mountain Park, CO	136	110	144	132
San Gorgonio Wilderness, CA	266	101	127	140
Shenandoah Park, VA	70	68	24	54
Superstition Mountains, AZ	234	185	153	166
Theodore Roosevelt Park, ND	198	93	133	120
Voyageurs Park, MN	75	99	172	106
Weminuche Wilderness, CO	98	112	150	137
Wind Cave Park, SD	209	119	152	147
Yellowstone Park, WY	55	24	96	63
Yosemite Park, CA	56	75	69	68
Zion Park, UT	184	156	172	164

All data are included except for observations of snow-covered targets. (Winter and spring estimates are problematic at many western sites due to treatment of data obtained during conditions with snow-covered targets.) Seasonal values represent averages of all quarterly medians available. Quarterly medians are based on regressions fit to cumulative frequency plots. Source: Trijonis et al., 1990.

ered. In the West, the airport results reported by Husar are problematic because of the lack of sufficiently distant targets at some sites and also because of the limited statistical resolution in the reported results; interpretation also is difficult because the results are based on data from a mixture of urban and rural locations (Trijonis et al., 1990). Definitive results must await a revised analysis of the NPS camera data and airport visibility data or, better yet, an analysis of the transmissometer data from the new NPS monitoring system (see Chapter 4).

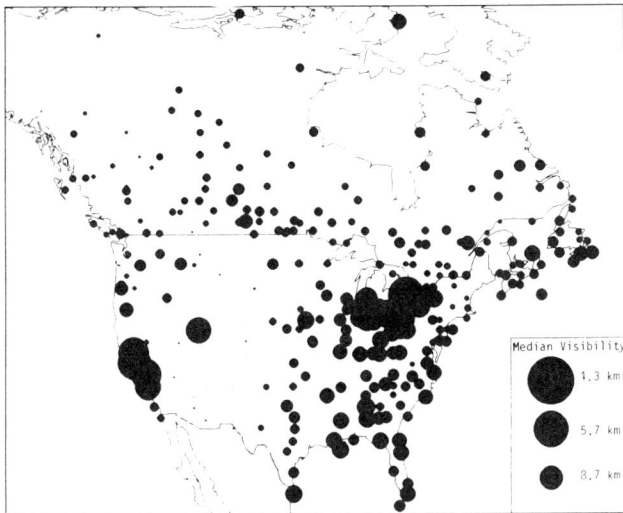

FIGURE 2-8a   Median airport visual range for January, averaged over
1979-1983. Larger circles correspond to lower visual range (greater light
extinction). Diameter of circles is proportioned to light extinction.
Source; R.B. Husar, pers. comm., Washington University, St. Louis,
Mo., 1989 (From Trijonis et al., 1990).

Despite these limitations, a few patterns emerge. For example, it is
well established that in Arizona and southern California, visibility is
lowest during the summer (Trijonis, 1982a; Air Resources Specialists,
1988; Trijonis et al., 1988; and Husar, 1989, pers. comm., Washington
University, St. Louis, Mo.). In contrast, for the northern two-thirds of
California, the Pacific Northwest, and the northern mountain states,
minimal visibility tends to occur during the fall and winter (Trijonis,
1982b; Air Resources Specialists, 1988; and Husar, 1989, pers. comm.,
Washington University, St. Louis, Mo.). The winter and fall visibility
minimum is especially obvious in the central valley of California
(Trijonis, 1982b; Husar, 1989, pers. comm., Washington University, St.
Louis, Mo.).

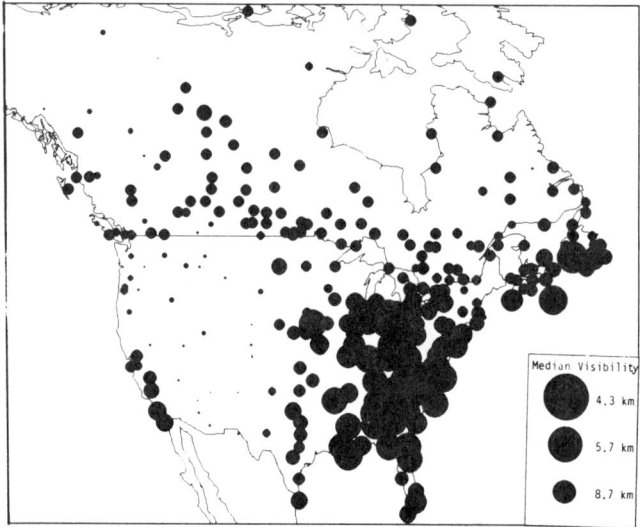

FIGURE 2-8b Median airport visual range for July, averaged over 1979-1983. Larger circles correspond to lower visual range (greater light extinction). The diameter of the circles is proportional to light extinction. Source: R. B. Husar, 1989, pers. comm., Washington University, St. Louis, Mo., 1989 (From Trijonis et al., 1990).

Statistical Patterns

The statistical distribution of visibility, especially with respect to worst-case values (greatest light extinction levels) is of interest. A comprehensive study of the statistical distribution of worst-case extinction levels was conducted by Gins et al. (1981), who used midday observations from 28 airports. Gins et al. sorted the data according to meteorological conditions, eliminating observations with fog, precipitation, blowing dust, clouds, and other obscuring factors; they estimated percentiles from airport data. They characterized the relationship between bad-case and median extinction according to the ratios of upper percentile extinction coefficients to median extinction coefficients. Table 2-2 summarizes the results for three percentiles (90th, 95th, and 99th), for the nation, for four regions, and for the four calendar quarters.

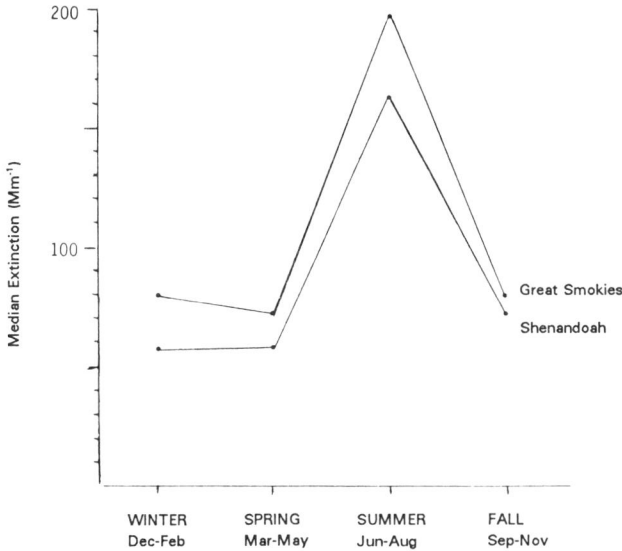

FIGURE 2-9 Seasonal pattern in median light extinction as measured by automated cameras at two eastern National Park Service sites, 1986-1988. Source: Trijonis et al., 1990.

The numbers in Table 2-2 are measures of the episodicity of light extinction levels; a high number indicates that extreme haze episodes are severe relative to the median. The four regions in Table 2-2 show considerable variation: the 99th percentile values exceed median extinction levels by a factor of about 3 to 8; the 95th percentile values by a factor of 2.5 to 4; and the 90th percentile values by a factor of 2 to 3. This last factor compares well with the results of several other studies for 90th percentiles (Trijonis et al., 1990).

Table 2-2 has some interesting features. For example, in the East, the annual episodicity is greater than the episodicity within individual seasons. This partly reflects the strong seasonality of light extinction levels in the East. Another interesting feature is that episodicity in the East is stronger in the summer than in the winter. This means that for worst-case conditions in the East (during the summer) the episodicity is even stronger than it is for average conditions. A similar situation might exist at many western sites, which tend to exhibit highest average extinction levels in the winter, as well as highest episodicity in the winter.

Historical Trends

Husar has analyzed historical visibility trends using airport data from the late 1940s through the early 1980s (1989, pers. comm., Washington University, St. Louis, Mo.). The 16 maps in Figure 2-11 represent four periods centered upon 1950, 1960, 1970, and 1980, grouped by calendar quarters (i.e., the four seasons).

During the first quarter, declining airport visual range shows evidence of decreased overall winter visibility in the Gulf states. Visibility decreased over the California-Oregon coast through the 1950s and 1960s, but it might have improved somewhat in the 1970s. The Northeast shows an improving trend in visibility.

In the second quarter, visual range decreased significantly over the Unit-

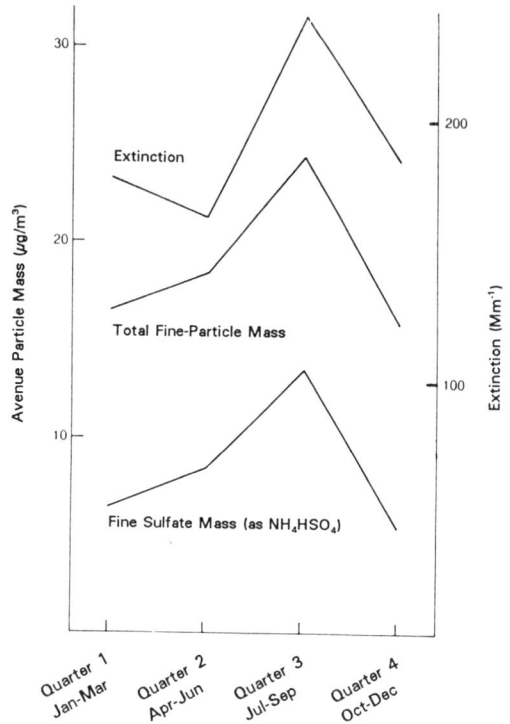

FIGURE 2-10 Seasonal patterns in SO_4^{2-}, fine-particle mass, and light extinction for rural areas of the eastern United States. The fine-particle mass and fine sulfate mass data are averages over five rural EPA dichotomous sampler sites (Will, Jersey, and Monroe Counties, IL; Erie County, NY; and Durham County, NC). The light extinction data are estimated from seasonal median visibilities (using a Koschmeider constant of 3.0) at eight rural airports. Source: Trijonis et al., 1990.

ed States east of the Rockies. Around 1950, the lowest-visibility region was confined to the Ohio and Mississippi valleys. By the 1980s, howev-

TABLE 2-2   Average Ratios of Selected Percentiles to the Median for Extinction Coefficients: 1974-1976

Period	Percent	Nation	Pacific	Rocky Mts.	Central	Eastern
Annual	90	2.4	2.5	2.3	2.0	2.9
	95	3.2	3.6	3.0	2.5	3.6
	99	5.4	6.5	5.9	3.6	5.8
Jan.-	90	2.5	2.7	3.0	2.0	2.3
Mar.	95	3.2	3.5	4.0	2.4	2.9
	99	5.6	6.3	7.7	3.6	4.5
April-	90	2.3	2.0	2.2	2.2	2.7
June	95	2.9	2.4	2.8	2.7	3.4
	99	4.4	3.1	5.2	3.8	5.2
July-	90	2.2	2.1	1.8	2.1	2.8
Sept.	95	2.8	2.6	2.3	2.6	3.6
	99	4.0	3.4	3.7	3.5	5.5
Oct.-	90	2.5	3.2	2.7	1.9	2.5
Dec.	95	3.2	4.2	3.6	2.2	3.0
	99	5.5	6.5	7.9	2.9	4.5

Source: Gins et al., 1981 (From Trijonis et al., 1990).

er, the lowest-visibility region extended over much of the eastern United States.

The most significant historical visibility declines occurred in the East for the third quarter, a phenomenon also reported by others (Munn, 1973; Trijonis et al., 1978; Husar, 1988; Husar et al., 1981; Sloane, 1982, 1983, 1984; Trijonis, 1982b). The visibility decline was especially great in the southeastern states.

The visual-range pattern for the fourth quarter is qualitatively and quantitatively similar to that of the first quarter; of note, visibility in the Northeast increased at the same time that visibility over the Southeast decreased.

These historical records of haziness show significant, coherent trends of visibility over broad regions of the country. However, the trend lines

44

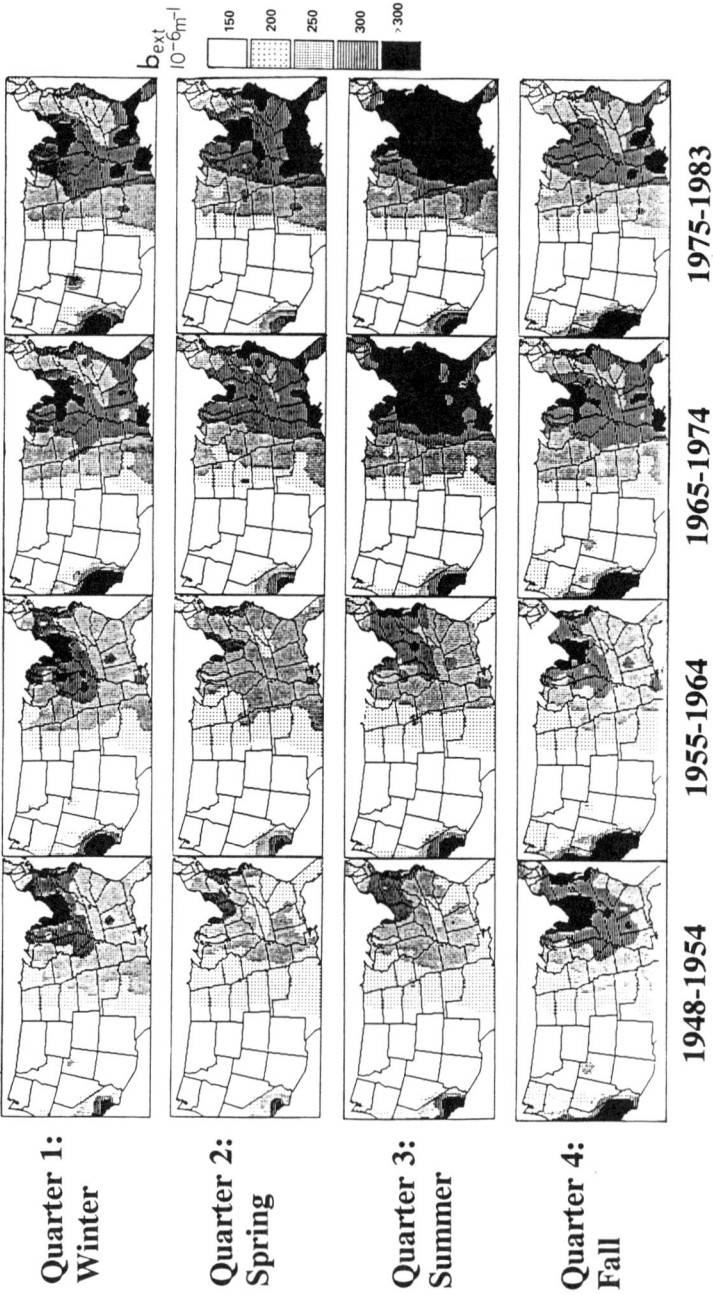

FIGURE 2-11 U.S. trend maps for the 75th percentile extinction coefficient, 1948-1983. Data are derived from airport visual range measurements for the calendar quarters: winter, spring, summer, and fall. Source: R.B. Husar, pers. comm., Washington Univeristy, St. Louis, Mo., 1989 (From Trijonis et al., 1990; Husar and Wilson, 1993).

are significantly different in the Northeast and Southeast. To emphasize this difference, regional and seasonal trends are displayed in Figures 2-12 and 2-13. These figures compare the aggregate for the Northeast —Indiana, Ohio, Pennsylvania, New York, New Jersey, Delaware, Maryland, Kentucky, West Virginia, and New England—with the aggregate for the Southeast—states south of the statistic above and east of the Mississippi. In the Northeast, the annual extinction coefficient is virtually unchanged (Figure 2-12), but the Southeast shows a 60% increase in annual haze over the 35-year period. In the Northeast, winter haze (Figure 2-13) shows a 25% decline, while the Southeast shows a 40% increase. The summer extinction coefficient in the Northeast shows a moderate overall rise. In the southeastern states the summer extinction coefficient increased by 80%, the change occurring mainly in the 1960s.

If sulfates are the major contributor to haze in the East, as many studies imply (see, for example, Table 6-1), then visibility could be expected to be related in some way to sulfur dioxide (SO_2) emissions. In light of the strong regional and seasonal differences discussed above, Husar et al. separately compared visibility trends with SO_2 emissions for the Northeast and the Southeast and for winter and summer (1989, pers. comm., Washington University, St. Louis, Mo.). Husar et al. compiled emission trends using the historical yearly emission data gathered by Gschwandnter et al. (1985) and Husar (1986) as well as the historical monthly emission data gathered by Knudson (1985).

Light extinction and sulfur dioxide (SO_2) emissions trends for the Northeast in the winter and summer are depicted in Figures 2-14a and 2-14b. The corresponding trends for the Southeast are illustrated in Figures 2-15a and 2-15b. Regional and seasonal trends are well correlated with historical light-extinction and SO_2 emission trends. Since the late 1940's, light extinction has declined moderately in the North during the winter and increased moderately during the summer; these trends are matched closely by the trends in SO_2 emissions. Over the same period, emissions and haze increased moderately during the winter in the Southeast, and a very strong increase was observed in both during the summer. The trends of increased emissions and increased light extinction show the greatest rise during the 1950s and 1960s and level off or decrease after the 1970s. The data suggest that, since the late 1940s, changes in light extinction over the eastern United States largely have been driven by changes in SO_2 emissions.

46

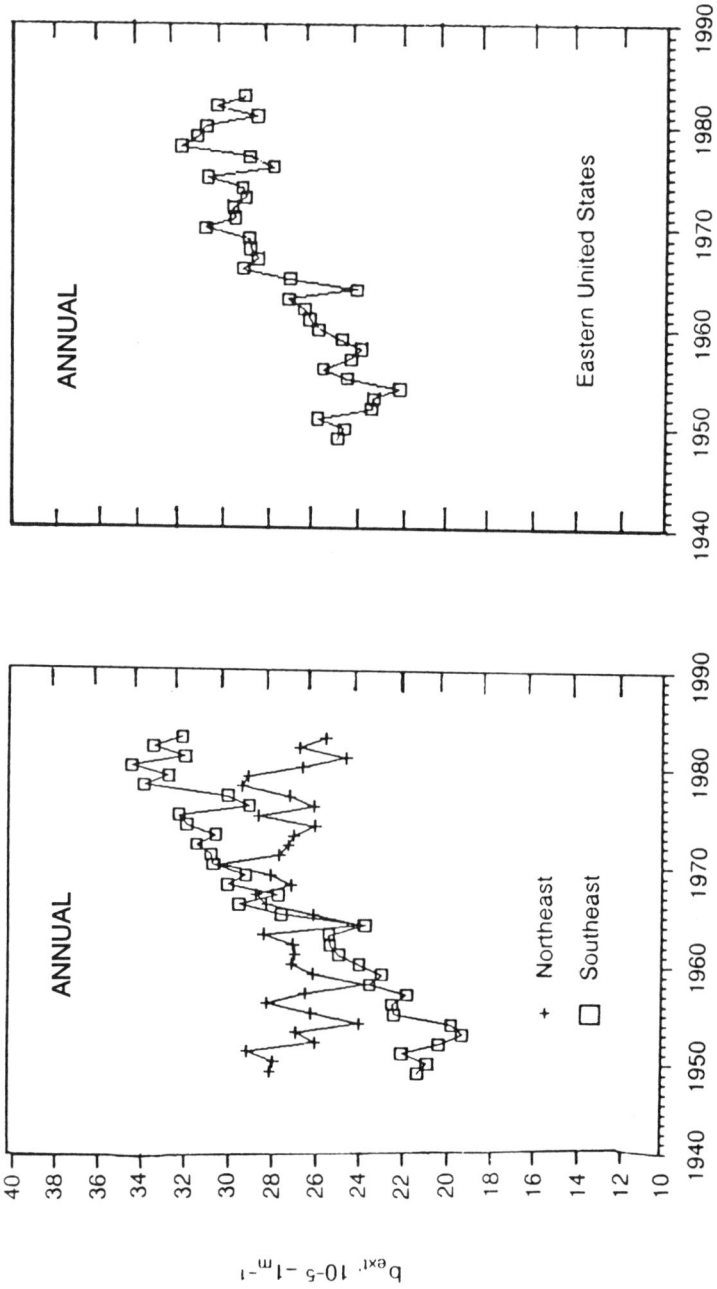

FIGURE 2-12. Annual trends for 75th percentile extinction coefficient for Northeast and Southeast United States and annual trend of average extinction coefficient for entire eastern United States. Source: Trijonis et al., 1990.

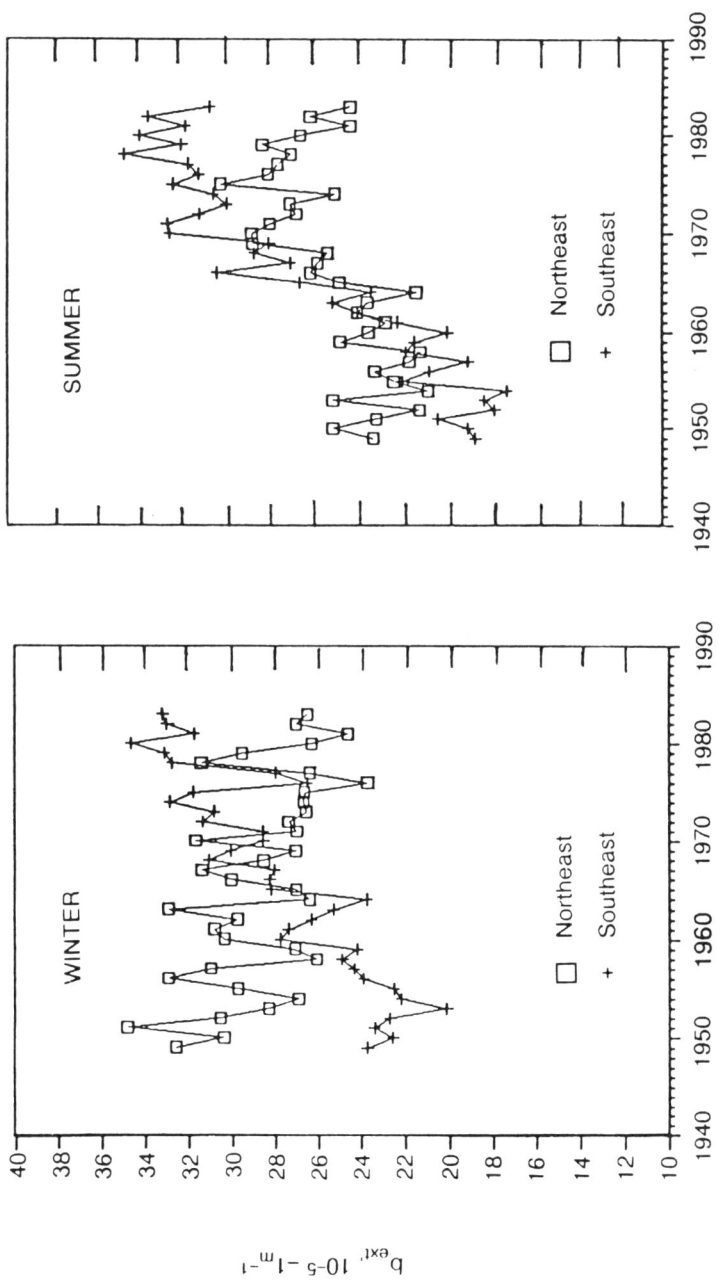

FIGURE 2-13. Trends for 75th percentile extinction coefficient for Northeast and Southeast United States, winter and summer. Source: Trijonis et al., 1990.

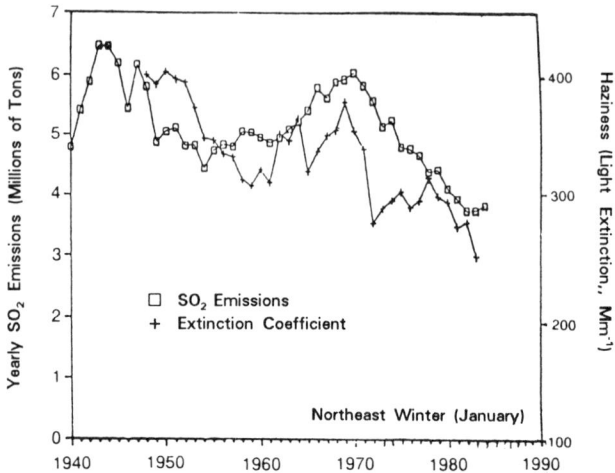

FIGURE 2-14a Comparison of SO_2 emission trends (□) and extinction coefficient (+) for the northeast United States during winter months. Source: R.B. Husar, pers. comm., Washington University, St. Louis, Mo., 1989 (From Trijonis et al., 1990; Husar and Wilson, 1993).

SOURCES OF VISIBILITY-IMPAIRING MATERIALS

Natural Sources

Trijonis (1982a,b) and Trijonis et al. (1990) have evaluated natural background visibility (i.e., visibility in the absence of anthropogenic pollution) and the types of materials present in the unpolluted atmosphere for the eastern and western United States. The evaluation incorporates data on the concentrations of the major components of airborne particles: SO_4^{2-}, organic matter, elemental carbon, nitrates, soil dust, and water. The estimates of natural aerosol concentrations are developed from three types of information:

• Compilations of data on emissions from natural and human sources;

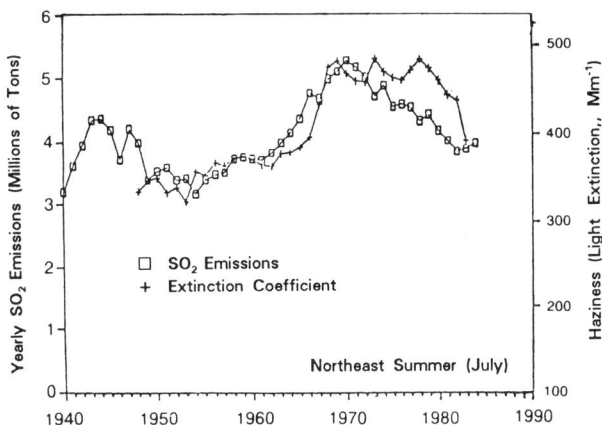

FIGURE 2-14b Comparison of SO_2 emission trends ($\square$) and extinction coefficient (+) for the northeast United States during summer months.
Source: R.B. Husar, pers. comm., Washington University, St. Louis, Mo., 1989 (From Trijonis et al., 1990; Husar and Wilson, 1993).

- Ambient measurements of aerosol species in remote areas (especially in the southern hemisphere);
- Regression studies of visibility against the concentrations of trace elements associated with anthropogenic and natural emissions.

The estimates exclude factors such as precipitation, fog, blowing snow, and sea spray. Because sea spray and fog are excluded, the results might not be representative of natural visibility in coastal areas, rather the natural background estimates are intended to be annual and spatial averages that exclude those areas.

For the East, the analysis yields an average natural background visual range of 150 $\pm$ 45 km. This is equivalent to a light extinction level of 26 $\pm$ 7 Mm^{-1}, or a little more than twice the level caused by Rayleigh (clear-air, "blue sky") scattering of light by air molecules. For the arid West, the average background visual range is estimated as 230 $\pm$ 35 km, a value equivalent to a light extinction level of 17 $\pm$ 2½ Mm^{-1}, which is about 1.5 times the level caused by Rayleigh scattering alone. (The error bounds are estimates of uncertainties in the spatial and annual averages.)

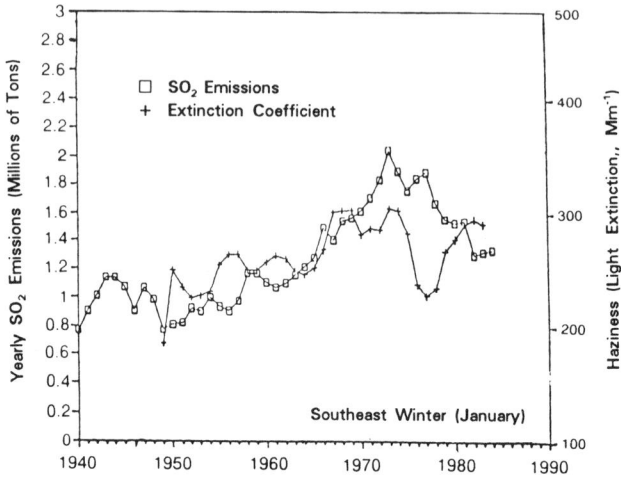

FIGURE 2-15a Comparison of SO_2 emission trends (□) and extinction coefficient (+) for the southeast United States during winter months. Source: R.B. Husar, pers. comm., Washington University, St. Louis, Mo., 1989 (From Trijonis et al., 1990).

Trijonis et al. (1990) estimated that the major contributors to natural extinction levels in the East are Rayleigh scattering (46%), organics (22%), water (19%), and suspended dust, including coarse particles (9%). The major contributors in the West are Rayleigh scattering (64%), suspended dust (14%), organics (11%), and water (7%). The contribution of water is due to the hygroscopic components in airborne particles. Water is the most uncertain contributor, and it might be overestimated, given the very small amounts of SO_4^{2-} and nitrates in the natural background particles (Trijonis et al., 1990). A significant fraction of the water in the natural background particles might be associated with organics because natural concentrations of organics are greater by a factor of 10 than those of SO_4^{2-} or nitrate.

Based on the above assessment, the most important sources of natural visibility-reducing particles are sources that emit organic materials and dust particles. Natural organic particles are produced as primary emissions (e.g., wildfire smoke, plant waxes, and pollen) and as a result of conversion from volatile organic compound (VOC) emissions (e.g., terpenes and other hydrocarbons). Natural mineral dust comes from the action of wind on soils.

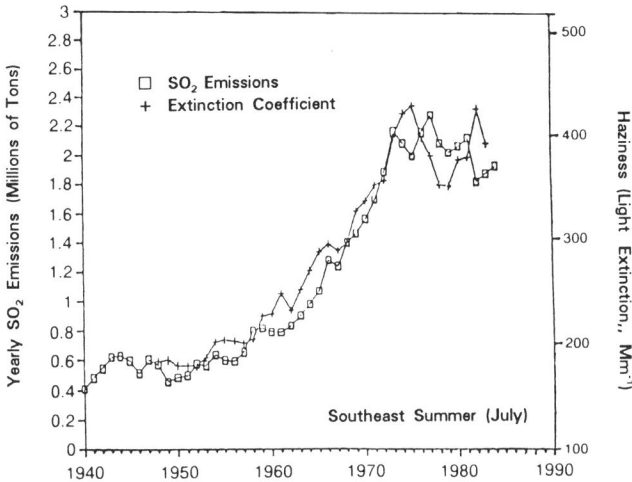

FIGURE 2-15b Comparison of SO_2 emission trends ($\square$) and extinction coefficient ($+$) for the southeast United States during summer months.
Source: R.B. Husar, pers. comm., Washington University, St. Louis, Mo., 1989 (From Trijonis et al., 1990).

On an annual average basis, the concentrations of natural particles are generally small compared with concentrations of anthropogenic particles. However, two natural particle sources—wildfires and windblown dust—are extremely episodic, and they can be the dominant cause of visibility reduction at certain times. As illustrated in Figure 2-16, most of the wildfire activity (in terms of acres) occurs in the Rocky Mountain states, the Pacific Coast, and the Southeast. The area most affected by intense dust storms centers around the Texas panhandle (see Figure 2-17). It should be noted that, for both of the above sources, the distinction between natural and anthropogenic influences is somewhat blurred.

Anthropogenic Sources

Fine particles are the primary cause of anthropogenic haze. Coarse particles (predominantly soil dust) and gaseous nitrogen dioxide (NO_2) can also play a significant role. Anthropogenic dry fine aerosol consists almost entirely of just five pollutants: sulfates, organics, elemental

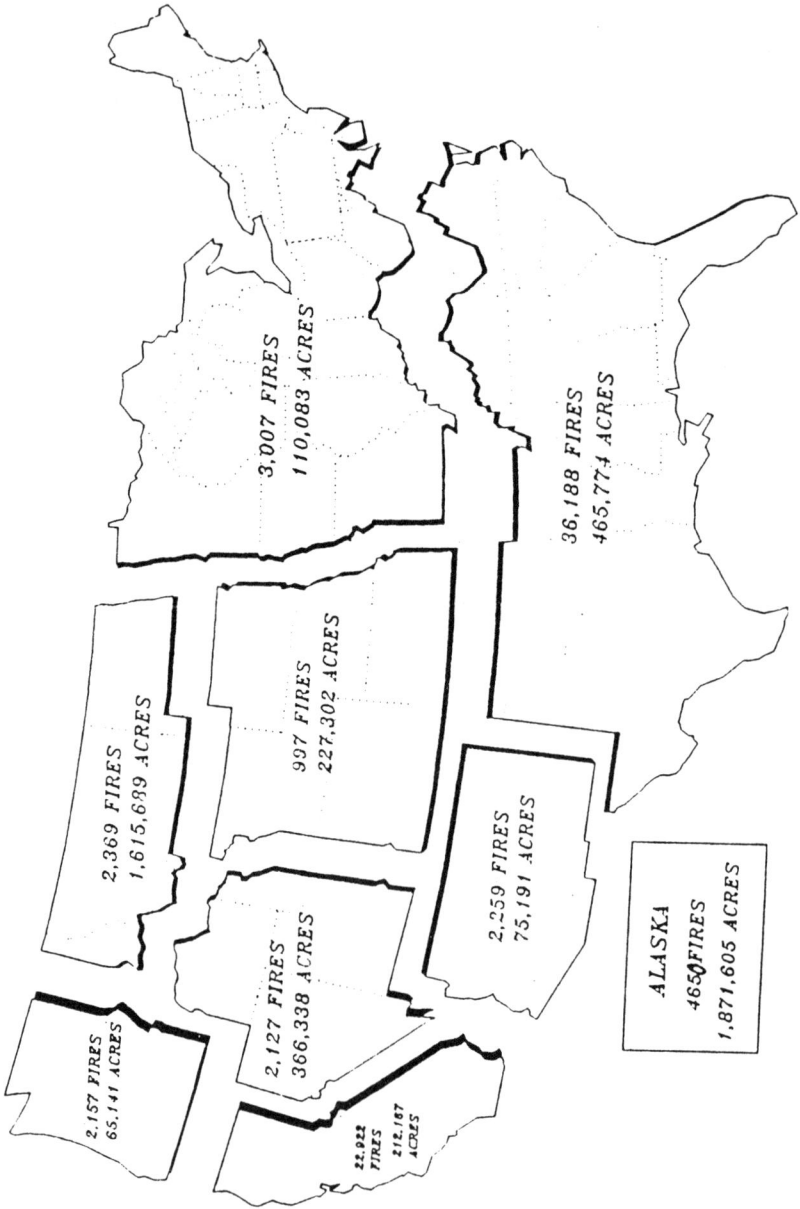

FIGURE 2-16 Frequency and extent of wildfires in the United States in 1988. Source: USDA, 1989.

FIGURE 2-17 Annual percentage frequency of dusty hours based on hourly observations from 343 weather observation stations that recorded dust, blowing dust, and sand when prevailing visibility was less than 7 miles (11 km). Shaded areas represent no observations of dust. Period covered is approximately 1940 to 1970. Source: Orgill and Sehmel, 1976.

carbon, soil dust, and nitrates. Consequently, from the standpoint of visibility, the most important anthropogenic emissions are SO_2 (precursor of SO_4^{2-} particles), primary organic particles, gaseous VOCs (precursor of secondary organic particles), primary elemental carbon particles, soil-derived material, ammonia (NH_3) (a precursor of ammonium nitrate), and nitrogen oxides (NO_x, which are precursors of nitrate aerosols and NO_2). Figure 2-18 illustrates the important anthropogenic sources of each emission in the United States. For some emissions, one source category stands out:

- SO_x: Electric utilities contribute about 70% of the total.
- NH_3: Livestock waste management operations are the dominant source.
- Elemental carbon: Diesel-fueled mobile sources account for about half of the emissions.
- Suspended soil dust: Vehicular traffic is presumably the predominant anthropogenic source.

For NO_x, three categories stand out: electric utilities, gasoline-fueled vehicles, and diesel-fueled mobile sources. VOCs and primary organic particles are emitted in significant quantities from a wide variety of sources; the most important are gasoline-fueled vehicles, residential wood burning, petroleum and chemical industrial sources, solvent evaporation, and burning for forest management.

SUMMARY

Visibility impairment episodes range in scale from local plumes to widespread regional haze. The most intense regional haze in the United States occurs in the East, where the median standard visual range (calculated from airport data) is generally less than 30 km. The best visibility in the country is found in the arid, mountainous Southwest, where median standard visual range exceeds 150 km. In the adjoining regions to the north and south, median standard visual range is also good, exceeding 100 km. Median visual range falls to less than 50 km along the northern Pacific coast, less than 30 km in the central valley of California, and less than 15 km in the Los Angeles basin. Visibility generally improves with altitude.

55

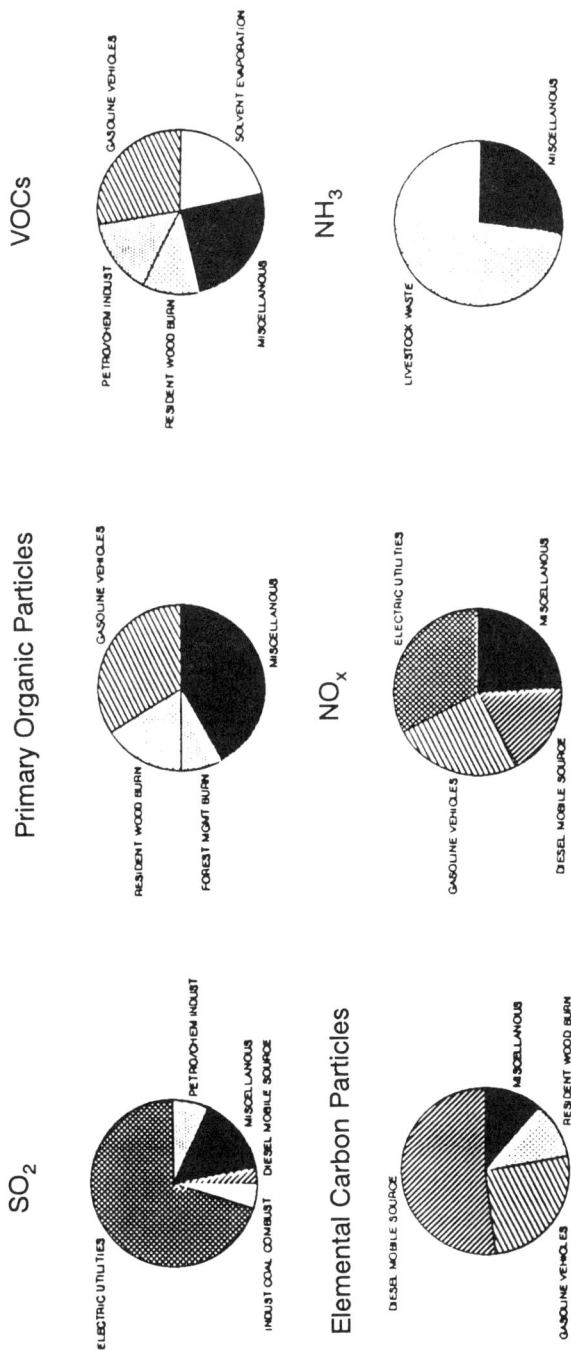

FIGURE 2-18 Anthropogenic inventory of visibility-related emissions for the United States. Based on the 1985 NAPAP inventory (Zimmerman et al., 1988a; Placet et al., 1990), except copper smelter emissions, which have been updated for 1988 (Trexler, pers. comm., DOE, Washington, D.C., 1990). A pie chart is not shown for suspended dust; vehicular traffic on paved and unpaved roads is assumed to be the predominate source.

National Park Service data show that sulfate particle concentrations are about six times greater in the rural East than in the rural West; elemental carbon and organic particle concentrations are about twice as great in the rural East as in the rural West; and concentrations of fine soil dust and nitrates are about the same in both regions. In the East, airport data and NPS camera data reveal a strong summertime visibility minimum that is strongly linked to a summertime maximum in sulfate concentrations.

Airport data from the late 1940s through the early 1980s show coherent visibility trends over large regions. In the Northeast, visibility has moderately improved during the winter and moderately decreased during the summer. Visibility in the Southeast has worsened moderately during the winter and substantially during the summer. The greatest increases occurred during the 1950s and 1960s. It is likely that changes in SO_2 emissions are largely responsible for these changes in visibility.

The average natural background visual range varies from about 150 km in the eastern United States to 230 km in the arid West. The major contributors to natural extinction are Rayleigh scattering, organics, water, and suspended dust. The main constituents of anthropogenic haze are sulfates, organics, elemental carbon, soil dust, nitrates, and water. The principal anthropogenic haze-causing emissions are sulfur dioxide (SO_2, a precursor of sulfate particles), primary organic particles, gaseous volatile organic compounds (VOCs, precursors of secondary organic particles), primary elemental carbon, primary crustal material, ammonia (NH_3, a precursor of ammonium nitrate), and nitrogen oxides (NO_x, precursors of nitrate particles and NO_2). Except for VOCs and primary organics, the anthropogenic emission inventories for these species are dominated by relatively few source categories.

3

Legal and Institutional Context

To be useful to policy makers, an evaluation of methods for determining the visibility effects of sources and for assessing alternative control measures must take account of the legal and institutional framework within which the methods will be implemented. This chapter describes that framework, as well as the history of the U.S. Environmental Protection Agency (EPA) and state efforts to use the Clean Air Act to protect and enhance visibility in national parks and wilderness areas.

The Clean Air Act Amendments of 1990 are intended to lead to an augmented program of visibility protection; this chapter describes various approaches such a program could take. Using this chapter in conjunction with Chapter 5, decision makers will be able to judge the compatibility of these techniques with one or more approaches.

PRESENT VISIBILITY PROTECTION PROGRAMS

The Clean Air Act

Relevant Provisions

The Clean Air Act (42 U.S.C. §7401-7671q) authorizes the EPA to establish National Ambient Air Quality Standards (NAAQS) at levels that protect public health with an adequate margin of safety (primary standards) and protect the public welfare from known or anticipated adverse effects (secondary standards). NAAQS currently exist for carbon monoxide, lead, ozone, nitrogen dioxide, particulate matter with a diameter of 10 μm or less (known as PM_{10}) and sulfur oxides (measured as sulfur dioxide).

The act establishes an array of programs intended to ensure that the standards are attained and maintained and that air quality superior to the standards is protected.

Each state must submit to EPA a state implementation plan (SIP) that demonstrates that primary NAAQS will be attained by the statutory deadlines and that secondary NAAQS will be attained as expeditiously as practicable. If a state does not do so, then EPA must prepare a federal implementation plan (FIP) for the state.

SIPs usually concentrate on existing sources in or near nonattainment areas—areas that violate one or more NAAQS. For many types of new sources, SIPs are supplemented by federal regulations that apply to clean air as well as nonattainment areas. For instance, EPA may promulgate new source performance standards for categories of stationary sources that contribute to air pollution. These standards require the use of the best demonstrated technology by new and modified sources in the category. Similarly, the act establishes emissions standards for new mobile sources and standards for fuel content; the act also authorizes EPA to establish additional regulations on those subjects.

Two programs under the act are concerned specifically with visibility in national parks and wilderness areas. One of these, the prevention of significant deterioration (PSD) program (§160-169), is directed primarily at new sources; the other, the visibility program (§169A-B), is aimed primarily at existing sources.

The PSD program requires that each applicant for a new or modified major emitting facility seeking to locate in a clean-air area (an area in which the NAAQS are met for one or more air pollutants) show that the facility will use the best available control technology (BACT) to minimize additional air pollution. (Every area meets the NAAQS for at least one pollutant; therefore, the PSD program applies nationwide.) BACT is defined as the maximum achievable degree of emission reduction, taking into account energy, environmental, and economic effects and other costs. It is determined on a case-by-case basis, but must be at least as stringent as any new source performance standard that applies to the facility's category.

The applicant must also demonstrate that the proposed new or expanded source will comply with air-quality "increments" that limit increases in air pollution. These increments specify the maximum permissible cumulative increases of several pollutants over the pre-existing baseline concentrations in each clean-air area. Increments currently apply to nitrogen dioxide, sulfur dioxide, and total suspended particles. (EPA is converting the increments for particulate matter so that, as with the ambient standards, only PM_{10} will be covered (EPA, 1989a)). The sizes of the increments vary with the area's classification; Class I increments are the smallest and Class III the largest. This system discourages sources from locating in areas that have restrictive increments (i.e. Class I areas) or that have almost used up their increments; it also encourages permit-writers to be strict in prescribing control technology to allow as much development as possible within the increments.[1]

The act classifies most clean air areas in the moderate-growth Class II category and gives states and Indian tribes the authority to redesignate these areas as desired. Few areas have been redesignated as Class I, and none as Class III.

Some parklands receive special protection: Large national parks and wilderness areas in existence when the PSD program was codified in 1977 are designated by Section 162(a) as mandatory Class I areas that may not be reclassified. This designation also covers additions to those areas made after 1977. Most of the 158 mandatory Class I areas are west of the Mississippi; nearly one-quarter of them are located in Utah, Arizona, Colorado, and New Mexico (Oren, 1989). Section 164(a) places other parklands in the intermediate category of "Class II floor areas"; these are initially classified Class II and can be redesignated Class I, but not Class III. Since no Class II area has been reclassified Class III, Section 164(a) has thus far been of little importance.

The PSD program includes a special air quality related values

---

[1]The above discussion focuses on the requirements that PSD imposes on new and modified major emitting facilities. The program can also potentially affect other sources. Once the first application for a PSD permit is received in an area, growth in non-major sources begins to count against the increment. If an increment violation occurs, the state must correct it. This could involve imposing controls on existing sources as well as on growth in sources not subject to the PSD program. There is, though, no known case to date in which an increment violation has occurred due to the interaction of growth of existing sources with permits to new and modified sources (see Oren, 1989). Thus minor and existing sources have been largely exempt.

(AQRV) test for evaluating a major emitting facility that might affect a Class I area. Section 165(d) charges the federal land manager (either the secretary of the interior, for lands managed through the National Park Service (NPS) or the Fish and Wildlife Service, or the secretary of agriculture for lands managed through the U.S. Forest Service (USFS)) with the responsibility to protect the AQRV (including visibility) of Class I areas. If the land manager demonstrates to the state that an area's AQRV would be affected adversely by a proposed new or modified major emitting facility within that state, the facility must be denied a permit; this is true even if the facility would not violate the Class I increments. On the other hand, a facility that would violate the increments can obtain a permit if the applicant shows that the source would not adversely affect an area's AQRV. In this way, the Class I increments amount to a device to assign the burden of proof on whether a proposed source should be allowed.

If the applicant shows that AQRVs in Class I areas will not be adversely affected, the facility is subject to a relaxed set of increments. If the applicant does not, the facility can be built only if it passes through a lengthy variance process that ultimately can require presidential approval for the permit.

The second program concerned with park visibility is established by Section 169A, which declares a national goal of preventing and remedying visibility impairment in mandatory Class I areas caused by manmade air pollution. Section 169A(a)(2) requires that EPA, in conjunction with the Department of the Interior, list mandatory Class I areas in which visibility is an important value. States that either include such areas or that contain sources that might contribute to visibility impairment in these areas are required by Section 169A(b)(2)(B) to include in their SIPs a long-term strategy for making reasonable progress toward the visibility goal. In addition, Section 169(b)(2)(A) requires states to revise their SIPs to require installation of the best available retrofit technology (BART) on any major stationary source placed in operation after August 7, 1962, if that source "emits any air pollutant which may reasonably be anticipated to cause or contribute to any impairment of visibility" in a listed mandatory Class I area. Under Section 169A(g)(2), BART is determined by taking into account, among other factors, the source's characteristics, such as the technological and economic feasibility of control, and "the degree of improvement in visibility which may reasonably be anticipated to result" from control.

The Clean Air Act Amendments of 1990 supplement the visibility protection program by enacting a new Section 169B. Section 169B(a) requires EPA, together with NPS and other appropriate federal agencies, to conduct research on visibility impairment. The research must include expansion of visibility-related monitoring, assessment of current sources of visibility-impairing pollution and clean-air corridors (areas from which unpolluted air flows to Class I areas), adaptation of regional air-quality models for the assessment of visibility, and studies of the atmospheric chemistry and physics of visibility. Interim findings are required by November 1993. Section 169B(b) also requires EPA to conduct by November 1992 an assessment of the improvements in visibility resulting from other provisions of the 1990 amendments (e.g., the acid precipitation control provisions). An assessment of visibility in Class I areas is required every 5 years thereafter.

Section 169B(c) authorizes EPA to establish a "visibility transport region" whenever EPA believes that interstate atmospheric transport of air pollution contributes significantly to visibility impairment in a Class I area. EPA must establish for each region a "transport commission" composed of the governors of significantly contributing and affected states and ex-officio representatives of EPA and federal land managers. Section 169B(f) requires a transport commission to be established by November 1991 for the region affecting visibility in the Grand Canyon National Park.

Under Section 169B(d), a transport commission must, within 4 years of its establishment, assess visibility effects from present and projected emissions in the region and recommend corrective action. The report from the commission must consider whether 1) clean-air corridors should be established in which emission limits would be especially stringent, 2) new sources in clean-air corridors should be held to the same requirements as new sources in nonattainment areas, and 3) EPA should promulgate regulations under Section 169A to control regional haze.

Section 169B(e) requires EPA, within 18 months after receiving a transport commission's report, to carry out its regulatory responsibilities under Section 169A, including the issuance of criteria for measuring reasonable progress toward the national goal of ending anthropogenic visibility impairment in mandatory Class I areas. It is not clear whether EPA may act before receiving the report. States would be given 1 year to revise their SIPs to carry out EPA's regulations. Under the section's timetable, this would occur by May 1998.

Visibility in mandatory Class I areas also will be affected by the new acid precipitation control program created by the 1990 Amendments. This program is intended to reduce sulfur dioxide emissions permanently by about 10 million tons per year below 1980 levels and to control nitrogen oxide emissions. By reducing the precursors to sulfates, nitrates, and nitrogen dioxide—all causes of visibility impairment—the new program is expected to improve visibility in the eastern and midwestern United States, where the bulk of the emission reductions are anticipated to occur. Projections for EPA by ICF, Inc. indicate that Title IV will halve, but not prevent, expected growth in sulfur dioxide emissions from electric utilities in 11 western states between now and 2010 (ICF, 1991). ICF currently projects an increase of as much as 180,000 tons in annual emissions, a 33% increase over 1985 levels, in these states. This would partly negate the reductions in sulfur dioxide emissions in the West that have been brought about through the control of nonferrous smelters.

The act also authorizes a program of research on air-pollution-related issues. The amount of funding actually available, however, depends on the annual appropriations. In addition, the federal land management agencies have supported research on visibility issues pursuant to their own statutory missions.

Basic Principles

The complex Clean Air Act provisions summarized above have several common themes. The most important is that preventive regulatory action is allowed despite scientific uncertainty. For instance, EPA is not required to prove that a substance poses a risk of harm before setting an ambient standard for that substance. This has been explicit since the enactment of Section 401 of the Clean Air Act Amendments of 1977, which altered the act's regulatory authorities to provide that EPA may regulate a substance or its emitters if, "in [the administrator's] judgment," the substance "may reasonably be anticipated" to endanger public health or welfare. The House Committee that developed this language explained that it intended to emphasize the preventive or precautionary nature of the act and to authorize the administrator to weigh risks and make reasonable projections of future trends (U.S. Congress, House of Representatives, 1977). This language was intended to adopt the interpretation of the act by the D.C. Court of Appeals in *Ethyl Corp. v. EPA* (541 F.2d 1 (D.C. Cir.) (*en banc*), cert. denied 426 U.S. 941 (1976)), which held that EPA need not prove airborne lead to be a hazard to restrict lead content in gasoline.

The visibility protection program outlined in Section 169A appears to be based upon this philosophy. Section 169A(b)(2) prescribes that BART be required for certain sources that "emit any air pollutant which may reasonably be anticipated to cause or contribute to any impairment of visibility" in a listed mandatory Class I area. By using the words "may reasonably be anticipated" in Section 169A, Congress apparently intended that the philosophy of precautionary action should apply to visibility protection as it does in other areas.[2] Similarly, Section 169A(c)(1) puts the burden of proof on a source to show that its effect on visibility within a Class I area is not significant, either by itself or in combination with other sources. In this way, the risks of scientific ignorance are borne by the source rather than by the public.

The PSD program also is based on a precautionary approach. In advocating codification of the program in 1977, the House and Senate committees argued that clean-air areas need to be protected because air-quality concentrations within the ambient standards might later be found to endanger health or welfare. The committees stated that the PSD program would minimize this danger by controlling the growth of pollutant concentrations (U.S. Congress, House of Representatives, 1977; U.S. Congress, Senate, 1977).

The primary means of control under the PSD program—BACT and increments—therefore apply regardless of whether a particular source can be shown to endanger health or welfare. A showing of risk is necessary only when a land manager wishes to prevent the issuance of a permit to a source that complies with BACT and the increments.

Thus EPA may base regulation not only on what present evidence shows, but also on a reasonable evaluation of what future evidence is

---

[2]Unlike other parts of the Clean Air Act, Section 169A does not include explicit reference to the judgment of the administrator. That reference had been included in the other authorities to make plain that EPA need not make a factual finding that harm could reasonably be anticipated but instead could rely on judgment (U.S. Congress, House of Representatives, 1977). The exclusion of this language from Section 169A might indicate less commitment to precautionary action in visibility than in other areas.

likely to show about the effects of current emissions and the agency's policy views about the relative risks of overprotection and underprotection. Similarly, the agency may use emerging techniques so long as it takes reasonable account of the possible flaws in those techniques. This principle is important to the committee's evaluation of source attribution methods because EPA is not required to show with certainty the suitability of a given method. The committee, therefore, has not required such certainty in its evaluation of source attribution methods.

Eclecticism is another important feature of the act. To a large extent, the act links the stringency of regulation to the degree of environmental risk that a source poses. The most familiar example of this principle is the SIP process, in which a source located in a high-pollution area may be treated differently from an identical source located in a low-pollution area. Similarly, the BART requirement of Section 169A differentiates among sources based on their contribution to visibility degradation in a mandatory Class I area; it does so by requiring that the BART determination incorporate consideration of "the degree of improvement in visibility which may reasonably be anticipated to result from the use of [the proposed] technology." This language emphasizes the importance of techniques to assess the effect on visibility from a given control measure.

The sponsors of the act recognized, however, the shortcomings of basing a regulatory system entirely on environmental risk. Such a system can result in underregulation when information on risk is lacking; moreover, it can give areas that meet risk-based standards a substantial and arguably undesirable advantage in attracting new sources. The act therefore requires a technology-based minimum of control on new sources, and, to a lesser extent, on existing sources. This differentiation between new and existing sources reflects the relative ease of designing controls into a new source and the relatively long life expectancy of new sources, which make it desirable to use technology-based controls as a means of ensuring against gaps in knowledge about the risks a particular new source may pose.

The act's diversity extends as well to the means of implementing environmental standards. For the most part, the act relies on a traditional regulatory approach under which state and federal regulators prescribe emission limits for sources. The 1990 amendments, however, incorporate market and other economic incentive programs to an unprecedented degree and encourage further experimentation with these devices.

The committee has been conscious that a given technique for apportioning visibility impairment among emission sources might be used in a broad array of programs. It has attempted, therefore, to provide information that would assist policy makers in assessing the compatibility of a technique with alternative program designs.

Implementation of the
Visibility Protection Programs

As the foregoing discussion shows, the Clean Air Act offers several tools for protecting visibility in national parks and wilderness areas. For instance, EPA could proceed by establishing national secondary air-quality standards at levels sufficient to protect visibility. This is legally possible under Section 302's broad definition of welfare (protected by secondary standards), which includes effects on visibility. Nonetheless, EPA has not found this an attractive route. Although the agency in the 1970s set a secondary standard for sulfur dioxide that was intended in part to protect visibility, the standard was judicially remanded to EPA for further explanation; EPA abandoned the effort due to the lack of quantitative information relating ambient sulfur dioxide concentrations to visibility effects (EPA, 1973).

Similarly, EPA has taken only the first steps to protect visibility by moving toward establishing a secondary standard to cover particulate matter smaller than 2.5 μm in diameter (EPA, 1987a). The nationally uniform nature of the NAAQS may explain the agency's apparent reluctance: a particulate standard sufficient to protect visibility in the "Golden Circle" of parks in the Southwest would require a reduction of pollution concentrations below natural background levels (those that exist in the absence of pollution) in the East (EPA, 1985a).[3] However, the NAAQS have made an indirect contribution to visibility improvement and maintenance, because the pollutants that impair visibility are subject to primary and secondary NAAQS established to protect health and other

---

[3]On the other hand, as discussed later in the chapter, an environmental quality standard approach might have several advantages, especially if it led to PSD increments for small particles that are most responsible for visibility impairment.

forms of welfare; these standards have led to reduced emissions of pollutants when the emissions interfere with attainment and maintenance of an ambient standard. For instance, enforcement in the West of the NAAQS for sulfur dioxide, principally through requiring control of nonferrous smelters, brought about a reduction of three million tons in annual emissions of this pollutant, which is a precursor to visibility-reducing sulfates.

The PSD program has had a greater effect. The program has benefitted visibility by reducing growth in atmospheric loadings of air pollutants that contribute to regional haze. According to a series of analyses performed for EPA in the early- to mid-1980s by TRW and Radian, Inc., the program has resulted in lower emissions from new sources in clean-air areas than would otherwise have been the case (EPA, 1982, 1985b, 1986a; Oren, 1988). Those studies found that the required case-by-case BACT determinations appears to be resulting in control technology requirements for new sources that are more stringent than EPA's categorywide new-source performance standards or, for sources in categories not subject to those standards, new-source requirements in SIPs.

The effects of BACT might have increased since then. When the analyses were done, EPA's guidance for BACT allowed the permit applicant to formulate the BACT alternatives and placed on the permitting authority the burden of showing the achievability of more stringent control options; EPA since has advised that BACT should be set at the most stringent limit achieved by a similar source, unless the applicant can show that level to be unachievable. (The agency is considering whether to formally adopt or abandon this policy, known as the "top-down" approach.) EPA has not updated the data base for the PSD program permits since 1984, and so the extent of the PSD program's effects is not clear.

There is also some evidence that the existence of the increments encourages permit-writers to be more stringent in setting control technology requirements for new sources that might otherwise be the case. This occurs because the increments limit the margin of clean air available for development. There is also anecdotal evidence that the restrictive Class I increments may be discouraging prospective sources from locating near national parks (U.S. Congress, House of Representatives, 1981).

But the increments are of limited effectiveness. First, the restrictive

Class I increments apply only to large federal parks created before 1977; many other scenic areas are Class II and receive no special protection. It is not clear even that the Class I increments ensure effective protection against new sources that might cause visibility impairment (NRC, 1981). One reason is that the increments do not distinguish between particles in the 0.1-1.0 μm range—which have the greatest potential to degrade visibility—and larger particles. In addition, the increments, like the ambient standards, focus on the concentration of pollution at a given time and place; but visibility impairment depends on the total magnitude of fine particulate matter between an object and an observer (Sloane and White, 1986).

Moreover, the AQRV mechanism for individual assessments of sources that might impair visibility in parks has been little-used, although the Department of the Interior (DOI) has promulgated criteria for such assessments (DOI, 1982). Only recently did a federal land manager attempt to persuade a state to deny a permit to a proposed source that complied with the Class I increments (DOI, 1990, 1992). As a result, the Class I increments, despite their shortcomings, have been used as the exclusive criterion for protecting parks from new sources.

A recent study by the U.S. General Accounting Office found that the present PSD program is not satisfactorily protecting the parks. The report points out that sources accounting for up to 90% of pollutants emitted near five Class I areas are exempt from PSD requirements. The reason is that PSD applies only to the construction and modification of major emitting facilities; in effect, baseline concentrations of pollutants are ignored. As a result, existing sources and small new sources are ignored by the program. Even for sources covered by the program, the GAO found deficiencies. Its report concludes that the AQRV process for review of proposed permits by the federal land managers has not been fully implemented. The GAO found that permit applications are not always forwarded to the land managers for review, and, even when the applications are forwarded, the land managers do not always review them and provide comments to permitting agencies. Moreover, according to the GAO, land managers do not feel they have sufficient information on the AQRVs in their areas to evaluate proposed permits (GAO, 1990).

The visibility protection program of Section 169A has been especially slow to take hold. In 1979, EPA promulgated a list of mandatory Class

I areas in which visibility is an important value; this list includes all but 2 of the 158 mandatory Class I areas (EPA, 1979). The following year, a court order caused EPA to establish visibility protection rules that set requirements for the SIPs of the 36 states that contain listed Class I areas (EPA, 1980a).[4] These rules take a phased approach to implementing Section 169A that concentrates on visible plumes—impairment attributed to smoke, dust, or layered haze, sometimes known as plume blight—rather than on regional haze. Under the rules, states (or EPA, when a state defaults) must require BART to remedy impairment in mandatory Class I areas that can be traced to a single existing stationary facility or small group of existing stationary facilities. The BART requirement, however, covers only sources whose contribution to visibility impairment is "reasonably attributable" through "visual observation or any other technique the State deems appropriate." The regulator, therefore, need not use modeling or any other sophisticated analytical method. As a result, regional haze has largely been ignored.

The rules also require the 36 states containing listed Class I areas to revise their SIPs to include

• A long-term strategy to make reasonable progress towards the national visibility goal of eliminating present and preventing future anthropogenic visibility impairment in mandatory Class I areas. The strategy must include steps to protect any "integral vista"—a view looking out from the listed mandatory Class I area that is important to an observer's visual experience of the area—identified by the federal land manager for the area. (Federal land managers were given until the end of 1985 to designate such vistas.)

• Procedures to ensure consideration of visibility effects within listed Class I areas from proposed new and modified major stationary sources. These rules implement the PSD program's requirement that a prospective source's effect on visibility in Class I areas be considered. In addi-

---

[4]EPA has the legal authority to extend the program to states without listed areas, but which contain emission sources that may affect visibility in such areas. However, EPA has not done so. This is a corollary of the agency's decision, as described in the text, to focus largely on plume blight caused by sources near parks.

tion, the procedures apply to proposed sources in nonattainment areas not subject to the PSD program and proposed sources that might affect integral vistas. In both instances, a state is free to balance the visibility effects of the proposed source against the energy and economic effects of denying a permit to construct.

• Development of state visibility-monitoring programs to assess impairment, strategy effectiveness, and trends.

• State and federal land manager coordination regarding visibility programs and PSD program permit review activities.

EPA has not initiated a second phase of rulemaking to focus on regional haze, and for procedural reasons, the courts have declined to order EPA to do so (see *Maine v. Thomas*, 874 F.2d 883 (1st Cir. 1989), affirming 690 F. Supp. 1106 (D.Me 1988)). One reason for EPA's failure to act is dispute over appropriate methods of linking individual sources or groups of sources to visibility impairment within Class I areas.

The implementation even of the limited program promulgated by EPA has been slow. Neither the Department of Agriculture nor DOI designated integral vistas for protection by the regulatory deadline. Moreover, the change in administrations shortly after promulgation of EPA's rules in December 1980 created uncertainty about the direction of the program and the schedule for SIP submittals. Many states delayed action on their SIPs until EPA in April 1984 announced an implementation schedule pursuant to settlement of a law suit brought by the Environmental Defense Fund (EPA, 1984). Even now, only ten states (Georgia, North Carolina, Arkansas, Louisiana, Colorado, North Dakota, Utah, Wyoming, Alaska, and Oregon) have revised their SIPs to meet the requirements of the 1980 rules. EPA has promulgated FIPs incorporating the 1980 rules for an additional 16 states, and the remaining 10 states have a combination of SIPs and FIPs (D. Scott and D. Stonefield, pers. comm., Office of Air Quality Planning and Standards, U.S. EPA, Aug. 5, 1991).

Those plans have had very little regulatory impact. In promulgating its visibility rules, EPA projected that "few, if any, existing stationary sources will have to retrofit controls"; this has proven to be accurate (EPA, 1980b). Typically, the control strategies for ensuring reasonable progress rely on emission reductions being achieved through other regulatory programs and do not include any new programs specifically aimed at reducing visibility impairment. Except for the recently announced agreement to install scrubbers at the Navajo Generating Station near the Grand Canyon National Park, BART has not affected any source.

The most recent comprehensive examination by federal regulatory agencies of the options for regulating regional haze is the 1985 report of the Interagency Visibility Task Force. The task force consisted of representatives from EPA, the Department of Agriculture, and DOI and was formed to develop long-term strategies for remedying regional haze and to recommend a long-range program to mitigate it. The task force recommended that research funding be enhanced and that visibility effects be taken into account in designing regulatory programs for other purposes; however, it did not propose immediate steps to control regional haze (EPA, 1985a).

Funding Issues

Federal funding for visibility research and regulation has been inconsistent. EPA funding for visibility-related programs, including research, dropped markedly after 1980 (Figure 3-1). However, the NPS Air Quality Division more than tripled its funding for visibility research between 1981 and 1987; since then, NPS funding largely has followed EPA funding (Table 3-1). In addition, the funds designated by the USFS and the Department of Defense for visibility research were halved from the mid-1980s through fiscal year 1990.

Such inconsistent funding has hampered fundamental research. However, compensating factors should be noted. First, research funding in the past decade aimed at understanding environmental issues such as global climate change and acid rain indirectly contributed to advances in visibility research, as have other efforts focused on airborne particle and associated gas-phase processes and radiative transfer. Funding by the National Aeronautics and Space Administration (NASA), the National Oceanic and Atmospheric Administration (NOAA), and the National Science Foundation (NSF) has also supported such research. Second, the private sector, such as the Electric Power Research Institute and the American Petroleum Institute, has contributed substantial resources to a variety of visibility and related studies. Nonetheless, the overall low level of support is a matter of concern.

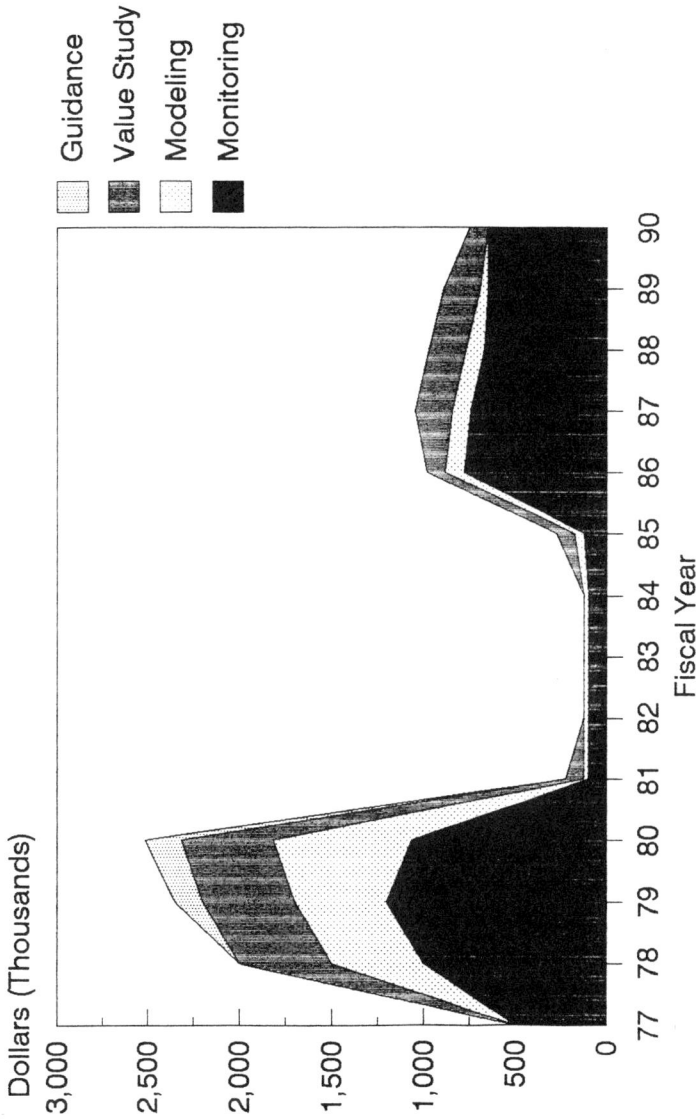

FIGURE 3-1. EPA funding for visibility research, including OPPE (Office of Policy Planning and Evaluation) and IMPROVE (Interagency Monitoring of Protected Visual Environments) funding. Does not include regional modeling or minor work on chemistry and physics. Source: pers. comm., EPA, Office of Air Quality Planning and Standards, Ambient Standards Branch, 1992.

TABLE 3-1　Visibility Research Funding Summary, 1971-1991

Year	Dispersion/ Visibility Modeling	Visibility Monitoring	Visibility Effects Research	Total
1979	$ 150,000	$ 416,000	$ 0	$ 566,000
1980	55,000	250,000	170,000	475,000
1981	64,200	492,537	100,000	656,737
1982	192,300	706,000	460,000	1,358,300
1983	150,000	1,130,000	335,000	1,615,000
1984	218,000	1,249,900	205,000	1,672,900
1985	221,000	1,128,000	387,000	1,736,000
1986	292,000	1,320,664	325,000	1,937,664
1987	272,000	1,915,700	500,000	2,687,700
1988	40,000	2,370,800	85,000	2,495,800
1989	50,000	2,351,000	50,000	2,451,000
1990	30,000	2,285,340	30,000	2,345,340
1991	100,000	2,173,500	230,500	2,504,000
Total	1,834,500	17,789,441	2,877,500	22,501,441

Figures not adjusted for inflation.
Source: M. Scruggs, pers. comm., NPS, Denver, Colo., 1992.

In response to the drastic cut in EPA's support of visibility-related research in the early 1980s, NPS launched its own effort. The shift in visibility research funding in the early 1980s from EPA to NPS caused inevitable discontinuities in progress. In some cases, NPS failed to take advantage of scientific expertise that had been accumulated over previous decades. The resulting costs, such as lapses in data collection, have been high (see Chapter 4). Moreover, by focusing its monitoring in national parks, NPS implicitly has minimized the broader regional character of visibility impairment. Nonetheless, NPS deserves credit for working effectively with limited resources and for taking the initiative to establish a national visibility-monitoring program at a time when EPA was disbanding its effort.

State Visibility Protection Programs

Few states have adopted programs that go beyond federal law to protect visibility in Class I areas. Three such programs are described below.

Oregon

A significant portion of visibility impairment in Oregon's Class I areas is caused by forestry slash and agricultural burning, and EPA's visibility program, which requires control only of large point sources, would have done little to improve visibility.

Therefore, Oregon established its own program that includes the following key elements:

• Implementation of a visibility monitoring program at seven sites adjacent to Class I areas. Optical, aerosol, meteorological, and observer data are collected to support source attribution studies as well as to determine the frequency, duration, and intensity of impairment. The state monitoring program is performed in cooperation with the USFS and NPS.

• Adoption of short- and long-term control strategies that have resulted in restrictions on forestry and agricultural burning. Forestry burning has been prohibited during the July 4-Labor Day period, when Class I areas are most heavily used by the public. These strategies have reduced the frequency of substantial impairment by about 80% in the summertime relative to the precontrol period of 1982-84.

• Active coordination with the federal land managers through the Oregon Visibility Advisory Committee. That committee provides recommendations to the Oregon Department of Environmental Quality on a wide range of issues, including the determination of impairment conditions, the adequacy of control strategies, and the scope of the protection program. The committee meets regularly to review the program and to consider new issues as they arise.

Washington

Washington's visibility protection program includes the elements of Oregon's program (State of Washington, 1983). The state monitoring program is more limited than Oregon's; therefore, the regulatory program relies heavily on monitoring conducted by NPS and the USFS.

Washington's control strategies call for a 30% reduction in emissions from forestry slash burning and a reduction in impacts during summer weekend days. As a result of improved forestry smoke-management programs, reductions in slash burning, and increased use of residues, visibility conditions in Washington's Cascade Mountains (a Class I area) has improved considerably during the summer months.

Vermont

Visibility impairment in the Lye Brook Wilderness, Vermont's only Class I area, is caused by sulfate-dominated regional haze episodes rather than individual sources (State of Vermont, 1986). Vermont's visibility program attempted to control regional haze caused by sources in Ohio, Pennsylvania, West Virginia, Indiana, Illinois, Michigan, Kentucky and Tennessee. The state adopted a 2 $\mu g/m^3$, 24-hour air-quality standard for sulfates to protect visibility within the Lye Brook Wilderness. Vermont requested that EPA approve the program as a revision of the state's SIP, and order the eight upwind states to revise their SIPs to ensure compliance with the Vermont standard. EPA refused to do either (EPA, 1987b); its decision was upheld by the U.S. Court of Appeals in *Vermont v. Thomas*, 850 F.2d 99 (2d Cir. 1988).

ALTERNATIVE REGULATORY APPROACHES

EPA has considered a variety of means to protect visibility in Class I areas (EPA, 1985a). A program might use, singly or in combination, any of four categories of approaches.

Air Quality Management Approach

Under this approach, the regulating authority sets a desired level of environmental quality, and emissions limits for sources are based on achieving that level. One example of this approach is the ambient standard/SIP process under the Clean Air Act. For instance, EPA could seek to protect visibility through setting an ambient standard for particles smaller than 2.5 μm. The Clean Air Act calls for the establishment of the desired environmental level could be expressed in other ways. For instance, California has long had a state ambient standard specifying that the concentration of light scattering and absorbing particles should be low enough that visibility is 10 miles (16km) or greater (30 miles in the Lake Tahoe area) when relative humidity is less than 70% (California Administrative Code, tit. 17, §70200). Similarly, Colorado recently adopted a standard that limits maximum permissible light extinction; special control measures, for example, limits on wood burning in the Denver area, go into effect on days when this standard is violated. A similar approach has been proposed on a national scale by one commentator, who suggests that each state should be required to reduce gradually the extinction coefficient caused by sources in that state (Pitchford et al., 1990).

Among the advantages of the air-quality management approach are that

• It can be cost-efficient in ensuring the achievement of a desired standard at a specific point, because the amount of control imposed on a source is linked to expected environmental improvement;

• It can encourage research in developing improved control technologies, if the mandated emission limits for sources call for more control than is possible using current technology;

• An environmental quality standard is an easily understood measure of the effectiveness of control.

The primary disadvantage of an air-quality management approach is that it requires at least a rough quantitative understanding of the relationship between emissions and air quality. This, in turn, requires detailed information on emission levels and accurate air-quality models for estimating the effect of a reduction in emissions. Such information and models have been difficult to develop for secondary pollutants, such as sulfates and ozone, that are not emitted directly from any source but are rather the result of atmospheric transformation of source emissions. Moreover, the air-quality management approach ignores the possible benefits of maintaining an area at a level superior to the desired standard. In addition, the environmental quality standard means that identical sources in different areas of the country will be treated differently, depending on the extent to which an area violates the air-quality management standard. Although this might minimize compliance costs, it can be regarded as posing issues of equity between regions.

Technology-Based Approach

A technology-based approach sets emission limits for sources at a level judged technologically and economically feasible. The new-source performance standard program under the Clean Air Act is an example. The advantages of this approach are its relative simplicity, uniformity, and lack of need for information tying the emissions of individual sources to air-quality levels (Latin, 1985). Its disadvantages are the following:

• It may not result in enough control to eliminate the environmental problem.
• It does not encourage the investigation of ways to reduce emissions below the level that is currently technologically feasible. (It should be noted, however, that source owners still have an incentive to find the least expensive way to meet the technologically feasible level; this might result in the development of controls that could meet still stricter standards.)
• It might be cost-inefficient in that the level of control is not tied to the level necessary to attain satisfactory air quality.

Nondegradation Approach

This approach focuses on preventing new environmental problems. A cap that limits emissions to a given level (e.g., 8.9 million tons of sulfur dioxide in the Clean Air Act Amendments of 1990) is one example of this approach. But the nondegradation objective can be expressed in terms other than emissions. The PSD program, which measures degradation in terms of increased pollutant concentrations, is an example.

Like a technology-based approach, a pure nondegradation strategy is simple in concept and implementation. Like the present PSD program, it can curb the growth of emissions. But such an approach takes existing levels of air pollution as a given and will not remedy existing environmental problems. For instance, the PSD program of park protection, because it relies largely on a nondegradation approach, has been criticized for disregarding existing high levels of pollution in parks (Oren, 1989). Moreover, it often is difficult to measure whether and how much degradation is occurring.

Market-Based Approaches

Market-based approaches are alternative techniques for implementing the three approaches described above. Most U.S. environmental programs have been implemented through a traditional regulatory approach for example, when a state specifies emissions levels to be achieved by each source under its auspices. However, without a great deal of information on specific facilities, such programs can be economically inefficient and expensive to administer.

Interest in market-based approaches has been increasing (Breger et al., 1991). Those approaches can take various forms, such as the following:

* EPA and some states have been moderately successful in establishing programs to allow individual sources to find the least-expensive method to achieve required reductions in emissions (Liroff, 1986).
* In a fundamental departure from the traditional regulatory structure, the acid rain provisions of the 1990 Clean Air Act Amendments assign emission levels to large sulfur dioxide sources but permit trading of reduction credits between sources. Under this plan, a source might find its least-expensive option for emissions reduction would be to pay another source to achieve compensating reductions in emissions. This would encourage sources to seek out the cheapest means of reduction, increase economic efficiency of the reduction program, and foster innovation in the development of cost-effective controls.
* Taxes can be imposed on emission of particular pollutants. This would be an incentive to eliminate all emissions for which the cost of

control is less than the amount of the tax. Chlorofluorocarbon production already is taxed, and the 1990 amendments to the Clean Air Act establish "excess emissions" fees.

The primary advantage of market-based approaches is their potential for cost-efficiencies in regulation. One disadvantage, however, is that these approaches might not have the same element of moral condemnation as do conventional regulatory approaches; to that extent, use of these approaches might lessen public concern about pollution. An immediate disadvantage is that all market-based approaches require that emissions be monitored precisely. The tax scheme has the further disadvantage that it is hard to predict in advance the level of fee necessary for reducing emissions to the desired level. A scheme based on trading emissions allowances can work only if the number of regulated facilities is great enough to constitute a market; it will be hindered if regulated facilities have local effects that preclude allowing trades with outside sources. The acid rain and chlorofluorocarbon problems are relatively independent of the precise location of a source. The same might be true of regional haze, but not of visible plumes.

The Present Visibility Program Compared With
Possible Future Approaches

The present program is a blend of the first three approaches: air-quality management, technology-based, and nondegradation approaches. Market-based approaches are included to the extent that the new acid rain program improves visibility in the East. Such a combination of approaches is not unusual; the paradigms are often blended because they have different strengths and weaknesses.

The PSD program uses the technology-based approach in that large new sources in clean-air areas are required to install BACT to limit their emissions. The program also uses a modified version of the nondegradation approach by limiting, (but not forbidding) increases in air pollution in clean-air areas.

The visibility protection program of Section 169A is also a hybrid. For example, technological feasibility and the likely improvement in visibility are considered in setting BART for large existing sources. In this way, BART blends the air-quality management and technology-based approaches.

Aside from its BART requirement, Section 169A gives EPA substantial flexibility in designing a visibility protection program. Although Section 169A requires that states must revise SIPs to make reasonable progress toward the national goal of eliminating visibility impairment in Class I areas, it does not specify the strategy to be used in achieving reasonable progress. EPA could adopt any of the four regulatory approaches identified above or some combination as the basis for further visibility regulation in conjunction with BART. This flexibility does not appear to be altered by the Clean Air Act Amendments of 1990. New Section 169B, for instance, calls for consideration of the regulation of regional haze, but does not specify how EPA might do so. For this reason, the committee believes it appropriate to evaluate source apportionment methods for their consistency with all four approaches identified above.

SUMMARY

The Clean Air Act establishes several mechanisms that could be used to reduce visibility impairment in national parks and wilderness areas, but the effectiveness of each has been limited. EPA has been reluctant, for policy and technical reasons, to use ambient air-quality standards as a means of controlling visibility impairment. The PSD program has decreased atmospheric loadings from new sources and has safeguarded some large parklands from new sources. The amount of protection, though, is limited by the lack of correlation between visibility effects and the Class I increments for maximum permissible increases, the primary measure of whether a new source will be allowed near a Class I area. To date, EPA and other implementing agencies have not established supplemental mechanisms to protect Class I areas. The visibility protection program of Section 169A has had little effect, largely because of EPA's decision to confine the implementation of the provision to impairment that is "reasonably attributable" to sources through the use of simple techniques.

Some states have responded to the limits in the federal effort by creating their own programs to curb visibility impairment in Class I areas.

In addition, interest has increased in exploring possible ways to alter the current federal programs to provide more effective protection. Even under current law, these could fit into any of the four paradigms identified above.

4

Haze Formation and
Visibility Impairment

To develop an effective strategy for ameliorating the effects of human activities on visibility, the complex processes that form haze and impair visibility must be understood. The primary visibility attributes—light extinction, contrast, discoloration, and visual range—can be quantitatively measured, and despite some limitations in knowledge about visibility, changes in those attributes can be related to changes in the chemical and physical properties of the atmosphere.

This chapter presents the current scientific understanding of the processes involved in haze formation and visibility impairment. In this chapter we discuss

- Some of the fundamental factors that relate to haze and visibility;
- The role of meteorological processes in haze formation;
- Experimental strategies for monitoring visibility;
- The modeling of the relationship between aerosol properties and visibility;
- Issues related to quality assurance and quality control.

The measurement techniques used to characterize the components that affect visibility are reviewed in Appendices A and B; Appendix B discusses techniques used to relate the human perception of visibility degradation to physical measurements.

FUNDAMENTALS OF VISIBILITY AND
RELATED MEASUREMENTS

Fundamental Processes in Visibility

If an observer is to see an object, light from that object must reach the observer's eye. The perceived visual character of the image depends on the light emitted from or reflected by the object and on the subsequent interaction of that light with the atmosphere. When an observer views a distant object, the light reaching the observer is weakened by two processes: *absorption* of energy or *scattering* by gases or particles in the atmosphere. These two processes are referred to collectively as extinction and are depicted in Figure 4-1.

Transmitted light is not the primary factor that determines visibility. The visibility of a distant object also is affected by light from extraneous sources (e.g., sun, sky, and ground) that is scattered toward the observer by the atmosphere (Figure 4-1). This extraneous light is referred to as air light. The air light behind an object provides backlighting and causes the object to stand out in silhouette (iv in Figure 4-1); the air light between the observer and an object tends to reduce the contrast of the object and to mute its colors (v in Figure 4-1).

Air light can be an important element of a view; it can have a positive as well as a negative effect on perception. The appearance of the daytime sky is due the scattering of sunlight by gases and particles in the atmosphere. If there were no scattering (or if there were no atmosphere), the daytime sky would be black, allowing the stars to be seen during the day. Air light also provides diffuse light to the surface below; without air light, objects viewed on Earth would have the deep shadow effects seen in photographs of the Moon.

Haze affects the quality and quantity of air light because absorption and scattering are wavelength dependent. That dependence accounts for the deep blue color of the sky in pristine areas, as well as the gray color of smog. Air light is proportional to extinction and, like extinction, depends on particle concentrations. Unlike extinction, air light also depends on viewing angle; particles scatter preferentially in forward directions, so that haze tends to appear brighter in the direction of the sun.

The extinction coefficient, b_{ext}, is a key measure of atmospheric trans-

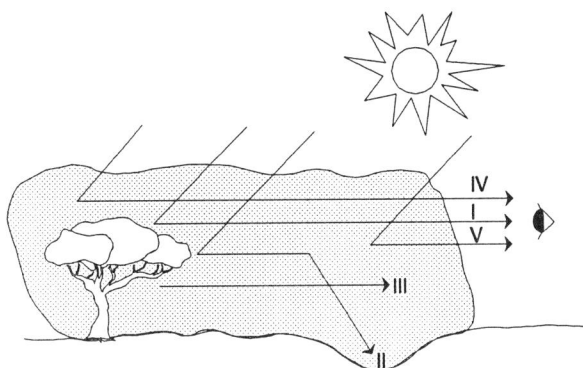

FIGURE 4-1 Elements of daytime visibility. The atmosphere modifies an observer's view of a distant object, in this case a tree illuminated by the sun. The paths illustrate: (i) light from the target that reaches the observer; (ii) light from the target that is scattered out of the observer's line of sight; (iii) light absorbed by gases or particles in the atmosphere; (iv, v) light from the sun or sky that is scattered by the intervening atmosphere into the observer's line of sight; process iv causes the object to stand out in silhouette while v reduces the contrast of the object. Source: EPA, 1979.

parency and is the measure most directly related to the composition of the atmosphere. It is a measure of the fraction of light energy dE lost from a collimated beam of energy E in traversing a unit thickness of atmosphere dx: $dE = -b_{ext}E dx$. The extinction coefficient has dimensions of inverse length (e.g., m^{-1}). The extinction coefficient comprises four additive components:

$$b_{ext} = b_{sg} + b_{ag} + b_{sp} + b_{ap}$$

where
b_{sg} = light scattering by gas molecules. Gas scattering is almost entirely attributable to oxygen and nitrogen molecules in the air and often is referred to as Rayleigh or natural "blue-sky" scatter. It is essentially unaffected by pollutant gases.

b_{ag} = light absorption by gases. Nitrogen dioxide (NO_2) is the only common atmospheric gaseous species that significantly absorbs light. b_{sp} = light scattering by particles. This scattering usually is dominated by fine particles, because particles 0.1-1.0 μm have the greatest scattering efficiency. Many pollutant airborne particles are in this size range.
b_{ap} = light absorption by particles. Absorption arises nearly entirely from black carbon particles.

The extinction coefficient usually is given in units of Mm^{-1} or km^{-1}. The extinction coefficient for visible light in the ambient atmosphere can range from as little as 10^{-2} km^{-1} in pristine deserts to as much as 1 km^{-1} in polluted urban areas.

The behavior of light in the sky is a complex process that depends on many factors. It is because of this complexity that the sky presents such a fascinating spectacle to the observer. However, this complexity also makes it difficult to characterize the visual environment, especially when human perceptions are involved. Nonetheless, techniques are available to characterize the optical properties of the atmosphere and to identify and quantify the determinants of visual air quality that are directly affected by pollutant emissions.

Visibility Measurements

There is no standard approach to measuring and quantifying optical air quality. Instruments for these purposes are commercially manufactured specialty items and are not widely available. EPA has no instrument standards, and uniformity is lacking in field measurements. Consequently, the regulatory community is uncertain which methods should be used. Visibility instruments usually measure either: the energy scattered out of the direct path of the beam or the energy that remains in the beam after it passes through the atmosphere. The nephelometer shown in Figures 4-2a and 4-2b is based on the measurement of scattered light; the transmissometer measures transmitted light (see Appendix B).

These two instruments are fundamentally different not only in what they measure, but also in the way data are obtained and can be used.

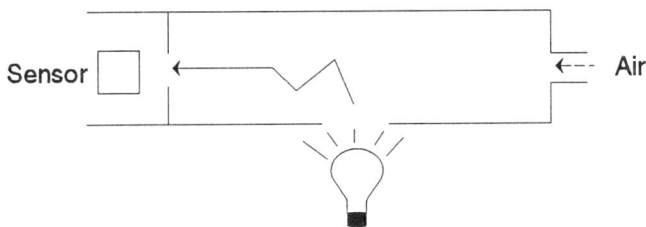

FIGURE 4-2a Approaches to the measurement of extinction. The nephelometer consists of a light-tight container that is fitted with a light source and a photodetector (represented by the sensor in the figure). The interior of the instrument is painted black and contains baffles so that the detector is not directly illuminated by the source; the detector only sees the light that is scattered from the light path. Ambient air is drawn through the instrument; the increase in the signal from the detector (compared to the signal obtained with clean, filtered air) is proportional to the scattering component of the extinction coefficient.

The nephelometer provides a point measurement, and the data obtained with it can be compared directly with other physical and chemical measurements made at the site (e.g., gas and aerosol concentration and composition and particle-size distribution). In contrast, transmissometers measure over long path lengths, at least several km (in clean air, typically 15 km), thereby yielding measurements of the mean transmittance over a long distance. Because of heterogeneities in the atmosphere, it is difficult to relate transmissometer data to chemical and physical measurements, which usually can be made only at one point or, at best, a few points.

Relationship Between Particle Concentrations and Visibility

Visibility impairment is approximately proportional to the product of airborne particle concentration and viewing distance (Figure 4-3). Consequently, relatively low particle concentrations can affect visibility substantially, as shown in the following example. A dark mountain at a distance of 100 km may be clearly visible in clean air, assuming an

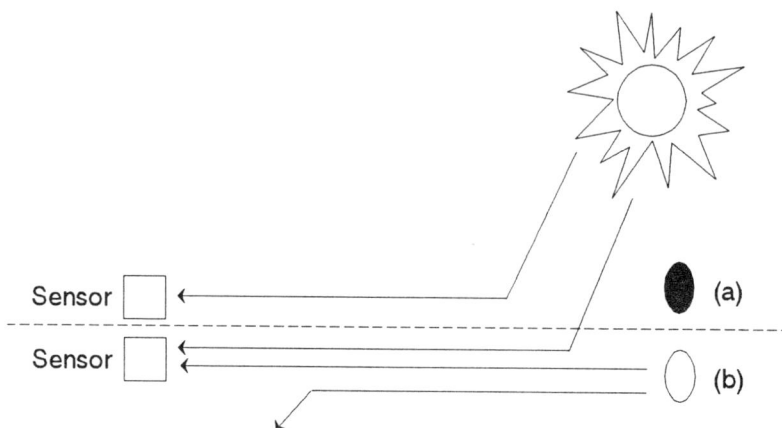

FIGURE 4-2b Radiance difference techniques are based on the teleradiometric measurement of adjacent bright and dark targets located several km or
more from the detectors (here represented by a sensor). In transmissometry, the target radiance difference is commonly obtained by using a single,
switched light source which serves as both the bright (on) and dark (off)
targets. With the light off (a), the detector sees only air light; with the
light on (b), the detector sees, in addition to air light, the light transmitted
through the atmosphere from the source, some of which is scattered in
transit. The difference in the measured radiances depends solely on the
transmittance of the intervening atmosphere. (The radiances of both targets will be affected to the same degree by air light.) The average extinction coefficient (absorption and scattering combined) can be calculated if
one knows the radiance difference of the targets.

average extinction coefficient of about 0.015 km^{-1}; under such conditions, the mountain's contrast with the background sky will be about
20%. If the particle concentration increases sufficiently to increase the
extinction by 0.015 km^{-1}, the contrast will fall below the threshold for
detection (about 5%) and the mountain will no longer be visible. An
extinction increment of this magnitude can be produced by a relatively
small concentration of fine particles, about 3-5 μg/m^3 of particles with
diameters between 0.1 and 1.0 μm.

Concentrations of a few μg/m^3 are not unusual, even in remote regions. At these concentrations, particles usually constitute only a small

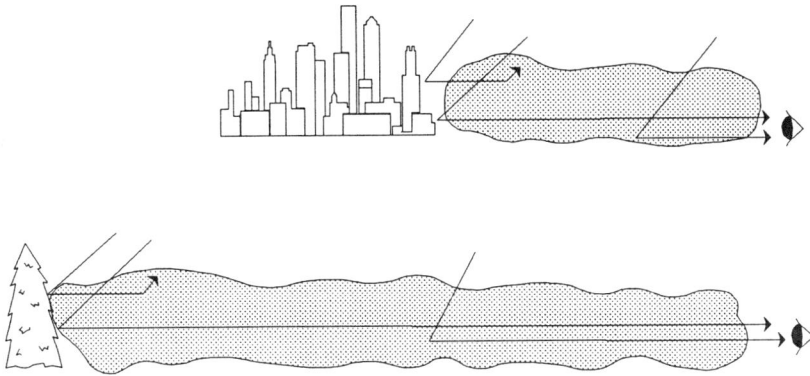

FIGURE 4-3 Visibility impairment and contaminant concentration.
All other things being equal, visibility impairment depends on the product
of path length and average concentration along the sight path, not on the
average concentration. In this figure the two sight paths have the same
total particulate mass (represented by shaded areas) along the sight path.
The upper portion of the figure represents a hazy urban environment,
while the lower represents a relatively clean natural setting. A photon
from the target has the same chance of reaching the observer in either of
the two situations; also the same number of extraneous photons (air light)
will be scattered into the observer's line of sight. The radiation received
by the observer is therefore the same whether the intervening atmosphere
is deep and relatively clean or shallow and relatively turbid. In practice,
the "other things" of our qualifier are seldom all equal, because different
extinction coefficients usually arise from differing proportions of particles
and gases which could have differing scattering and absorption characteristics. Nonetheless, the simple dependence of visibility on the product of
concentration and distance is a useful approximation. Among other things,
it explains why the most transparent atmospheres are the most sensitive to
contamination.

fraction of the total trace materials (gases and particles) found in the
atmosphere, even in relatively clean regions. Sulfate (SO_4^{2-}), nitrate
(NO_3^-), and organic carbon are usually the most important airborne
particle fractions on a mass basis, and they are the trace materials that
usually reduce visibility the most. The sulfur in 1 $\mu g/m^3$ of ammonium
sulfate aerosol is equivalent to 0.2 ppb of sulfur dioxide (SO_2). This

SO_2 gas-phase equivalent is low compared with the concentration found in a typical urban region. (The National Ambient Air Quality Standards permit 24-hour-average SO_2 concentrations of up to 140 ppb. If this quantity of SO_2 were converted to ammonium sulfate aerosol, the resulting concentration would be 700 $\mu g/m^3$.) Indeed, the equivalent gas phase SO_2 concentration calculated in this example for 1 $\mu g/m^3$ of SO_4^{2-} is below the detection threshold of the instruments normally used to monitor compliance with SO_2 air-quality standards (see Appendix B). In contrast, transmissometers easily and accurately can measure the light extinction produced by several $\mu g/m^3$ of particles while nephelometers can do the same by fractions of a $\mu g/m^3$ (see Appendix B). Similar conclusions hold for the gas-phase equivalents of typical nitrate and organic carbon particle concentrations.

Empirical Relationships Between Airborne Particles and Visibility

The components of extinction (i.e., particle and gas scattering and absorption) and their relationship to visibility have been well characterized in a wide range of environments. These empirical relationships are shown in Figures 4-4 a-d. The relationship between visual range and the scattering coefficient (as measured with an integrating nephelometer) is shown for an urban area (Seattle) in Figure 4-4a and for an area near Shenandoah National Park in Figure 4-4b. If atmosphere and illumination are uniform, visual range and extinction in theory are inversely proportional. Because scattering is almost always the dominant component of atmospheric extinction, visual range should be inversely related to scattering as well. Figures 4-4a and 4-4b empirically confirm this expectation for hazy conditions, where sightpaths are relatively short and air masses are fairly uniformly mixed.

Figures 4-4c and 4-4d illustrate the relationship between atmospheric light extinction, as measured with nephelometers, and particle concentrations. Figure 4-4c (for Seattle) and Figure 4-4d (for an area outside of Shenandoah National Park) show that scattering is approximately proportional to total particle mass.

Figure 4-5 shows the fraction of the non-Rayleigh extinction attributable to the various components of scattering and absorption. In all

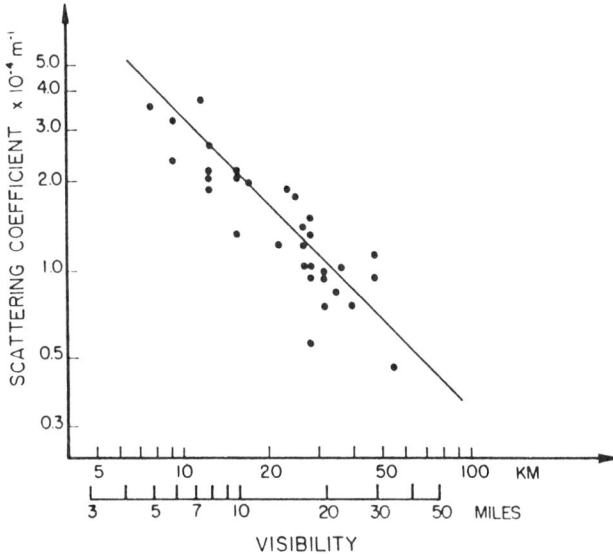

FIGURE 4-4a Empirical visibility relationships. Light scattering coefficient
  as a function of prevailing visibility (visual range attained over at least half
  of the horizon circle), in Seattle. Data are from summer 1968 at humidities below 65%. Visual range was determined by human observer; light
  scattering was measured by unheated nephelometer and is corrected to a
  photopic spectral response. Both scales are logarithmic; sloping line indicates the theoretical Koschmieder (1924) relationship $V = 3.9/b_{ext}$ for a
  contrast threshold of 0.02. Source: Reprinted from *Atmospheric Environment* 3:543-550, H. Horvath and K.E. Noll, "The relationship between
  atmospheric light scattering coefficient and visibility," 1969, with permission from Pergamon Press Ltd., Headington Hill Hall, Oxford, OX3
  OBW, UK.

cases, fine-particle scattering is the dominant contributor to light extinction; this is especially true for eastern locations. In the West, coarse-
particle scattering (usually soil dust) and particle absorption also contribute significantly.

  In all regions, gases have a minor role. The only atmospheric trace
gas contributing to visible extinction is nitrogen dioxide (NO_2), which
has a broad absorption band at the blue end of the spectrum; consequently, when NO_2 concentrations are high, the atmosphere has a distinct

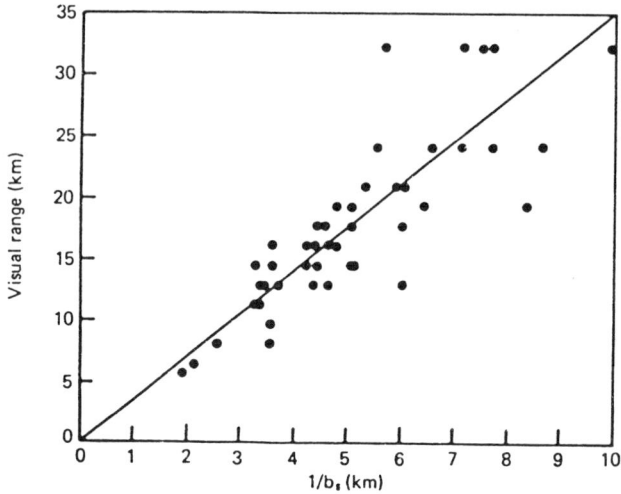

FIGURE 4-4b Visual range as a function of the light scattering coefficient, just outside Shenandoah National Park. Data were obtained during non-overcast conditions when the atmosphere was well mixed during mid-summer 1980. Visual range was determined by human observation of mountain peaks aligned to the southwest of the site; easily identified peaks were available at 2-3 km intervals from 5 to 24 km. Light scattering was measured by unheated nephelometer and is plotted on a reciprocal scale, in accordance with the Koschmieder (1924) relationship $V \propto 1/b_{ext}$. Source: Adapted from Ferman et al., 1981.

red-brown color. However, NO_2 is relatively reactive, and its concentration is generally small, except in urban areas near emissions sources. Therefore it usually is a small contributor to regional optical air quality.

Aerosol Chemistry and Particle Size Distributions

The optical effects of atmospheric aerosols depend on the chemical composition and size distribution of the airborne particles. Particle size distributions in the atmosphere change with time; the size distribution is determined by the characteristics of the particles emitted directly by a

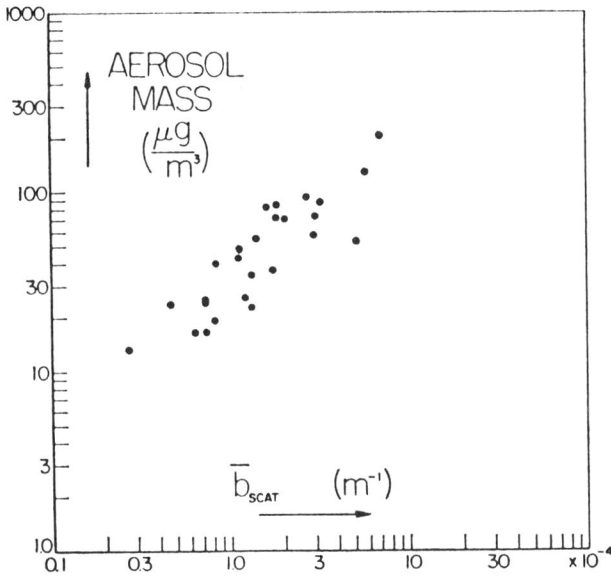

FIGURE 4-4c   Particle mass concentration as a function of light scattering
coefficient, in Seattle. Data are from winter 1966, averaged over periods
of 2 hours to 24 hours. Light scattering was measured by unheated nephelometer, mass by a total filter located in the nephelometer outlet. Both
scales are logarithmic. Source: Reprinted from *Atmospheric Environment*
1:469-478, R.J. Charlson, H. Horvath, and R.F. Pueschel, "The direct
measurement of atmospheric light scattering coefficient for studies of visibility and pollution," 1967, with permission from Pergamon Press Ltd.,
Headington Hill Hall, Oxford, OX3 OBW, UK.

source, the subsequent formation of airborne particles by reactions of the
emitted gases (especially SO_2), and processes that remove the particles
and gases from the atmosphere. Those processes are sensitive to variations in the composition of the emissions and to meteorological conditions, including sunlight intensity, temperature, humidity, and the presence of clouds, fog, or rain.

    Primary airborne particles are those emitted directly from a source—
for example, soot, fly ash, and soil dust, but a major portion of the fine-
particle mass fraction (particles with diameters between 0.1 and 1.0 μm)
usually is formed in the atmosphere by the conversion of species emitted

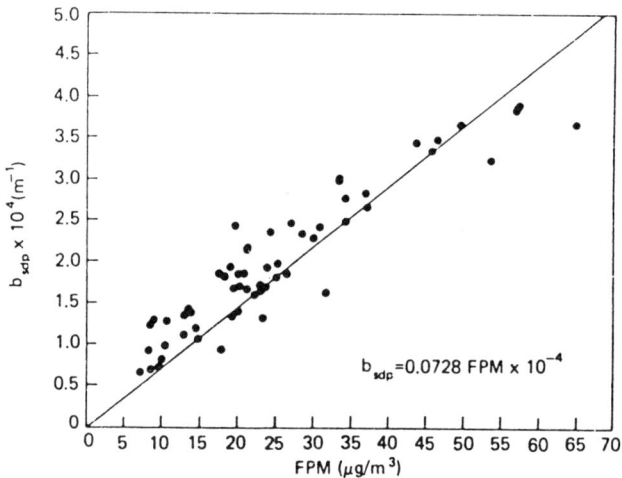

FIGURE 4-4d Light scattering coefficient as a function of fine particle mass concentration, outside Shenandoah National park. Data are 8-hour averages from midsummer 1980. Light scattering was measured by heated nephelometer and mass by a co-located filter sampler behind a cyclone separator. Source: Adapted from Ferman et al., 1981.

as gases (see Appendix A). These secondary particles include SO_4^{2-}, NO_3^-, and organic compounds. Figure 4-6 gives a qualitative overview of the processes by which secondary particles are formed in the atmosphere. The figure shows the relationships between trace gas molecules and the secondary particles that are generated from them. The generation of secondary particles begins with the generation of oxidants such as OH and O_3 in the presence of sunlight and proceeds through the interaction of various reactive, transient species with various pollutant molecules (e.g., SO_2, NO_2, and VOCs). Heterogeneous chemistry in clouds and fog are important in many of these processes. For further discussion and a more quantitative explanation of the transformations of gases into airborne particles, see Appendix A.

The effects of primary particle emissions and chemical transformations on atmospheric particle size distributions are illustrated in Figure 4-7 (see Appendix B for particle size measurement techniques). Nucleation-mode particles (particles with diameters < 0.1 μm) can be emitted

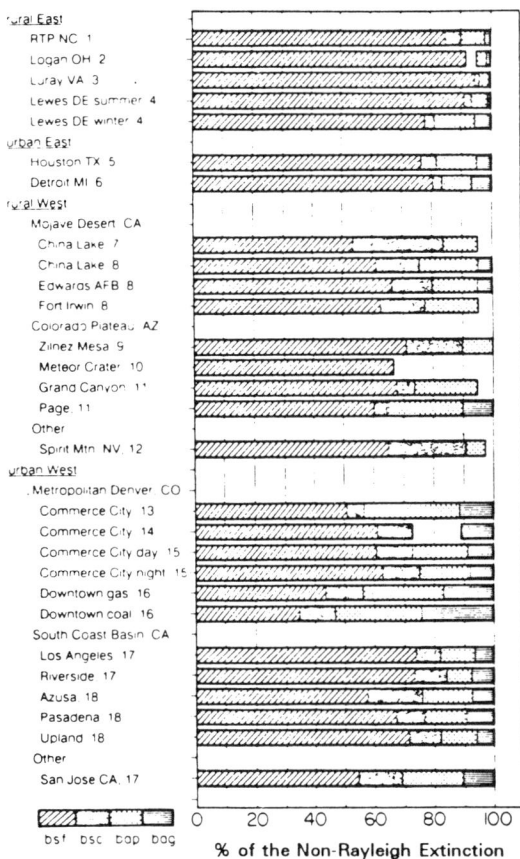

FIGURE 4-5 Additive components of extinction. This summarizes the relative contributions of components of extinction, excluding Rayleigh extinction, as determined in various field studies. The extinction attributable to each component is expressed as a fraction of the total non-Rayleigh extinction. The components are:

b_{sf}, scattering by fine (diameter < 2.5 μm) particles;

b_{sc}, scattering by coarse (diameter > 2.5 μm) particles;

b_{ap}, absorption by particles; and

b_{ag}, absorption by gases.

The figure clearly shows that fine particle scattering is the dominant component at all locations. The relative amounts of extinction components vary considerably showing clear regional trends. Source: Reprinted from *Atmospheric Environment* 24:2673-2680, W.H. White, "The components of atmospheric light extinction: A survey of ground-level budgets," 1990, with permission from Pergamon Press Ltd., Headington Hill Hall, Oxford, OX3 OBW, UK.

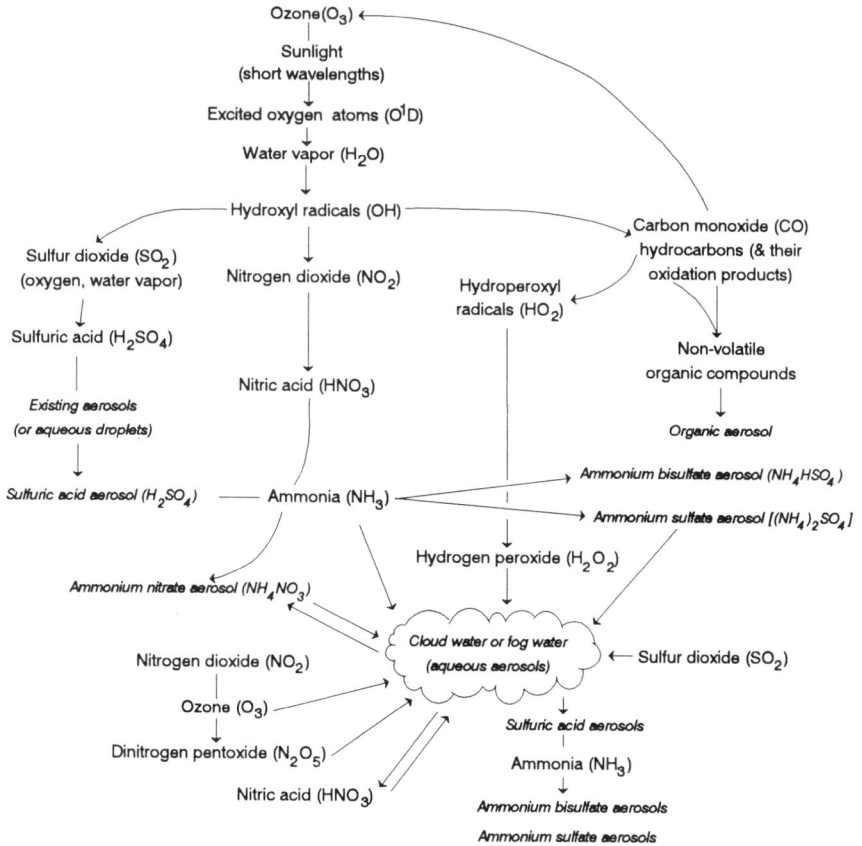

FIGURE 4-6 Pathways to secondary particle generation in the atmosphere. The diagram shows the relationships between the trace gas molecules (italicized) and the secondary that are generated from them. (See Appendix A for further discussion).

directly into the atmosphere from combustion sources or formed in the atmosphere by homogeneous nucleation. Coarse particles (particles with diameters $> 1~\mu m$) include wind-blown dust, plant particles, sea spray, and volcanic emissions. Secondary reaction products (especially NO_3^-) also are found on the surfaces of coarse particles (e.g., Savoie and Prospero, 1982; John et al., 1990). Accumulation mode particles (particles with diameters between 0.1 and 1.0 μm) can be primary or secondary particles; the latter are usually dominant.

95

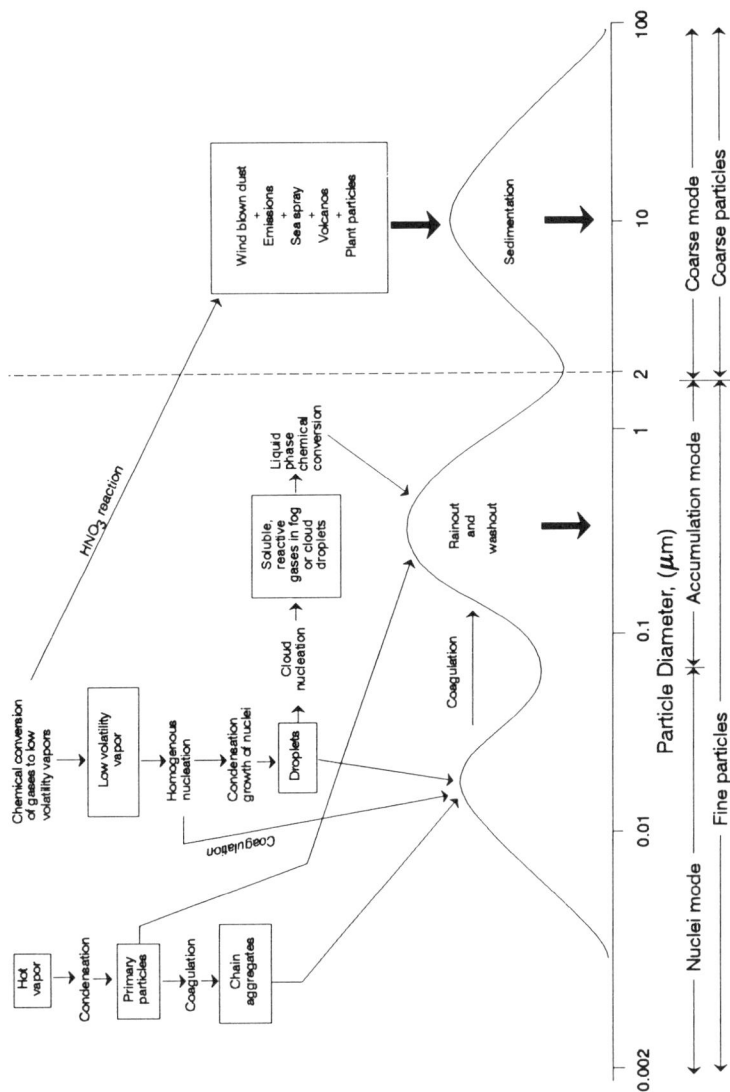

FIGURE 4-7 Trimodal size distribution. Schematic diagram of an airborne particle surface area distribution showing the three modes, the main source of mass for each mode, the principal processes involved in inserting mass into each mode, and the principal removal mechanisms for the modes. Source: Adapted from NRC, 1981.

The physics and chemistry of atmospheric particle formation results in a trimodal size distribution. Those size differences have a profound effect on the physical and optical properties of the resulting aerosol. The characterization of aerosol effects is further complicated by the highly dynamic nature of particle size distributions. Particle size is affected not only by a wide range of formation and transformation processes but also by atmospheric removal processes. Particles are removed from the atmosphere by wet processes (precipitation, cloud, and fog) and dry processes (gravitation, diffusion, and impaction).

Large particles settle toward the earth, with sedimentation velocities proportional to d_p^2 (Friedlander, 1977). Loss rates are proportional to sedimentation velocities, and therefore the rates increase rapidly with increasing particle size. In contrast, small particles are highly mobile and are lost primarily through attachment to other particles that they encounter during their random motions. The resulting loss rates for the smallest particles are proportional to their diffusion coefficients, yielding a dependence on particle size in the range d_p^{-2} to d_p^{-1} (Friedlander, 1977). Other diffusional losses (e.g., within control devices or to vegetation) display a similar dependence on particle size.

The efficiency of wet removal processes is also highly size-dependent (see the section below, The Role of Meteorology). The net result of the various wet and dry removal processes is that particles in the nucleation- and coarse-modes tend to have a relatively short residence time while those in the accumulation mode have a relatively long residence times (see Figure 4-8). In the next section, we show that these long-lived accumulation-mode particles have the greatest effect per unit mass on visibility and that it is for this reason that visibility impacts can often extend over large regions.

Particle Optics and Visibility

The optical properties of airborne particles are affected by several factors, among them particle size. Figure 4-9 shows the scattering efficiency of ammonium sulfate particles as a function of their diameter for monochromatic 530 nm light. The oscillations that are shown correspond to size-dependent resonances in scattering. For white light, a smooth curve that peaks at about the same particle size is found. The

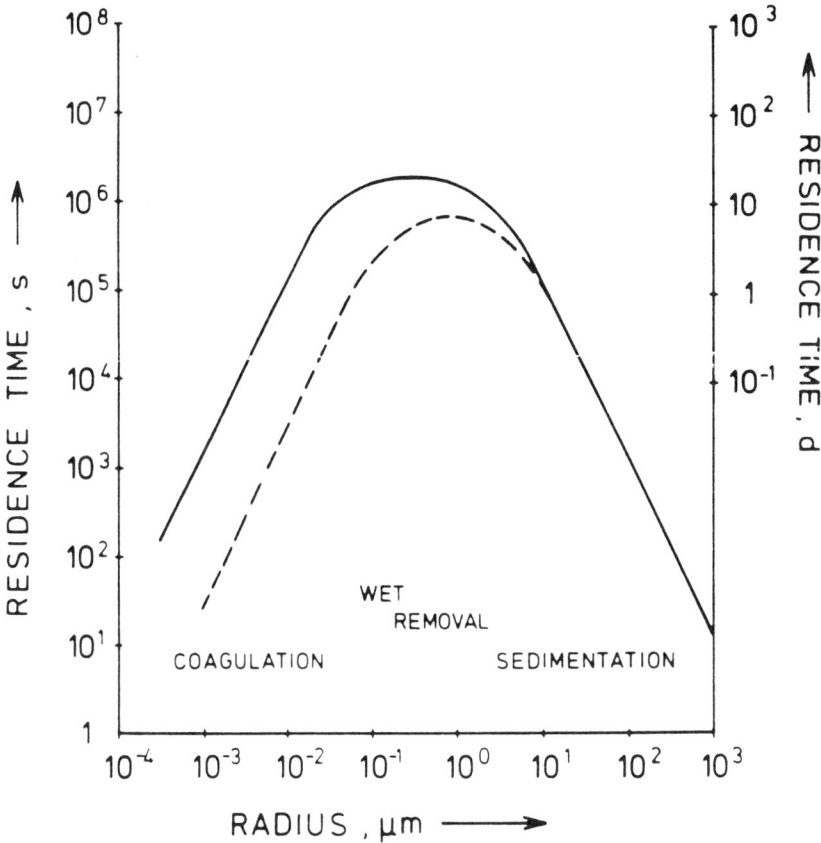

FIGURE 4-8 Particle loss rates as functions of particle size. Particle loss mechanisms are dependent on particle diameter. By comparing this figure with Figure 4-9, which shows particle scattering efficiency as a function of particle size, one sees that none of the common removal mechanisms is very effective in removing particles that are the most efficient scatterers of light. This fact accounts for the accumulation and persistence of large hazy air masses. Source: Jaenicke, 1980.

size-dependent scattering efficiencies of other types of airborne particles are qualitatively similar. A given mass concentration scatters most effectively when it is distributed among particles having diameters comparable to the wavelength of the illumination (Friedlander, 1977).

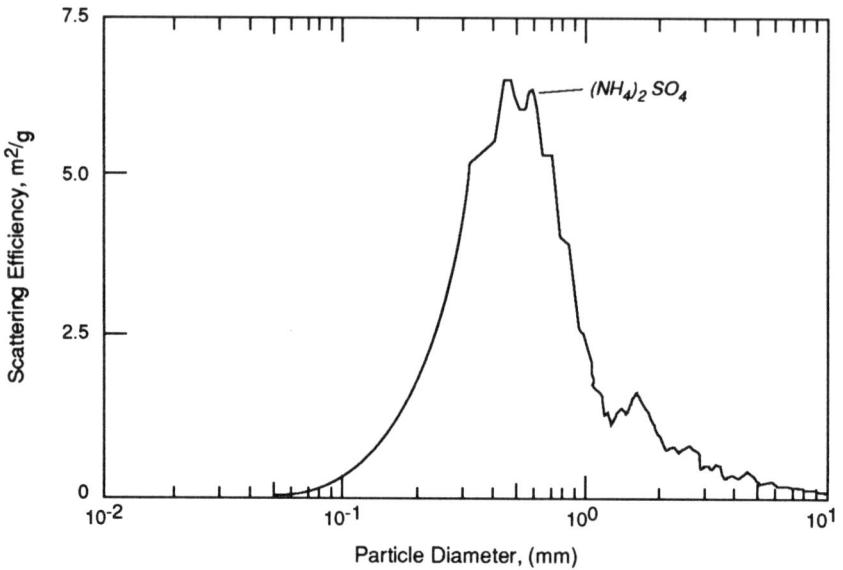

FIGURE 4-9 Dependence of light scattering on particle size. The scattering efficiency of particles depends on several factors, among them particle size, wavelength of the light, and optical properties of the aerosol material. The theoretical scattering efficiency can be computed as shown here for ammonium sulfate particles at wavelength 530 nm (the curve labeled $(NH_4)_2 SO_4$). Particles scatter light most effectively when their diameters are comparable to the wavelength of the illumination (Friedlander, 1977). Therefore, airborne particles, which tend to accumulate in the size range 0.1 to 1.0 μm diameter, can have a large effect on visibility. Particles much larger than the wavelength of visible light show a sharply decreasing scattering efficiency with increasing particle size. These particles scatter energy in proportion to their geometric cross-sectional area; that is, they "cast shadows" just as larger objects do. However, for particles (and gas molecules) much smaller than the wavelength of visible light, the scattering efficiency increases rapidly with increasing particle (and gas molecule) size (a few tenths of a μm or less); these particles scatter energy in proportion to the square of their mass. In the size range of about 0.3-1.0 μm, the scattering behavior of particles is very complex. Source: Adapted from Ouimette and Flagan, 1982.

Secondary particles can have a strong effect on visibility and visibility impairment because these particles tend to accumulate in the size range of 0.4-0.7 μm, the region of the range for the visible spectrum.

Figure 4-10a shows the production rates of excess aerosol sulfur (i.e., the SO_4^{2-} aerosol concentration in excess of that in the regional background aerosol) as a function of solar age of the St. Louis urban plume. The solar age is the equivalent number of hours of exposure to clear-sky midday solar radiation. The data show that secondary sulfur particle production varies roughly linearly with solar age. The data are consistent with the conversion of SO_2 to SO_4^{2-} at average rates that range from 2% to 4% per hour. Figure 4-10b shows (for the same data set) the change of the excess aerosol light-scattering cross section as a function of solar age. It is important to note that nearly all the light scattering is due to secondary particles and that the emissions from sources in St. Louis have their greatest effects a considerable distance downwind.

Figures 4-7, 4-8, and 4-9 illustrate the relationship between the visibility problem and pollution:

• Sources emit gases that are converted to secondary airborne particles,

• Those particles are produced primarily in the 0.1-1.0 μm diameter range,

• The 0.1-1.0 μm size range is where the light scattering per unit mass is greatest and where removal rates are slowest.

Because the secondary aerosol particles have long lifetimes, they can be carried great distances by winds; consequently, visibility impairment is usually a regional problem. Further, these particles are small, and they have a large effect on visibility per unit mass. It follows that visibility impairment is a sensitive indicator of air pollution, and visibility can be significantly improved only by addressing the problem on a regional scale.

Some Experimental Difficulties
in Aerosol Chemistry Studies

Because of the relationship of visibility to various airborne particles

PARTICULATE SULFUR

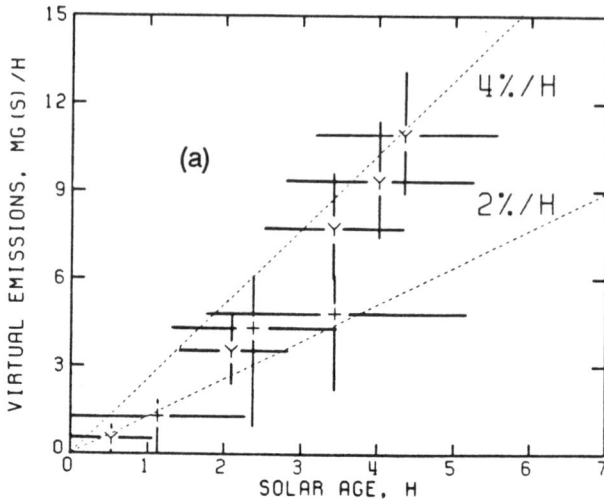

FIGURE 4-10a, b Increase in light-scattering aerosol with distance from source. This figure show measured flow rates of excess (a) sulfate particles (Mg sulfur/hour) and (b) light scattering coefficient (km^2/hour) downwind of metropolitan St. Louis. The data were obtained from a detailed mapping of the urban plume by instrumented aircraft, together with intensive pilot balloon determinations of wind. Observations are plotted as a function of 'solar age', in equivalents of exposure to clear-sky midday solar radiation. These data show that most of the sulfate and light-scattering aerosol attributable to St. Louis is formed in the atmosphere hours downwind of the city; the sulfate is generated by the photochemical oxidation of SO_2 that is emitted from sources in St. Louis and carried away by the winds. Dotted lines in (a) indicate sulfate flow rates that would be expected for SO_2 conversion rates of 2% and 4% per hour. Source: White et al., 1983.

and their size distributions, it is important that aerosol composition and particle-size distributions be measured accurately in visibility programs. In practice, such measurements are difficult. (A more detailed discussion is found in Appendix B.)

One problem is that most of the important gas and particulate phase species are highly reactive. Of particular concern is the conversion of gas-phase species within a sampling system; gas concentrations can be hundreds of times greater than those of the ambient airborne particle

HAZE

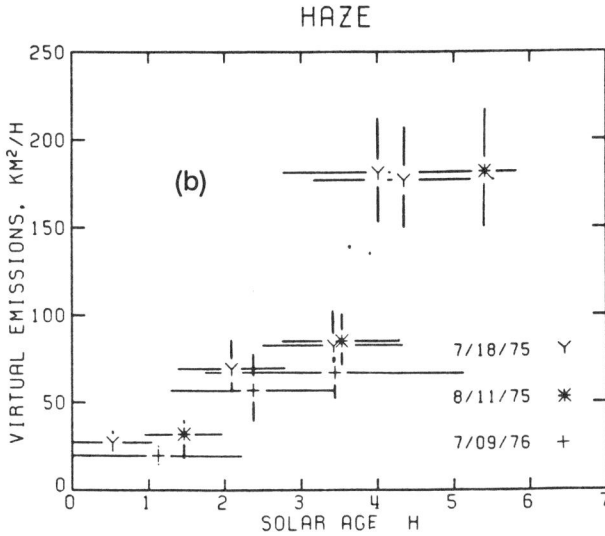

FIGURE 4-10a,b (continued)

equivalents. For example, glass fiber filters were used for many years to collect particulate matter. In the late 1970s, it was found that these chemically basic glass fibers efficiently captured gaseous nitric acid, and yielded erroneously high values for particulate NO_3^-. As discussed in Appendix B, there are many other cases where gas-phase species react with the sampling medium to yield erroneously high particle concentrations.

The reverse process can occur as well. Aerosol substances can react in a sampling system to produce gaseous materials that are lost to the sampling stream. For example, Teflon filters, which are chemically inert, yield erroneously low NO_3^- aerosol values because of nitrate volatilization. When ambient NO_3^- particles (which must have been neutral or basic to exist in the atmosphere) are collected on the filter surface, they can react with acidic SO_4^{2-} particles or with SO_2 in the air stream; as a result, NO_3^- is converted to nitric acid, which subsequently evaporates.

One solution to these problems is to strip the gas-phase species such as nitric acid from the air stream before passing it though the filter. This is accomplished by a combination of gas denuders and filter packs (Figure 4-11). These sampling systems yield reliable measurements of

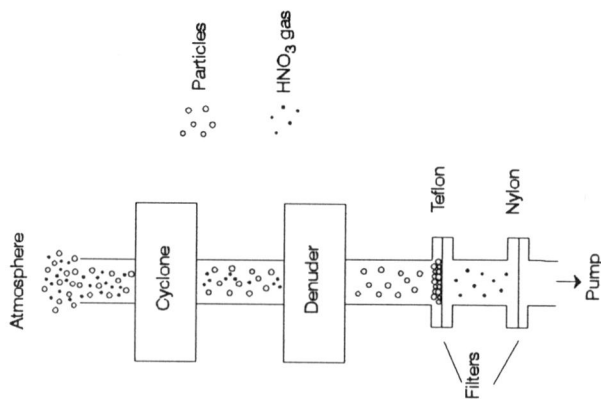

FIGURE 4-11 Phase-preserving particle sampling train. A combination of denuders and filter packs can be used to obtain a sample of particles and gases that is representative of the ambient atmosphere, because it minimizes interaction between various reactive species. The sample is drawn through an inertial separator (a cyclone) to remove large particles ($> 2.5 \mu m$ diam.) that would be subject to losses in the denuder. The air stream is then drawn through a denuder to remove gaseous nitric acid, which might form artifact particulate nitrate through reaction on the filter. The denuder typically consists of parallel plates or concentric cylinders separated by a few mm and coated with a reagent; the path length through the denuder is typically a few tens of cm. The denuder exploits the fact that gases are much more diffusive than particulate phases. After passing through the denuder, the air stream is drawn through a filter pack. The pack consists of a Teflon filter to collect particles, followed by a reactive filter (usually nylon) that will capture any nitric acid released from the particles collected on the Teflon filter. The sum of the nitrate collected in the cyclone and on the front and backup filters yields the total concentration in the particulate phase. The nitrate in the denuder is assumed to be derived from the gas-phase nitric acid. Additional denuder stages can be added in series to collect other gaseous substances (e.g., SO_2, NH_3 etc.) according to the reagent coating employed in the denuder. For additional information, see Appel et al. (1981); Forrest et al. (1982); Shaw et al. (1982); Eatough et al. (1988); Wiebe et al. (1990).

ambient fine-particle NO_3^- concentrations and of the gaseous nitric acid. With the addition of various denuders in series (or with two or more denuder/filter pack systems in parallel) these systems can be used to collect other gas-phase species (e.g., SO_2, NH_3, and HCl).

Water presents another problem. Because most of the aerosol particulate mass consists of hygroscopic materials (e.g., sulfuric acid, ammonium sulfate, ammonium bisulfate, and ammonium nitrate), the size of the airborne particles depends on the relative humidity. For a change in relative humidity from 30% to 90%, the size of an ammonium sulfate particle increases by a factor of 5, while a sulfuric acid particle grows by a factor of more than 3 (Figure 4-12). The air in the lower atmosphere usually contains a substantial amount of water (typically several g/m^3) even in the arid Southwest. Because the concentration of water vapor is millions of times greater than that of airborne particles, the conversion (condensation, sorption, or reaction) of even a minute fraction of this water to the particulate phase can have a major effect on visibility. Hygroscopic particles can take up water at humidities well below saturation. It follows that particle-borne water can play a major role in optical extinction at high relative humidities ($> 70\%$) (see Figure 4-13).

Even though water itself is a natural constituent of the atmosphere, in the context of visibility impairment, the water associated with anthropogenic SO_4^{2-} and NO_3^- must be regarded as a contaminant since it condenses to further degrade visibility in the presence of hygroscopic particles. The direct measurement of particulate water is a formidable challenge. Because the water associated with particles constitutes only a small fraction of the total water, it cannot be collected using denuders and filter packs. As discussed further in Appendix B, techniques are needed to quantify particle water content.

THE ROLE OF METEOROLOGY

Meteorology can play a dominant role in visibility degradation. For example, wind speed either can reduce or increase the likelihood of visibility degradation. At low wind speeds, the ventilation of emitted pollutants is reduced, thus increasing the concentration of pollutants in the atmospheric boundary layer. High winds can also degrade visibility

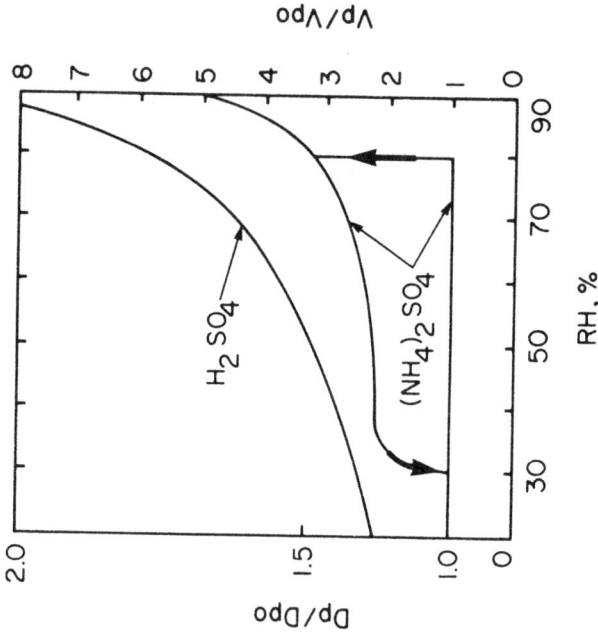

FIGURE 4-12 Dependence of particle size on composition and relative humidity. Under many common ambient conditions, much scattering associated with hygroscopic particles could be attributable to water held in the liquid phase. The interaction with water vapor can be complex and vary strongly with chemical composition. This figure shows the difference between behavior of H_2SO_4 and $(NH_4)_2SO_4$ particles, two common pollutant aerosol species. H_2SO_4 droplets pick up (and lose) water readily over a wide range of RH values. $(NH_4)_2SO_4$ has much more complex behavior. When an $(NH_4)_2SO_4$ particle is exposed to increasingly moist air, it does not pick up water vapor until RH reaches 79.5%, the deliquescence point for the salt; at this point, the droplet grows rapidly (indicated by the upward-pointing arrow). When droplets of $(NH_4)_2SO_4$ solution are exposed to increasingly dry air, the salt retains water to RH well below the deliquescence point, and the solution becomes supersaturated. Clearly, it is not enough to know the concentration of sulfate in the atmosphere; speciation must be known, because the difference in the behavior of H_2SO_4 and $(NH_4)_2SO_4$ particles is large, especially at low RH values. Also, note that the deliquescence point is at an RH value that is common in many regions, especially during the summer. Source: Adapted from Tang, 1980.

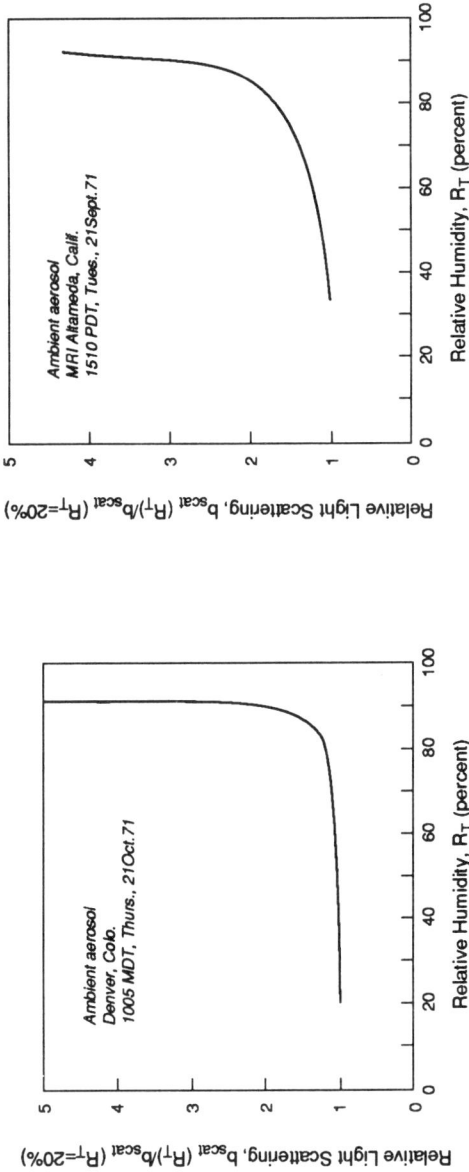

FIGURE 4-13 Atmospheric measurements of light scattering as a function of relative humidity. The observed dependence of particle light scattering on relative humidity (RH) at two locations, Altadena, CA, and Denver, as measured with an integrating nephelometer. The relative humidity of the air stream entering the nephelometer is varied from 20% to over 90% and the change in scattering is recorded. The figure presents the results as the ratio of the scatter measured at a specific RH to that measured at 20% RH. At Altadena the scatter increased continuously over the entire range of RH (especially above 80%) in a manner similar to that for H_2SO_4 in Figure 4-12. Light scatter increases about a factor of two between 20% and 85% RH. This suggests that the Altadena particles are very hygroscopic and that visibility effects vary greatly with RH. In contrast, Denver particles are relatively non-hygroscopic; there is very little increase in scatter until the RH exceeds 90%. These data demonstrate that scattering is very sensitive to RH and that the behavior of airborne particles with respect to RH can differ with location, depending on the composition of the particles. Source: Covert et al., 1972.

locally by picking up and carrying dry soil. When low wind speeds are associated with low temperatures (as is common in the western United States during winter), stagnation occurs and pollutants accumulate. Pollutants may build up further during periods of low temperature due to increased heating requirements (e.g., increased power-plant emissions and wood smoke).

The following section discusses some examples of the role of meteorology in visibility. Appendix A discusses meteorological factors in more detail.

Transport

The transport of atmospheric pollutants depends strongly on meteorological conditions. For example, high SO_4^{2-} concentrations in the Adirondack Mountains most often are associated with transport from polluted regions to the south and southwest of New York (Galvin et al., 1978). In the Shenandoah Valley, 78-86% of the light extinction is attributed to anthropogenic airborne particles, most of which originate in the Midwest (Ferman et al., 1981).

Presently there is good understanding of the meteorological factors that affect regional haze transport in the eastern United States. However, knowledge about meteorological effects on visibility in the West is less advanced. It is known that in the western and northwestern United States, the types of meteorological conditions associated with decreased visibility change seasonally. Incidents of reduced wintertime visibility in the Southwest usually are associated with low wind speeds and high relative humidities (NPS, 1989). In pristine regions, where the visual range can be great, small increases in SO_4^{2-} aerosol concentrations can lead to readily apparent decreases in visibility. Consequently, visibility in those areas is sensitive to meteorological conditions that maximize the effect of local emissions or transport emissions from distant sources.

In the Southwest, the winter episodes of decreased visibility usually occur during mesoscale meteorological events. Few data are available with which to delineate the areal coverage of these haze episodes. Nonetheless, evidence from the WHITEX study suggests that the spatial extent of haze during that study was small compared with the size of haze episodes in the East. Southwestern episodes are, however, too large to be accommodated by the predictive techniques used for plume blight; plume models typically are restricted to sources within 50 km of the receptor. Thus, adequate plume models are not available for dealing with the western haze episodes.

During the summer in the Southwest, decreased visibility is associated with a wide range of meteorological conditions. Winds can carry heavily polluted air from southern California eastward to regions of the desert Southwest, including the Grand Canyon National Park. Similar conditions occur in the San Joaquin Valley, where winds carry pollutants from the San Francisco Bay area toward the national parks of the Sierra Nevadas.

Meteorology also plays an important role in the chemical transformation of pollutant gases to particles. Conversion of SO_2 to SO_4^{2-} is greatly accelerated in the presence of water droplets, in particular fog or cloud droplets (see Appendix A). Figure 4-14 shows the ratio of particulate sulfur (i.e., SO_4^{2-}) to total sulfur (primarily SO_2 and SO_4^{2-}) as a function of plume age. The figure summarizes data obtained in measurements from a variety of sources by many different investigators. Figure 4-14 shows that from <1% to >10% of the SO_2 can be converted to light-scattering SO_4^{2-} aerosol particles after several hours. Thus, the effect of a particular source on a receptor region can vary tremendously, depending on ambient atmospheric conditions.

It is clear that the meteorological conditions associated with reduced visibility in national parks and wilderness areas in the West are different from those in the East. Consequently, a range of meteorological analysis options needs to be devised to attribute haze to sources in the two areas.

Dispersion

The stability of the atmosphere affects the amount of vertical mixing that takes place during plume transport. Enhanced mixing reduces pollutant concentrations near the sources (i.e., within a few tens of kilometers). The transport path of a plume is determined by the wind speed and direction combined with the effects of vertical mixing. Strong vertical mixing dilutes the plumes, making them less dense and less visible.

FIGURE 4-14 Extent of gas-to-particle sulfur conversion as a function of emissions age for daytime (top) and nighttime (bottom). S_p/S_t is the ratio of particulate sulfur (mostly SO_4^{2-}) to total sulfur (mostly SO_4^{2-} and SO_2). The data are collated from measurements made in the plumes of 12 different point sources, as reported by 8 different organizations. The concentrations S_p and S_t have been corrected for background and primary particle contributions. Different symbols distinguish the data from different reporting organizations. Data corresponding to plume ages of less than 1 hour are not shown. The lines are labeled for different nominal conversion rates (%/hour). Note that daytime rates are considerably greater than nighttime rates. Source: Reprinted from *Atmospheric Environment* 15:2573-2581, W.E. Wilson, Jr., "Sulfate formation in point source plumes: A review of recent field studies," 1981, with permission from Pergamon Press Ltd., Headington Hill Hall, Oxford, OX3 OBW, UK.

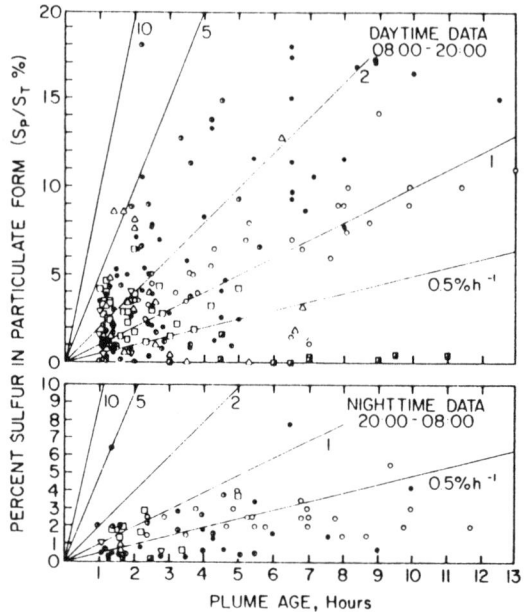

Vertical mixing also can break up surface layers into which plumes can become entrained. Vertical mixing subsequently redistributes the pollutants into a thicker layer of more homogeneous haze. In contrast, reduced mixing combined with low wind speeds can increase the likelihood of formation of valley fogs; valley fogs are often an important factor in pollution episodes, because many industrial sources are located in valleys, near water resources.

Deposition and Resuspension

The transport of particles to and from the Earth's surface has an important effect on haze. Wet and dry deposition processes cleanse the air, while the resuspension of soil dust and plant material by winds can be a significant source of visibility-impairing particles. In this section we briefly describe the role of these surface-exchange processes in visibility.

Wet Deposition

Wet deposition is the result of processes that occur within and below clouds. Field studies have shown that within clouds most (>90%) of the visibility-impairing particles are incorporated into cloud droplets (Brink et al., 1987) along with a large fraction of the reactive gases. Cloud droplets have two fates: they can be removed as precipitation or they can evaporate. It has been estimated that only about 10% of the cloud cover dissipates through precipitation; the other 90% evaporates (McDonald, 1958). When clouds evaporate, the particles that were incorporated into cloud droplets are re-released to the atmosphere. However, the composition of the resulting aerosol changes because of the aqueous-phase chemistry that takes place in cloud droplets, principally because of reactions with atmospheric gases such as SO_2 and HNO_3. Airborne particles are likely to cycle through clouds many times before they are removed by precipitation, and the composition and size of the particles changes with each cycle. This process accounts for the fact that individual particles are often found to consist of internal mixtures of a wide range of chemical species (Andreae et al., 1986)

It is generally believed that, because of the very large surface-to-volume ratio of cloud droplets, in-cloud processes are more effective than the below-cloud processes at cleansing the atmosphere of visibility-impairing particles and their gaseous precursors. Particles and gases also can be removed below cloud level by falling precipitation; however, the mass transfer to precipitation is relatively inefficient due to the large droplet sizes.

Dry Deposition

Dry deposition is the deposition of particles and gases to surfaces in the absence of precipitation. Dry removal rates of particles from the atmosphere depend on particle size, small-scale meteorological processes near the surface, and the chemical and physical characteristics of the receiving surface. Studies suggest that dry deposition typically accounts for roughly 20% to 40% of the sulfur removal from an airshed (Shannon, 1981). Young et al. (1988) estimated that dry deposition contributes about half of the acid deposition in mountainous regions of the western United States, although it might be less important than wet deposition in the eastern United States (Galloway et al., 1984). Dry deposition has been reviewed by McMahon and Denison (1979), Sehmel (1980), Hosker and Lindberg (1982), and Davidson and Wu (1989).

Theory shows that dry deposition rates of particles are lowest for particles in the 0.05-0.5 μm diameter range. Therefore, particles in this size range have a longer atmospheric lifetime than smaller or larger particles. This is significant from the standpoint of haze formation, because particles of this size are relatively effective at light scattering and absorption.

Resuspension of Soil Dust

The resuspension of soil dust is an important source of coarse atmospheric particles. Although those particles have relatively short atmospheric lifetimes, they can reduce visibility considerably under some conditions. Soil dust is resuspended by dust devils, wind erosion, agricultural tilling, and vehicular travel on paved and unpaved roads. Gillette and Sinclair (1990) found that resuspension of soil dust by dust devils is comparable in significance to other sources of that material. Vehicular travel is an important anthropogenic source. All estimates show that emissions of soil dust are higher in the arid Southwest than in other parts of the United States.

STRATEGIES FOR VISIBILITY MEASUREMENT PROGRAMS

The preceding sections discussed visibility measurement techniques without regard for the manner in which they might be integrated into a visibility study. In practice, such studies are performed in either a research or an operational mode:

In a research (or intensive) mode, a large array of measurements is made to understand the factors affecting visibility. Intensive studies often involve a large cooperative effort by scientists from academic, government, and private organizations. Such studies normally take place over a short period—weeks to a few months.

In a monitoring mode, measurements are made routinely over an extended period—usually many years—to detect and characterize patterns in visibility impairment and to identify the causes of such patterns. Standardized instrumentation is used in such studies and the procedures must be simple enough to be carried out by personnel without highly specialized training.

This section focuses on systems and procedures used in field measurement programs and different strategies for establishing a visibility monitoring program.

Criteria for Monitoring Programs

In monitoring programs, the optical properties measured are those that are closely related to human visual perception. Regulatory agencies with monitoring responsibilities design optical measurement programs on the basis of several practical considerations:

- The measurement methods should be inexpensive, reliable, and simple to operate under field conditions. Because extinction coefficients are likely to vary widely for monitoring programs that cover a wide geographical area, these methods should be capable of measuring the extinction coefficient over several orders of magnitude.
- The extinction data should be coordinated with measurements of the concentrations of atmospheric aerosols that cause the extinction so that source-apportionment analysis based on aerosol chemistry can be linked to extinction.
- The measurements should reflect visibility conditions as perceived by human observers, and the measured parameters should be presented in units that are understandable to decision makers and the public.
- Because the Clean Air Act and regulatory programs focus on anthropogenic rather than natural sources of visibility impairment, the

method should be insensitive to extinction caused by rain, fog, snow, and other weather conditions.

• Data-averaging times should be linked to the public perception of visibility (e.g., a 24-hour averaging period is of little or no help when regulators are concerned with visibility impairment only during the daylight hours).

These criteria can be difficult to meet. Visibility as perceived by a human observer cannot be fully replicated by any instrumental technique (see Appendix B). Because no single method can satisfy all of these criteria, regulatory agencies (which usually have very limited funding) must often rank their monitoring needs.

Monitoring meteorological variables in support of an assessment of regional haze should be conducted with these considerations in mind:

• At a minimum, the field program should be based on an analysis of the climatology of low-visibility episodes. The analysis should involve data on the wind flow, humidity, and atmospheric stability conditions most often associated with low-visibility episodes.

• Meteorological instruments should be sited to represent the air flow at suspected or proposed sources or source areas, at key receptor areas, and at intermediate locations. Wind measurements should represent the wind flow at the height of the emission plume, which usually requires aloft measurements of winds aloft.

Examples of Visibility Measurement Programs

The Interagency Monitoring of Protected Visual Environments Program

In response to Section 169A of the 1977 Clean Air Act Amendments, EPA promulgated regulations for a visibility monitoring strategy for Class I areas for states that have not incorporated such strategies in their state implementation plans (SIPs). The federal strategy called for the establishment of an interagency program with the cooperation of EPA and several federal land management agencies, including the National Park Service (NPS), the Fish and Wildlife Service (FWS) and the Bureau of Land Management (BLM) of the U.S. Department of Interior, and the Forest Service (FS) of the U.S. Department of Agriculture. The Interagency Monitoring of Protected Visual Environments (IMPROVE) program has been operating since March 1988 to satisfy the regulatory requirements.

The objectives of IMPROVE are (1) to characterize background visibility so as to be able to assess the effects of potential new sources, (2) to determine the present sources of visibility impairment and to assess the amounts of impairment from these sources, (3) to collect data that are useful for assessing progress toward the national visibility goal, and (4) to promote the development of improved visibility monitoring technology and the collection of visibility data (Pitchford and Joseph, 1990).

Twenty sites now operate in the IMPROVE network. Additional sites that employ similar measurement methodologies are operated by NPS in the TERPA (Tahoe Regional Planning Agency) and NESCAUM (Northeast States for Coordinated Air Use Management) networks. Figure 4-15 gives the locations of the 48 sites using the IMPROVE sampler. These networks are operated by Cahill and co-workers at the University of California at Davis.

The IMPROVE measurement protocol involves aerosol, optical, and view monitoring. Four particle samples are collected simultaneously over 24 hours on Wednesdays and Saturdays each week. The samplers include one PM_{10} filter sampling system (which collects particles smaller than 10 μm diameter) and three $PM_{2.5}$ filter systems (for particles smaller than 2.5 μm). One of the $PM_{2.5}$ samplers is preceded by a potassium carbonate denuder to remove acidic gases so as to facilitate the measure ment of particulate nitrates. Table 4-1 indicates the measured quantities and the analytical techniques used for each filter type. All of the filter analyses are done at the University of California at Davis, except for ion chromatography (IC) and thermal optical reflectance (TOR), which are subcontracted to other laboratories (Pitchford and Joseph, 1990). Temperature and relative humidity measurements are made with rotronic model MP-1007 humidity temperature meteorological probes. According to manufacturer's specifications, these sensors record RH to "within a few %RH over the temperature operating range of the probe." The operating temperature range is -20°C to +55°C.

The IMPROVE sampling strategy provides information on major and trace particulate species. The TOR measurements of organic and ele-

114

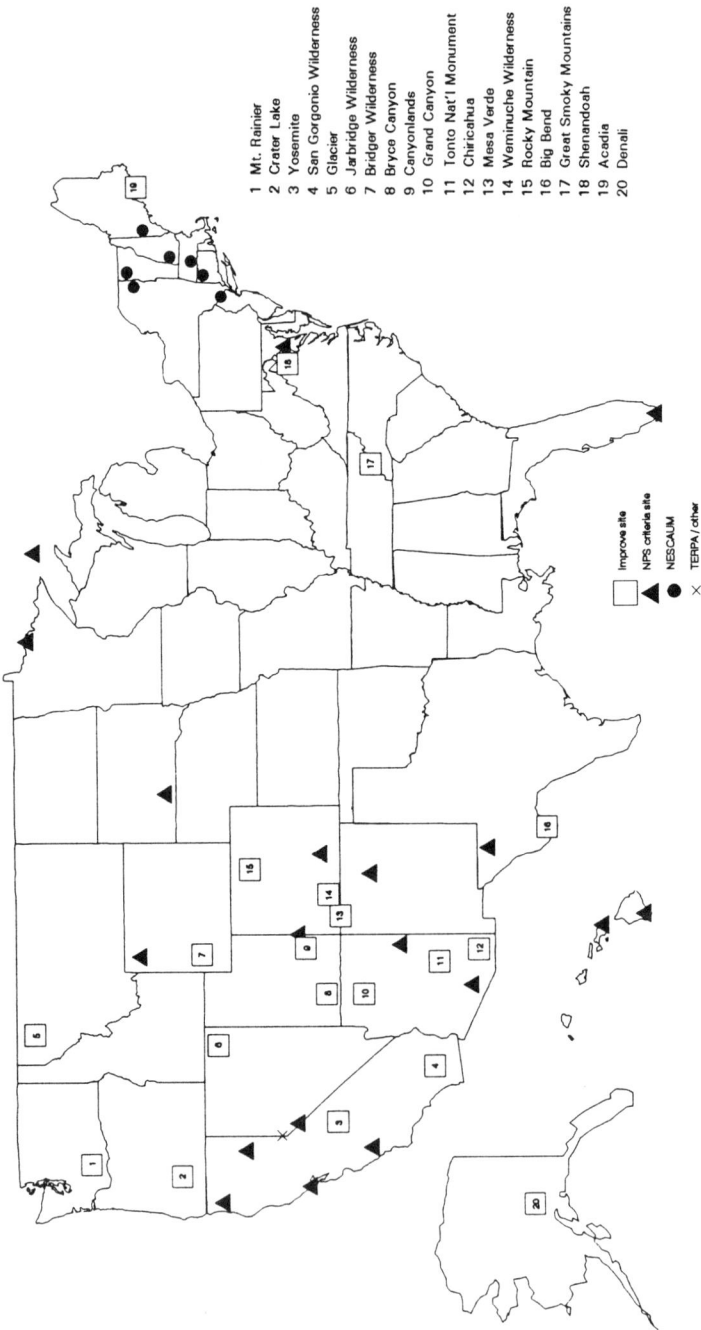

FIGURE 4-15 The Interagency Monitoring of Protected Visual Environments (IMPROVE) background visibility monitoring network (triangles) and IMPROVE "look-a-like" sites (see legend). The locations of the IMPROVE sites are listed in the figure. Source: Pitchford and Joseph, 1990.

TABLE 4-1 Particle Measurements Made with the Interagency Monitoring of Protected Visual Environments (IMPROVE) Sampler[a]

	Filter Type	Quantity Measured	Analytical Technique
Module A	Fine teflon filter[b]	Mass of collected particles	Gravimetric analysis
		Optical absorption (b_{abs})	Laser integrating plate method
		Elements Na to Pb	Particle-induced X-ray emission
		Hydrogen	Proton elastic scattering
Module B	Sodium carbonate denuder and fine nylon filter[b]	Chloride, nitrite, nitrate, sulfate	Ion chromatography
Module C	Fine quartz filter[b]	Organic and elemental carbon	Thermal optical reflectance
Module D	PM_{10} filter[b]	Mass of collected particles	Gravimetric analysis
	Impregnated quartz filter[c]	SO_2	Ion chromatography

[a]The IMPROVE Modular Aerosol Monitoring Sampler consists of four independent filter modules, a control module, and a pump house containing a pump for each module. Each module collects two 24-hour filter samples per week. The filters are collected weekly and shipped to laboratories for analysis.

[b]The three fine filters collect particles of diameter less than 2.5 μm. The PM_{10} filter collects particles of diameter less than 10 μm.

[c]This filter, which collects gaseous SO_2, is used in samplers at National Park Service Criteria Pollutant Monitoring sites. Samplers at these sites are otherwise identical to samplers at IMPROVE sites. Sources: Eldred, 1988; Pitchford and Joseph, 1990; R. Eldred, pers. comm., University of California at Davis, July, 1992.

mental carbon particulate concentrations are likely to be the least accurate measurements. An independent estimate of particulate carbon concentrations is obtained from proton elastic scattering analysis (PESA)
measurements of hydrogen (e.g., Eldred et al., 1989) using the nonsulfate hydrogen technique. In this procedure, the amount of hydrogen
associated with SO_4^{2-} is subtracted out by assuming that the SO_4^{2-}
consist of pure ammonium sulfate; for samples collected at Great Smoky
Mountains and Shenandoah, the SO_4^{2-} is assumed to be 75% ammonium sulfate and 25% sulfuric acid (e.g., Eldred et al., 1989). It is further assumed that hydrogen associated with nitrates or water is lost when
the sample is brought to vacuum during PESA analysis. The resulting
hydrogen concentration is converted to an equivalent carbon value by
assuming that hydrogen constitutes 9% of the organic mass. These
assumptions yield surprisingly good correlations between TOR and
PESA estimates of carbon concentrations. It would be far preferable,
however, to measure particulate carbon more accurately and directly, for
reasons to be discussed below.

Transmissometers are used to measure extinction coefficients in the
IMPROVE network. Temperature and relative humidity are also measured continuously on site. Data from these instruments are radio-trans-
mitted via satellite to a central computer for daily retrieval; thus, mal-
functions can be discovered quickly and remedied.

There is a major concern about the quality of the data obtained in the
IMPROVE network. Because of limited resources, comprehensive quality assurance evaluations have not been carried out by independent audi-
tors. However, intercomparisons with various measurements from other
groups have been done in conjunction with several intensive field programs. Also, outlier points can be identified through comparisons
among interrelated variables (Pitchford and Joseph, 1990). Nonetheless,
it is essential that the quality of these data be characterized and clearly
documented so that long-term trends can be evaluated.

One example is the concern about the quality of the organic carbon
concentration data estimated by the nonsulfate hydrogen technique. Data
for average concentrations across the United States from June 1984 to
June 1986 are approximately a factor of two lower than average concentrations measured between March 1988 and February 1989 (Cahill et al.,
1989; Eldred et al., 1990). The data during the earlier period were
collected with the sequential filter unit (SFU), a predecessor to the IM-

PROVE sampler that was used in the more recent measurements. It is not known whether the differences are due to an actual change in ambient concentrations (which would be surprisingly large for so short a time), to differences in the sampler operating characteristics, or to some other factor. It is essential that the reasons for such discrepancies be documented clearly.

State Programs

In this section we present some examples of state visibility monitoring programs. This discussion is not intended to be a comprehensive survey. In presenting these examples, the committee does not endorse or condemn either the design strategy for the cited programs or the manner in which they are implemented. The following examples show some approaches to state monitoring needs.

Sequential filter samplers are used for aerosol monitoring by Oregon and Washington at remote sites near Northwest Class I areas (Core, 1985). The sequential filter sampler first was developed during the Portland Aerosol Characterization Study (PACS) (Watson, 1979) and later adapted for use in the Sulfate Regional Experiment (SURE) (Mueller and Hidy, 1983). The current design (with a PM_{10} inlet) has been designated by EPA as an equivalent method for PM_{10} monitoring in Oregon.

In this system, 12-hour sampling periods provide adequate analytical sensitivities. Timers control the sampling time and intervals. As many as 12 filter sets can be loaded into the sampler at any one time, thereby minimizing the number of site visits needed to maintain continuous operation. The filters are contained in cassettes to minimize possibilities of contamination and are routinely analyzed for gravimetric mass. Selected samples are analyzed by x-ray fluorescence (XRF), IC, and TOR to provide aerosol composition data for receptor modeling and extinction budget analysis.

The state regulatory agencies of Washington, Oregon, and California measure extinction as part of their visibility monitoring programs. In each case, the states have chosen measurement methods simpler and less costly than those that would normally be used in field research programs.

The California Air Resources Board (CARB) reviewed several optical methods for measuring visibility impairment throughout the state (CARB, 1989). These included measurements of transmittance using active and passive transmissometers, of extinction caused by light scattering based on scanning and integrating nephelometers, and of light absorption as measured by the integrating plate and coefficient of haze (COH) methods. Indirect methods of measuring extinction, including contrast ratio measurements from densitometer analysis of 35-mm color slides and teleradiometry, also were evaluated, as was modeling of extinction from aerosol chemistry measurements.

Following an extensive consideration of the costs, the relative advantages and disadvantages, and the ease of implementation of the various methods, the CARB Committee on Visibility adopted three measurement methods: 1) integrating nephelometry (MRI 1550 B with heated inlet) for measuring dry particle light scattering at a nominal wavelength of 550 nm; 2) COH tape sampler measurements as an indicator of light absorption in urban areas; and 3) ambient air hygrometer measurements of relative humidity. The humidity measurements are used to flag observation periods when relative humidity exceeds 70%.

In developing monitoring protocols, California took a pragmatic point of view, opting for simple, reliable measurements of parameters that could be related to anthropogenic sources of impairment and thereby support regulatory programs. Automated cameras and densitometric radiometry were recommended as an alternative approach to document scene quality at specific levels of measured extinction.

The Oregon State Department of Environmental Quality also chose to use MRI 1550 B integrating nephelometers equipped with heated inlets as their primary measure of extinction. Important considerations in making this selection were that the method is insensitive to extinction caused by weather conditions and that the instrument is reliable. Unlike California's program, all of Oregon's monitoring is conducted at remote sites near Class I areas where absorption is a minor component (typically 8%) of the total extinction (Beck and Associates, 1986). As a result, the Oregon program does not include routine measurements of light absorption. The state's visibility goals are expressed in terms of reductions in the frequency of impairment (defined in the SIP) as measured by integrating nephelometry.

In areas where commercial power is not available, 35-mm cameras are used to document scene quality three times daily. Standard visual range measurements are then made from the slides by densitometric radiometry.

The Washington State Department of Ecology carries out a visibility monitoring program similar to Oregon's, except that MRI 1590 integrating nephelometers (with unheated inlets) are used (principally because the equipment was available in this configuration). Like Oregon, Washington has adopted a working definition of visibility impairment (as measured by nephelometry) of 40 Mm^{-1}, exclusive of Rayleigh scattering (State of Washington, Department of Ecology, 1983).

Washington also uses automated 35-mm cameras to document scene quality. Visual range is estimated from the slides by densitometric radiometry.

As a result of these measurement programs, Oregon and Washington have adopted restrictions on sources that impair visibility in their Class I areas.

Intensive Programs

This section describes an intensive experiment designed to evaluate the visibility effects of a particular point source, the Navajo Generating Station (NGS). In providing this example, the committee does not imply that it endorses or condemns either the design strategy or the manner in which the program was implemented.

NGS is a coal-fired power station; with a generating capacity of 2400 MW, it is one of the largest power plants in the western United States. NGS is located approximately 25 km from the nearest border of the Grand Canyon National Park and about 110 km from the Grand Canyon Village tourist area. This region experiences extended periods of stagnation during the winter; hazes are known to occur at such times. The NGS visibility study was carried out in 1990 to determine the extent to which wintertime visibility at the GCNP would be improved if NGS emissions of SO_2 were reduced. (For a more complete discussion of this study, see Richards et al., 1991).

Field measurements were made in an array of sampling sites in the vicinity of NGS over an 81-day period from January to March 1990; these included air quality and optical parameters and meteorological data. As one component of the program, the investigators injected perfluorocarbon tracers into the NGS stacks; at the monitoring sites, the concentration of this tracer was measured along with the other parameters. Various aspects of NGS emissions (concentrations of SO_2, NO_x, and particles, along with opacity) also were monitored continuously during the project. Instrumented aircraft were used to characterize the composition of the NGS plume and of the regional background during selected intensive operation periods. Several special experiments were conducted to characterize the aerosol in more detail than was possible with the routine filter data.

A summary of the meteorological measurements made during the NGS visibility study is given in Table 4-2; site locations are shown in Figure 4-16. These measurements were designed to provide data for forecasting intensive operation periods and to support diagnostic modeling analyses. Surface meteorology measurements included wind direction and speed, temperature, relative humidity, precipitation, and radiation. In addition, rawinsondes, airsondes, and tethersondes were used periodically at several sites to measure winds, pressure, temperature, and relative humidity as functions of altitude. Wind fields were mapped using radar profilers, Doppler sodar, monostatic sodar, and Doppler lidar.

Surface air quality measurements were made at 27 sites. Table 4-3 summarizes the measurements made at these sites; the locations of tracer and air-quality monitoring sites are shown in Figure 4-17. Not all of the measurements listed in Table 4-3 were made at all sites. However, certain measurements, such as SO_2 and SO_4^{2-} concentrations and fine particle mass were made routinely at most sites. Other measurements, such as organic and elemental carbon concentrations, size-resolved aerosol chemical composition, aerosol water content, cloud water chemistry, and aerosol optical measurements, were made at selected sites. Approximately 60,000 substrates were analyzed for particulate mass and chemical composition.

Two approaches were used to determine the effect of NGS emissions on haze. One approach was largely empirical: the data were examined to determine the relationships among NGS emissions, meteorology, air quality, and visibility during the study period. The second approach involved mechanistic modeling. The contributions of NGS and other sources to SO_4^{2-} levels in the study area were obtained by numerical

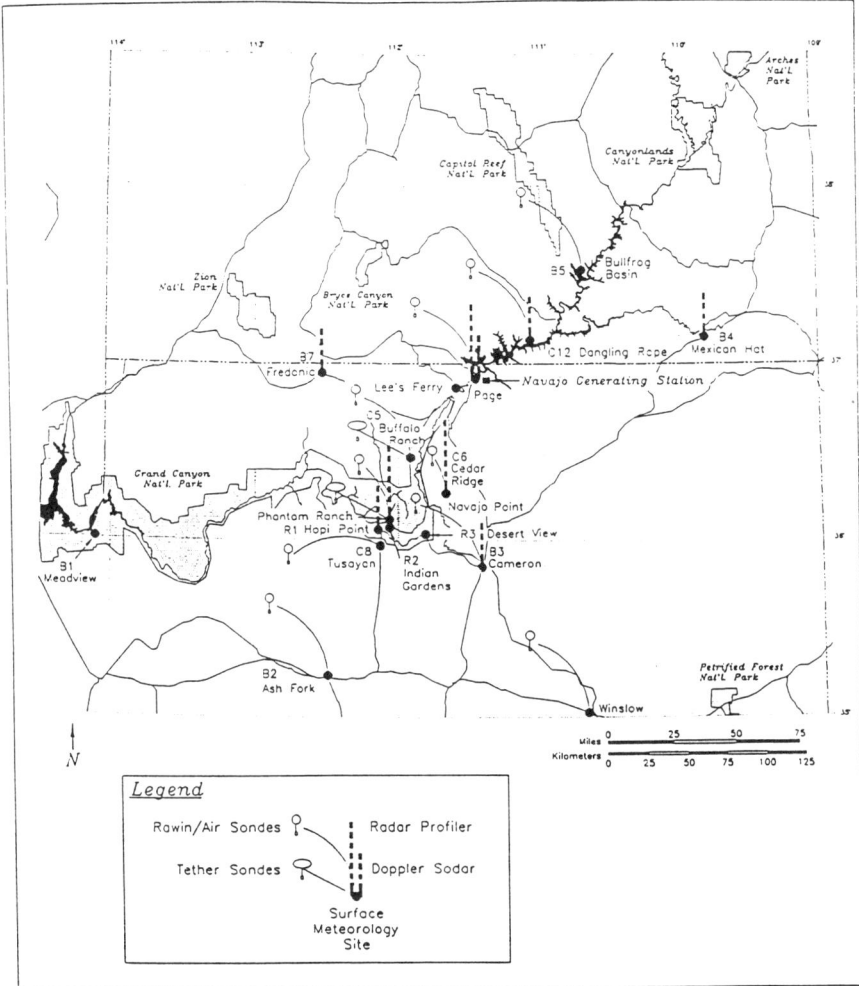

FIGURE 4-16 The locations of meteorological measurement sites in the experimental area for the 1990 NGS Visibility Study. Source: Richards et al., 1991.

The field program was designed so that measurements provided the required information for each analytical approach.

An intensive study of this kind can provide detailed information on factors that affect visibility. One major limitation of intensive programs is that they are confined to relatively short periods. Because year-to-

TABLE 4-2 Meteorological Measurements. Typical numbers of routine and intensive measurements/day are shown for non-continuous measurements.[a]

Site#	Site Name	Abbr.	Surface Meteorology	R'sondes/Day Routine/Int.	Airsondes/Day Routine/Int.	T'sondes/Day Routine/Int.	Radar Profiler	Doppler Sodar	Monostatic Sodar	Doppler Lidar
B2	Ash Fork		✓	2/2						
C5	Buffalo Ranch		✓		0/V[b]	0/V[a]				
B5	Bullfrog Basin	BUL	✓	2/4[c]				✓		
B3	Cameron		✓	0/4[c]						
C6	Cedar Ridge		✓		2/2[d]		✓	✓		
C12	Dangling Rope	DNG	✓	0/4						
R3	Desert View	DSV	✓					✓		
B7	Fredonia	FDN	✓							
	Grand Junction		✓	2/2						
R1	Hopi Point	HOP	✓						✓	

R2	Indian Gardens	ING	✓		
	Las Vegas		✓	2/2	
	Lee's Ferry	LEY	✓		
B1	Mead-view	MVW	✓		
B4	Mexican Hat	MEX	✓		✓
	Navajo Point				✓
	Page	PGA	✓	2/4	✓
	Phantom	PTN	✓	$0/V^a$	$0/V^a$ ✓
	Ranch				
	Tusayan		✓	$2/4^c$	
	Winslow		✓	2/2	

[a] ✓ means measurement was made at the site.

[b] V = Variable

[c] After 3/19 Soundings were performed on a routine schedule with 2/day at Cameron and Ash Fork, 4/day at Bullfrog Basin and Tusayan, and none at Dangling Rope

[d] Only temperature and RH were measured. Winds were measured by the radar profiler.

Source: Richards et al., 1991.

TABLE 4-3   Measurement Methods

Parameter	Sampling Method[a]	Sampling Frequency	Duration[b]
PM$_{2.5}$:			
Mass	Filter sampling	6/day	4 hours
		3/day	8 hours
		2/day	12 hours
Sulfate (total sulfur)	Filter sampling	6/day	4 hours
Nitrate	Filter sampling	3/day	8 hours
Carbon	Filter sampling	3/day	8 hours
Trace elements	Filter sampling	3/day	8 hours
Size fractionated Trace elements	DRUM	3/day	8 hours resolution
Size fractionated Multiple species	MOUDI	Daily	24 hours
Water	TDMA	Continuous	--
Size fractionated Sulfate	LPI	3/day	8 hours
Gases:			
SO$_2$	Filter pack	6/day	4 hours
NO/NO$_x$, SO$_2$, O$_3$	Automatic analyzer	Continuous	--
Tracer	Bag sampler	6/day	4 hours
		24/day	1 hour[c]
Cloud water chemistry	String sampler	Irregular	--
Extinction:			
b$_{ext}$	Transmissometer	Continuous[d]	--[d]
b$_{sp}$	Nephelometer	Continuous	--
b$_{ap}$	Filter sampling	6/day	4 hours
		3/day	8 hours
Visual records:			
Vista and sky conditions at R1 and R2	Photography	08, 09, 10, 11, 12, 13, 15, & 16 hours MST	--

TABLE 4-3 (continued)

Parameter	Sampling Method[a]	Sampling Frequency	Duration[b]
Vista and sky condition at C7	Time-lapse photography	Camera 1 at 08, 10, 12, 14, & 16 hours; Camera 2 at 09, 11, 13, 15 & 17 hours MST. Time-lapse at 75-s intervals, 07-17 hours MST	--
Vista and sky condition at R3	Time-lapse photography	Cameras[e] at 08, 09, 10, 11, 12, 13, 15, 16 hours MST; Time lapse[f] at 60-s. intervals 0630-1800 MST	--

[a]DRUM: Davis rotating-drum universal-size-cut monitoring sampler (a cascade impactor) (Raabe et al., 1988). MOUDI: Microorifice uniform deposit impactor (Marple et al., 1991). TDMA: Tandem differential mobility analyzer (McMurry and Stolzenburg, 1989). LPI: Low pressure impactor (Hering et al., 1979).

[b]Four-hour samples were changed at 0200, 0600, 1000, 1400, 1800, and 2200 MST. Eight-hour samples were changed at 0200, 1000, and 1800 MST. Twelve-hour SCISAS samples were changed at 1000 and 2200 MST. Twenty-four-hour MOUDI samples were changed at 1800 MST.

[c]Operated only at IOP days.

[d]Ten-minute sample each hour.

[e]From December 13 to December 27, 1989, single photographs were taken three times daily at 0900, 1200, and 1500 MST.

[f]Time lapse for Mt. Trumbull view was taken at 30-second intervals from 0600-1800 MST starting on January 4, 1990.

Source: Richards et al., 1991.

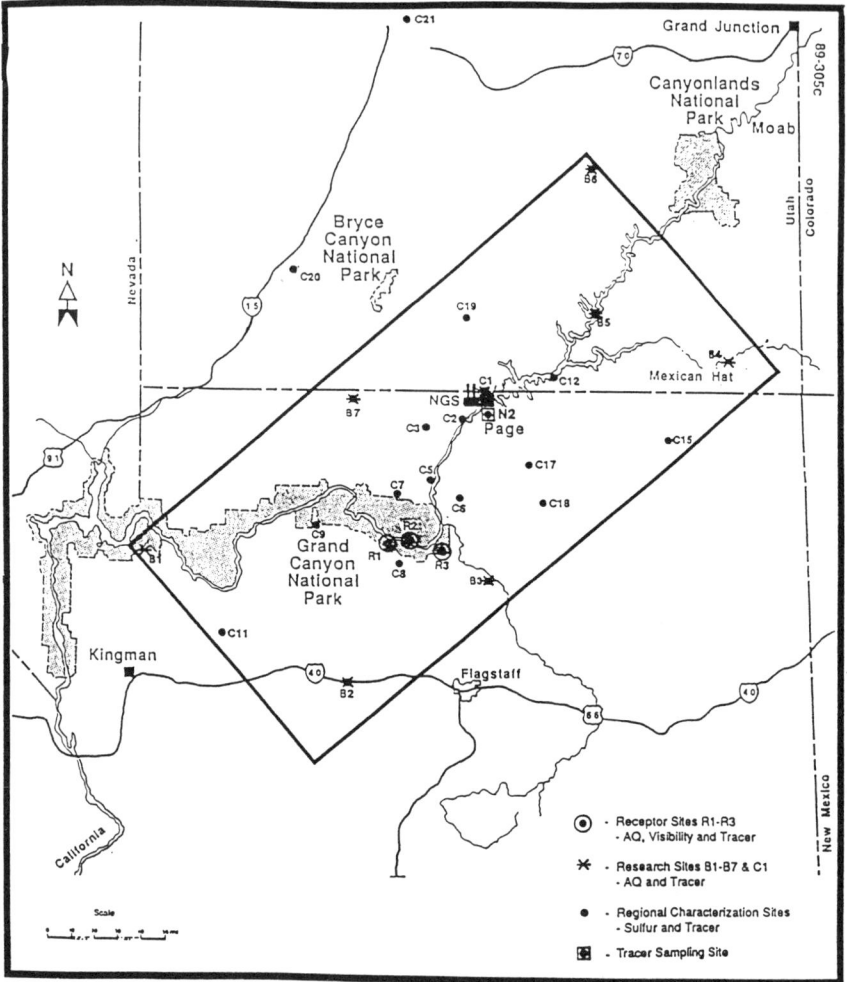

FIGURE 4-17 The locations of tracer and air-quality monitoring sites for the
   1990 NGS Visibility Study. Source: Adapted from Richards et al., 1991.

year meteorological changes substantially can affect wind fields and
chemical transformations, average or typical effects usually cannot be
inferred from measurements made during a particular year. As a result,
the insights obtained from intensive studies must be supplemented with
observations made during routine atmospheric monitoring.

It should be noted that such a program makes great demands on manpower and resources and is extremely expensive. The NGS study is estimated to have cost about $14 million (A.S. Bhardwaja, pers. comm., Salt River Project, Phoenix, Ariz., 1991). Because of the great cost, such large programs rarely are carried out. By way of comparison, the Oregon monitoring program costs about $20,000 per year to operate, and that in Washington costs about $10,000. The entire atmospheric chemistry program at the National Science Foundation had an annual budget of $12 million in 1991.

MODELING OF AEROSOL
EFFECTS ON VISIBILITY

The effect of airborne particles on the optical properties of the atmosphere is determined by the radiative properties (such as sun angle and solar intensity) as well as the chemical and physical characteristics of the particles. The physical relationships among these effects are fairly well understood and have been incorporated in several models described in this section.

In principle, theoretical models can provide information about the sensitivity of atmospheric optical properties to the concentration of selected airborne particle species. If the theory includes information about the dependence of particle size and composition on relative humidity, the models can also be used to quantify the role of adsorbed or condensed water. Thus, models could be used to evaluate the visibility benefits of various emission control strategies.

Optical Modeling (Mie Theory)

The scattering, absorption, and extinction coefficients for atmospheric particles can be calculated from measurements of the size-resolved chemical composition made at a given location and time. The procedure involves converting airborne particle measurements to number distributions; these are then multiplied by particle projected areas and by single particle scattering, absorption, or extinction efficiencies and integrated over the particle size distribution. The cross sections are determined by the optical properties of the particles. For chemically homogeneous spheres with a given complex refractive index, cross sections can be calculated from Mie theory (Mie, 1908; Bohren and Huffman, 1983).

Researchers have used this theoretical approach to investigate the contributions of various species of atmospheric particles to extinction (e.g., Ouimette and Flagan, 1982; Hasan and Dzubay, 1983; Sloane, 1983; Sloane 1984; Sloane and Wolff, 1985). In each of these studies, number distributions of airborne particles were calculated from cascade impactor measurements of size-resolved chemical mass distributions. In calculating the number distributions, particle densities (which are needed to convert distributions from aerodynamic to actual size) and refractive indices (which are needed for scattering cross sections) were determined from the measured size-dependent particle composition. These studies generally have been successful at reconciling measured and calculated scattering or total extinction.

There are two major difficulties in applying Mie-theory models. First, airborne particle characteristics have not been measured with sufficient detail to permit unambiguous modeling. The sensitivity of the scattering, absorption, or extinction coefficients to mass concentrations of a given species depends on the microscopic particle structure (White, 1986). Several particle properties can have an important effect on a chemical species' contribution to extinction, but have not been directly measured. These include:

- The distribution of chemical species among particles in a given size range (i.e., the degree of internal and external mixing);
- Particle density, including particle-to-particle variations;
- Particle complex refractive index, including particle-to-particle variations;
- Particle morphology; shape and phase composition;
- Hygroscopic and deliquescent behavior, including particle-to-particle variations.

All work to date has been based on cascade impactor measurements in which particles in a given aerodynamic diameter range are mixed together on the collection substrates. With such measurements, ad hoc assumptions must be made about internal and external mixing characteristics of different species of airborne particles. To reduce uncertainties, theoretical reconstructions of light scattering, absorption, or extinction by airborne particles should be based on data in which the above-mentioned particle properties are measured.

The second major difficulty with theoretical models is that atmospheric processes are not sufficiently well understood (Sloane and White, 1986). To determine the sensitivity of optical properties to changes in component concentrations, how the size distribution at the receptor will be affected needs to be known. This requires an understanding of how size distributions evolve. However, the current understanding of secondary atmospheric particles usually is inadequate to permit definitive calculations of secondary particle size distributions. Changes in size distributions can be estimated, but these estimates introduce uncertainties that are difficult to quantify.

Despite these limitations, theoretical models have the potential to provide definitive answers on the contributions of particular categories of airborne particles to atmospheric optical properties. For these methods to reach their full potential, improved techniques to characterize aerosols are needed, as is a more quantitative understanding of atmospheric processes.

Empirical Optical Models

When the size-resolved data necessary for the Mie-theory approach are unavailable, extinction usually is modeled as a linear function of aerosol composition:

$$b = \sum_k e_k c_k$$

where b is the extinction coefficient in units of m^{-1}, c_k is the concentration in g/m^3 of aerosol species k, and e_k is the extinction cross-section per unit mass in m^2/g of species k. (To account for the hygroscopicity of certain species, their concentrations may be scaled by $1/(1-RH)$ or some other function f(RH) of relative humidity.)

The unobserved coefficient e_k is usually referred to as the specific extinction or extinction efficiency of the kth species. It can be selected based on a literature review (NPS, 1989) or on Mie calculations for assumed particle size distributions. Alternatively, the coefficient e_k can be estimated by multiple linear regression of measured extinction on measured species concentrations over repeated observations (White and Roberts, 1977; Cass, 1979; Trijonis, 1979; Groblicki et al., 1981). Model simulations show this latter procedure to yield accurate results under favorable conditions (Sloane, 1988).

The linear functional form is commonly found to fit the data quite well, yielding multiple correlation coefficients sometimes approaching unity. However, a good fit cannot be interpreted as a confirmation of the coefficient values in a particular relationship; an empirical model that accurately describes the observed total extinction may be inaccurate in apportioning this total among individual species (White, 1986). The assumed linear relationship would be invalid, for example, if the mean particle size were found to depend on mass concentrations. Because the concentrations of most aerosol species strongly correlate with each other, quite different coefficients could yield fits that are nearly as good (Sloane, 1983); most of the observed variability in extinction is in fact associated with variations in total fine-particle mass.

Regression procedures provide standard estimates for the statistical uncertainty of results, but these tend to be somewhat optimistic, because they overlook the interdependence and heterogeneous variance of neglected factors (White, 1989a,b). Decisions on the many discretionary aspects of the analysis—whether to drop a marginally significant variable, for example, or to include an anomalous observation—are typically based in part on the plausibility of the outcome; this process makes the resulting apportionments less objective than they might appear.

Any empirical model can be distorted by factors extraneous to the optics-aerosol relationship. Errors in the aerosol measurements have systematic effects, even though the errors are themselves random; standard regression estimates tend to overstate the importance of precisely measured species such as sulfates, and understate those of poorly characterized species such as organics (White and Macias, 1987a,b). More fundamentally, concentration and extinction effectiveness may covary through a common dependence on air mass history and meteorological conditions; this yields empirical associations between pollutant concentrations and light extinction that have no basis in cause and effect (White, 1986).

An example of a linear scheme for allocating light extinction to atmospheric components is provided by the RESOLVE (Research on Operations Limiting Visual Extinction) study (Trijonis et al., 1987, 1988). Figure 4-18 presents an overview of the RESOLVE extinction allocation procedure, with boxes in the figure indicating the techniques used to determine and allocate the extinction contributions. For the RESOLVE data, average non-Rayleigh extinction (35 Mm^{-1}) consists of 1 Mm^{-1} from absorption by NO_2, 28 Mm^{-1} from scattering by particles, and 6 Mm^{-1} from absorption by particles. The contributions to total extinction of these three components, as well as those of fine and coarse particles, are known to be exactly linear and additive. Further contributions to particle extinction of different aerosol components (e.g., sulfates, organics, etc.) are approximated as linear by the use of extinction efficiencies. Table 4-4 presents the "consensus" fine-particle scattering efficiencies used in RESOLVE.

Perceptual Air-Quality Modeling

The preceding sections discussed models that deal with the physical processes involved in the interaction of light with particles. Although these models might predict how a specific visibility parameter might change as a result of changes in aerosol properties, these models cannot deal with the issue of visual air quality (VAQ), which is based on the human judgment of visibility (see Appendix B). Human judgments of VAQ often involve aesthetic values. Consequently, these judgments are not necessarily directly related to the measured physical properties of the atmosphere. Indeed, there is no widely accepted technique for measuring VAQ, nor are there satisfactory techniques for projecting the change in VAQ that might result from proposed policy actions.

It should be possible to establish predictive relationships for VAQ by linking changes in emissions to changes in human judgments of visibility and by characterizing the relationships among the important physical and perceptual visibility variables. There are well-established models based on physical and chemical principles that relate pollutant concentrations to emissions. Pollutant concentration patterns, together with descriptions of the visual environment (e.g., sun angle, sky conditions, observer location relative to the scene), can be used to calculate perceptual indices, such as contrast. Least understood is the relationship between

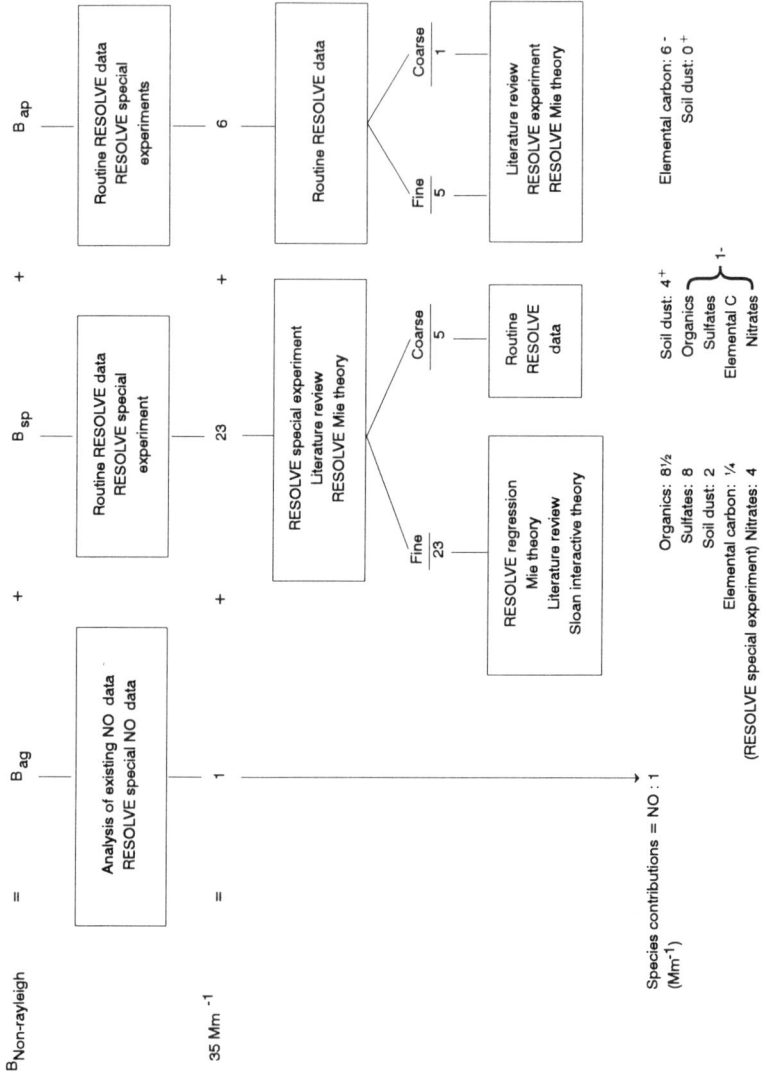

FIGURE 4-18 Allocation procedure for light extinction in the RESOLVE study. Data are averaged over three sites in the Mojave Desert. Source: Trijonis et al., 1987, 1988.

perceptual indices and human judgments of VAQ. Statistical analysis of observer data can be used to establish relationships between perceptual cue judgments (e.g., clarity of objects) and judgments of overall VAQ. The human judgments of cues and overall VAQ can be derived from field observations or from judgments of photographs as described in Appendix B.

Studies have attempted to establish relationships between judgments of the VAQ of natural scenes and various atmospheric and vista parameters, such as mountain/sky contrast, solar angle, extinction coefficient, sky color, and percent cloud cover (e.g., Malm and Pitchford, 1989; Malm et al., 1980; Malm et al., 1981; Latimer et al., 1981; Middleton et al., 1983a, 1984; Hill, 1990; Ely et al., 1991). Summaries of many of these study findings are given in Trijonis et al. (1990). A major implication of this research is that a small number of variables (e.g., sun angle, cloud cover, and scene composition) play a dominant role in judgments of overall VAQ or scenic beauty.

One example of this approach is given by Malm and Pitchford (1989), who suggested using a quadratic detection model to predict the change in atmospheric particle concentrations that would be required to evoke a just-noticeable-change (JNC) in the appearance of contrast-related landscape features in photographs. The change resulting from a new level of emissions could then be expressed as the number of JNCs between an earlier appearance and the appearance under current conditions. It should be emphasized that calculations of detection thresholds and JNCs are assessments of changes in information content in a scene and, as such, they are not necessarily good indicators of human judgments of overall VAQ. For instance, a change of 10 JNCs in a scene with low overall contrast might not be judged to have the same effect as a 10-JNC change in a high-contrast scene. Also, the relationships between JNC and human judgments have not been established under realistic field conditions; similarly the relationship of JNC to other optical parameters has not been studied.

The relationship between emission changes and visibility effects can be described for current conditions using the modeling framework described above. However, predictions of the effects of changes in emissions under a variety of atmospheric conditions are more difficult because the human response to visibility changes must be quantified. Perhaps the easiest way to document the effect of changes on a scenic resource is through photography. By making optical measurements concurrently with color photographs, it should be possible to establish a

TABLE 4-4 Fine-Particle Scattering Efficiencies Used in the RESOLVE Study

Methodology	Scattering efficiencies (m^2/g)			
	Organics	Sulfates	Elemental Carbon	Soil Dust
Multiple regression analysis (based on routine RESOLVE data at the three receptor sites)				
Ordinary least squares	3.7	5.0	0.6[b]	0.4
Corrected least squares[a]	3.8	5.1	-1.8[b]	0.5
Literature review (20 studies with the following adjustments for consistency)				
Nephelometer calibration Airport contrast = 5% Nephelometer λ = 530 mm Organics = 1.5 OC Relative humidity = 40%	2-3	3-6	(2-3)[c]	1-2¼

data base that would show pictorially the correspondence between measured values and the appearance of the scenic resource. Such a data base could capture a wide variety of atmospheric conditions; however, it would not necessarily reflect changes in emissions.

An alternative to taking photographs in conjunction with optical mea-

TABLE 4-4 (continued)

Methodology	Scattering efficiencies (m²/g)			
	Organics	Sulfates	Elemental Carbon	Soil Dust
Mie Theory (for Mojave Desert data)				
Ouimette and Flagan (1982)	2.5^d	3.2	NA	1.4
RESOLVE DRUM sampler data	NA	3.2	NA	1.4
Interactive Mie Theory (based on Detroit size distributions. Sloane, 1986)	3.8	4.7	(3.8)^c	1.3
Consensus	3¼	4¼	1½	1¼

[a]As in White and Macias (1987a), and White (1989b).
[b]Difference from zero not statistically significant.
[c]Elemental carbon grouped with organic carbon.
[d]Mie theory based on volume size distribution for all material, not just organics.
Source: Trijonis et al., 1988.

surements is to use image processing techniques (Williams et al., 1980; Malm et al., 1983; Larson et al., 1988). This method uses atmospheric optical models that simulate the effects of pollutants on a scene. With such an approach, the consequences of a variety of atmospheric conditions and emission scenarios can be represented pictorially; these pictures then could be judged by observers for their VAQ and acceptability. This approach is promising, but the ability of simulations to reproduce the effects obtained in real photographs has not been thoroughly tested (Larson et al., 1988).

EXPERIMENTAL DESIGN, QUALITY ASSURANCE, AND QUALITY CONTROL

The quality of data acquired in intensive or routine measurement programs depends, to a large extent, on the design of suitable measurement strategies and on the implementation of appropriate quality assurance plans. Issues that need to be considered in establishing measurement strategies include sampling periods and locations, sampling and analytical methodologies, choice of instrumentation, and coordination of activities among participants. Quality-assurance plans require that a significant portion of the budget for a given project be used for replicate analyses, collocated sampling, blanks (no sample collected), and the quantification of analytical capabilities including precision, accuracy, and detection limits.

Numerous factors have led to compromises in experimental design and quality assurance for visibility monitoring programs. For example, budgets are always limited and quality-assurance programs are expensive. If a significant portion of the available funds is invested in quality assurance, then the scope of work (number and diversity of sites, number of substrates to be analyzed, etc.) must be reduced. Because of limited funds, for example, independent system audits have not been incorporated in the IMPROVE network (Pitchford and Joseph, 1990). As a result, when an anomaly is detected, it is not clear if the trend is real or if it is an artifact of the techniques used; an important example, discussed earlier, is the sharp increase in the organic carbon concentration trends in the IMPROVE data.

Because of similarities in the sampling and analytical methods used in the current (1988-present) IMPROVE network to those employed at similar sites in the earlier Western Particulate Monitoring Network (1979-1986) and the NPS National Fine Particle Monitoring Network (1982-1986), an opportunity may exist to greatly extend the temporal coverage for certain variables and sites. However there is a concern about the compatibility of some measurements across the different monitoring programs. For example, during the past two decades visibility measurements have been made with nephelometers, telephotometers, cameras, and transmissometers. Because the data obtained with these various techniques are not equivalent, the existing record provides limited information on temporal and spatial trends. These examples underscore the importance of designing long-term measurement strategies around techniques that have been thoroughly tested and accepted by the research community. A greater effort should be made (with associated funding commitments) to test the compatibility (or lack thereof) of current and historical monitoring efforts through additional methods intercomparisons, filter analysis, and statistical assessments.

Priority should be given to establishing an independent scientific advisory committee with oversight responsibility for national visibility monitoring networks including IMPROVE. Such oversight could help to eliminate errors that may leave large gaps in the historical record, would help to ensure a consensus on measurement and sampling strategies, and would facilitate drawing on the broad experience that is available nationally.

The continuity of data records is vitally important. In this regard the committee expresses concern about the future of airport visual range measurements which are now made by human observers. As discussed in Chapter 2, these human observations have provided most of the information that has been used to assess long-term visibility trends and to relate these trends to changes in emissions. Despite the subjectivity and inherent variability of individual observers, the population of observers changes less over the decades than do the instrument technologies involved in other pollution-related measurements. However, these airport visual range measurements will soon be stopped and will be replaced by a different technique. As part of a comprehensive and long-planned modernization program (NRC, 1991a), the National Weather Service, Federal Aviation Administration, and Department of Defense intend to replace the present network of human observers by a network of instrumental visibility monitors. The new Automated Surface Observing System (ASOS), due to be in place by 1995, should provide better spatial coverage, higher time resolution, and improved standardization. Unfortunately, the new measurements will also be more narrowly focused on aviation needs.

The principal limitation of the ASOS measurements for haze studies is that they will not record variations in range during good visibility conditions. Only three visual range values in excess of 4 miles will be reported: 5 mi, 7 mi, and > 10 mi (J.T. Bradley, pers. comm., 1991, NOAA/NWS). However, in most regions the visual range is greater than 10 miles most of the time. Thus, this measurement technique will only provide useful data under extreme conditions such as those associated with the worst regional haze events in the East. Consequently, if there is any change in visibility conditions in the central or western parts of the nation (where most national parks and wilderness areas are located), the changes will not be observed until the degradation is severe.

The ASOS sensor will be a forward-scattering (40°) pulsed visible-light, open-air monitor. This instrument, like the candidates over which it was chosen, has been field tested against human observers and transmissometers (Bradley and Imbembo, 1985; Bradley, 1989) to assess its operational response to distinct weather classes (fog, rain, snow, and haze). There is as yet no commitment to continuing studies such as those currently planned for temperature and precipitation measurements (NWS, 1991); such studies are important for documenting the effect of the change-over on climate data continuity.

The nation's existing visibility monitoring network has been implemented largely by the National Park Service. Its approach has been aggressive and innovative. However, the NPS group is too small to have in-house expertise in all aspects of work pertaining to visibility monitoring, airborne particle sampling and analysis, and data interpretation. EPA personnel have a broad range of experience with the chemical measurements made in visibility sampling networks, and ought to be involved. Lamentably, EPA funding for visibility monitoring and research has been insignificant (see Figure 3-1) and, as a result, EPA's participation has been minimal.

In summary, the nation's visibility measurement program needs to establish a balance between innovation and standardization, and between the scope and the quality of work. It is important that routine monitoring networks provide data that are comparable over decades. In order to achieve these objectives, an independent science advisory panel with EPA sponsorship should be established. This would help to ensure a wider participation among the scientific community on important decisions regarding visibility monitoring and research.

We are particularly concerned that the historical record of visual range measurements made at airports by human observers will be interrupted by the new Automated Surface Observing System, which will not provide useful visibility information. We recommend that airports should be equipped with integrating nephelometers that are sensitive enough to measure the range of haze levels encountered in the atmosphere. Nephelometer data are closely correlated with visual range measurements made by trained observers. It is essential to have a continuous record of such data during the 1990s in order to determine the effect on visibility of the acid rain controls that will be implemented in response to the 1990 Clean Air Act Amendments.

SUMMARY AND CONCLUSIONS

Nature has contrived to maximize the effect of anthropogenic activities on visibility. As a result of physical and chemical processes in the atmosphere, a large fraction of anthropogenic primary and secondary airborne particles accumulate in the 0.1 to 1.0 μm diameter size range where removal mechanisms are least efficient. Because these particles have sizes comparable to the wavelength of sunlight, scattering and absorption are at a maximum. Thus, anthropogenic particles tend to accumulate in the size range that contributes most to haziness per unit mass.

As discussed in Appendix A, a great deal is known about the processes that produce visibility-impairing particles. However, there are some major gaps in the understanding of visibility impairment. For example, although organic particles can contribute significantly to visibility impairment, especially in the West, there is poor understanding of the concentration and composition of atmospheric organic materials in the particle and the vapor phase (Appendix B). There is also poor knowledge of the relative importance of primary and secondary organic carbon species and of their anthropogenic and natural sources (Appendix A). Furthermore, there are major problems associated with collecting representative organic carbon particulate samples. Also, important insights about atmospheric transport and transformations of visibility-impairing particles would be possible if instrumentation for measuring concentrations of particulate species on a continuous basis were available (Appendix B).

The current understanding of visibility is based on information from a variety of sources including historical data on airport visual range, data from routine state and national visibility networks, and intensive, short-term field programs. While these measurements have provided a good qualitative picture of visibility, there remain important gaps in measurement protocols. Inadequate attention has been paid to ensuring self-consistency between measurements using different sampling or analysis methods. As a result, it is sometimes difficult to determine whether the observed trends are real or are due to changes in measurement procedures. Also, there are no EPA-recognized performance standards for aerosol and optical sampling instruments. Many of the instruments used for visibility monitoring are expensive and are not readily available. This has led to a lack of uniformity in field measurements and to uncertainty within the regulatory community as to which sampling methods should be used. As discussed in Appendix B, adopting the integrating nephelometer as the instrument of choice for routine measurements of haze in monitoring networks would go a long way towards implementing one valuable standard for optical measurements.

The planned transition from human to automated airport visibility monitoring has unfortunate implications for visibility monitoring. Most existing information about historical haze trends is from airport data. The new automated instruments are designed to measure the very poor visibility conditions that are of primary concern for aviation safety, but will provide little or no information on haze under typical visibility conditions. We recommend that the proposed instrumentation be supplemented with integrating nephelometers which would permit measurements of light scattering coefficients under typical visibility conditions. Intercomparisons have shown that light scattering coefficients measured with integrating nephelometers are closely correlated with human observer visual range data. The addition of nephelometers to airport instrumentation would ensure that haze levels are monitored over a broad and representative geographic scale, thereby providing important information on spatial and temporal trends of regional haze. It is especially important that such trends be documented during the coming decade, so that the effect of acid rain controls on haze levels can be qualified. Emissions reductions being implemented to reduce acid rain provide atmospheric scientists and regulators with an unparalleled experiment of opportunity. It would be a serious and potentially costly error to fail to record key data.

A large fraction of submicron, visibility-impairing particles is produced in the atmosphere. This is a major obstacle to assessing the effects of new or existing pollution sources on visibility impairment. While laboratory studies have provided a great deal of information about rates and chemical mechanisms of gas-to-particle conversion, it is often difficult to apply this knowledge to the atmosphere. Atmospheric transformations can occur in clear air or in clouds. In-cloud processes depend on the availability of oxidants and on the frequency and duration of cloud processing. Clear air chemical transformations depend on sunlight intensity, and on the blend of NO_x and organic gases. Because these phenomena are so complex, they are difficult to characterize either empirically or theoretically. For example, if clouds or fog are involved and if all the H_2O_2 reacts with SO_2, then the conversion of a large fraction of the SO_2 would be a reasonable estimate because of the rapid oxidation in droplets; if clouds or fog are not involved, then little SO_2 would be oxidized in transit because of the slow rates of homogeneous oxidation. Clearly, a better understanding of atmospheric conversion processes is needed to link emissions adequately to their effect on visibility.

Visual air quality goals are usually stated in terms of some readily measurable quantity such as visual range or extinction coefficient. Ultimately, these criteria are, or should be, related to the human visual perception of what is desirable or acceptable. People base their judgments of visual air quality on a variety of perceptual cues, and the relative importance of these cues varies with the setting. A better understanding is needed of the factors that affect perception. It is also important to communicate visually the possible results of a visibility improvement program (or its absence). One promising technique is the use of computer visibility models to generate photographic representations of scenes under various conditions. However, more research is needed to establish the general validity of this approach.

The phenomena that lead to visibility impairment are reasonably well understood, particularly when compared with many other environmental issues which have much larger uncertainties. Nevertheless, a number of scientific and technical issues need to be resolved in order to reduce uncertainties in the understanding of the relationships between human activities and visibility.

5

Source Identification and Apportionment Methods

Two questions invariably present themselves to those who must devise ways to protect or improve visibility: "Which sources cause the visibility problem under study?" and "How large is each significant source's contribution of visibility-reducing particles and gases?" The first question, of source identification, must be answered to reach even a qualitative understanding of the problem. If emission controls are to be applied efficiently to the major sources, then one needs a quantitative understanding of each source's contribution to the visibility problem. The quantitative assignment of a fraction of an entire visibility problem to one or more sources is called source apportionment.

It usually is impractical to conduct a source apportionment study by experimenting on all the major air pollution sources in a large region—that would require an expensive control program just to observe its effects. Instead, analytical methods and computer-based predictive models have been developed to quantify the connection between pollutant emissions and changes in visibility. There are several major classes of methods and models. Speciated rollback models are relatively simple, spatially averaged models that take changes in pollutant concentrations to be directly proportional to changes in regional emissions of these pollutants or their precursors. Receptor-oriented methods and models infer source contributions by characterizing atmospheric aerosol samples, often using chemical elements or compounds in those samples as tracers for the presence of material from particular kinds of sources. Mechanistic computer-based models conceptually follow pollutant emissions from source to receptor, simulating as faithfully as possible the pollutants' atmospheric transport, dispersion, chemical conversion, and deposition. Mechanistic models are source oriented; they take emissions as given and ambient concentrations as quantities to be estimated. Because these models require pollutant concentrations only as initial and boundary conditions for a simulation they can therefore be used to predict the effects of sources before they are built.

The members of the committee do not aim to give advice on how to choose a single best source apportionment technique for analyzing a given visibility problem. Instead, the committee offers guidance on how to view the air quality modeling process. The way to view air quality models is that they provide a framework within which information about the basics of the problem can be effectively organized. This basic information includes data on the air pollutant emission sources, observations on meteorological conditions, data on the ambient air pollutants that govern visibility, and information on emission control possibilities. The quality of the outcome of the modeling process usually depends as much or more on the quality of the data used as inputs to the model than it does on the modeling method chosen, thus placing a premium on the accuracy with which the basic facts of the problem are known. The objective of the analyst is to capture the scientific relationships between emissions and air quality such that important decisions about the effect of emission controls or about the siting of new sources can be answered. Depending on the decisions to be made, there may be either a strict or a more relaxed requirement for technical accuracy or detail. Federal regulatory programs are permitted to make regulatory decisions in the face of continuing scientific uncertainty. Within many likely regulatory structures, attribution of contributions due to individual sources may be unnecessary—attribution to classes of like sources or upwind geographic regions would suffice. In those cases where approximate answers are satisfactory, there are many possible ways to approach answering questions about the relationship between emissions, air quality, and visibility.

When approaching the analysis of the causes of a particular visibility problem, the best strategy is generally to use a nested progression of techniques from simple screening through more complex methods. Simple methods can be used to screen the available data and find an approximate solution. Next, more complex methods can be applied to determine source contributions with greater resolution. Advanced methods are appropriate when a problem is scientifically complex or when control costs are high enough that more detailed or more highly resolved information is warranted. In general, the simpler methods use subsets of the data required by the more complex methods; this nesting of data requirements yields a natural progression of techniques. Receptor-oriented methods, for example, form a progressive series, where each additional measured variable contributes new information. When it is necessary to collect data to support a more complex method, simpler methods often can be applied inexpensively using these same data. Even when the simpler methods fail to produce sufficiently specific findings, the information they offer can be valuable because it is easy to grasp and to communicate to policy makers.

The cause of visibility impairment can take two general forms: widespread haze and plume blight, and source apportionment models for them must account for markedly different physical, chemical, and meteorologic processes. The committee evaluated models applicable to both kinds of impairment but focused on source apportionment models for multiple sources that contribute to widespread haze, because regional haze is the main cause of reduced visibility in Class I areas. A later section of this chapter contains a section on single source and plume blight models, and that section is prefaced by a description of the differences between widespread haze and plume blight.

First, we provide criteria for evaluating the relative merits of source identification and apportionment methods in the context of a national program to protect visibility. We then evaluate various methods, roughly in order of increasing resources required for their application: simple source identification methods; speciated rollback models; receptor models, including chemical mass balance models and regression analysis; models for transport only and for transport with linear chemistry (these are simplified mechanistic models that are either receptor or source oriented); advanced mechanistic models; and hybrid models. These methods are described in Appendix C. Appendix C should be read before encountering the critique of modeling methods that follows, as that appendix contains the definition of certain uncommon modeling methods (e.g., the *speciated* rollback model) and important text related to air quality models based on regression analysis. We generally describe models that predict source effects on atmospheric pollutant concentrations only, and not on visibility itself; these are known as air-quality models. It is understood that once pollutant concentrations are apportioned among sources, the source contributions to light extinction can be calculated by the optical models discussed in Chapter 4. We then discuss the selection of apportionment methods to assess single source siting problems and air-quality problems other than visibility.

CRITERIA FOR
EVALUATING SOURCE IDENTIFICATION
AND APPORTIONMENT METHODS

A national visibility protection program could employ many alternative modeling methods. Source apportionment studies are generally best conducted through the successive use of simple screening models followed by more precise methods. At each stage of this process, one must decide whether further analysis and investigation by more complex methods is warranted. How can one judge the merits of an investment in more sophisticated analysis? Will a particular source apportionment approach yield results of acceptable accuracy? Is that approach consistent with resource constraints and legal requirements?

This section sets forth criteria for use in comparing alternative methods of source apportionment. Some criteria might seem to reveal some methods as either adequate or inadequate, but the committee's intention is to provide standards for comparing methods across the board. Few, if any, source apportionment methods can be rated highly in all respects, and it can be expected that regulatory decisions will be based on imperfect models. Some of the desirable properties of source apportionment methods (technical validity and simplicity, for example) can in fact conflict with one another. However, the following criteria should help make more informed decisions about the suitability of a given method for application to a particular visibility problem.

Technical Adequacy

The first set of criteria concern the technical adequacy of source apportionment methods.

Validity

The methods for modeling air quality and visibility should have sound theoretical bases.

Air-quality models can be based on solving the atmospheric diffusion equation, which provides a mechanistic description of the atmospheric physics and chemistry of pollutant transport, transformation, and removal. Simplifying assumptions and approximations usually are made to speed the solving of these equations. In a particular model formulation, these assumptions should be made to capture the essence of the problem at hand rather than to oversimplify the problem to the extent that there is little assurance that source–receptor relationships are represented correctly. The same criteria apply to mechanistic models for predicting the optical properties of the atmosphere described by Mie theory. For example, the derivation of the calculation scheme must be understood, and the effects of any simplifying assumptions should be small enough that reasonably accurate results can be obtained.

Empirical models that relate emissions to air quality and air quality to visibility parameters also can be judged for their theoretical foundations. Some empirical models are derived directly from the differential equations that explain the physical phenomena of interest, and they have a well-understood theoretical basis. Other empirical models use the concepts of materials balances or concepts that require the whole to equal the sum of its parts. Finally, some empirical models are purely phenomenological, with little structural relationship to physical processes in the atmosphere. Source apportionment models should be examined for valid theoretical bases, and models that are not developed carefully and in the light of first principles should not be used.

Compatibility of Source and Optics Models

It should be possible to link the model for source contributions to pollutant concentrations to a model for pollutant effects on visibility. The assessment of source contributions to visibility impairment generally requires two types of calculations. First, source contributions to ambient concentrations of pollutants are computed; next, the effects of those pollutants on visibility are determined. The results of the ambient pollutant calculation should satisfy the input data requirements of the visibility model. Not all air-quality and visibility models are compatible; for example, a conventional rollback air-quality model probably will not provide the particle size distribution data needed to perform a Mie theory light-scattering calculation.

Input Data Requirements

The data required for application of a particular approach to source apportionment should be understood and obtainable in a practical sense. The input data requirements of the various methods for apportionment differ tremendously. A rollback air-quality model might require only tens to a few hundred observations on emissions source strength and air quality. A photochemically explicit model for secondary-particle formation, however, easily can require millions of pieces of spatially and temporally resolved emissions and meteorologic data, along with size-resolved and chemically resolved aerosol data to check the model's calculations. It should not come as a surprise that the more theoretically elegant techniques place the greatest demands for field experimental data.

Evaluation of Model Performance

The performance of the candidate source apportionment model should have been adequately evaluated under realistic field conditions.

Confidence in model performance builds up over time as a result of its successful applications. New and untested systems require thorough testing and evaluation before they can be recommended for use in a national visibility program.

Source Separation

The source apportionment method should distinguish the sources that contribute to a particular visibility problem with the level of accuracy required by the regulatory framework within which the model must operate.

Some source apportionment methods (receptor models) can attribute visibility impairment with considerable accuracy to generic source types (sources of sulfur oxides, for example) but cannot distinguish among different sources of the same type (they often cannot tell which power plants are contributing to the problem). Other modeling methods (such as speciated rollback models) could predict the effect of individual sources in a region on air quality at each receptor (the prediction would be that the atmospheric concentration increment is proportional to the fraction of the emissions contributed by that source to the air basin), but that prediction might not be accurate. The source separation achieved by a particular method should serve as a basis for an effective regulatory program, given that the amount of source separation needed depends on the legal framework within which that program must operate.

Temporal Variability

The source apportionment method should account for the temporal character of the visibility problem. Many models directly calculate pollutant concentrations over averaging times that range from a day to as long as a month or a year. However, reduction in visual range is instantaneous, and often it is impossible to explain short-term reductions in visibility from data on long-term average pollutant loadings. A model's averaging time can limit its usefulness in visibility analysis.

Geographic Context

The source apportionment approach should be suited to the geography of the visibility problem; the spatial characteristics of an air-quality model should be matched to the spatial character of the problem at hand. If the terrain of interest is complex (for example, the Grand Canyon), then models that assume flat topography might not capture the location of plumes that travel between the observer and the features of the elevated terrain. In the case of grid-based air-quality models, the spatial scale of the grid defines the smallest area for which air quality can be examined. If the grid system is too coarse, the essence of a source–receptor relationship can be lost.

Source Configurations

The source apportionment method should be suited to the physical arrangement of the sources. Some air-quality models, such as rollback models, assume that the locations of emissions sources will not change. Some receptor-oriented models can apportion emissions from existing sources but cannot readily predict emissions from new ones. Some mechanistic models are better than others at predicting the effects of changes in the elevation of emissions.

Error Analysis and Biases

The method's error characteristics should be known. No technique can be expected to be completely accurate in its attribution of an environmental effect to a particular source. The limits of scientific knowledge about the atmospheric dispersion of air pollution and the workings of chance in atmospheric processes prevent absolute certainty. Obviously, the greater a technique's expected error, the less useful it will be in a regulatory program. It is best to conduct a systematic analysis of the error bounds that surround the predictions made by a candidate method. It should be known whether the errors affect all source contribution estimates equally or whether biases are likely to distort the relative importance of different sources.

Attribution techniques are often skewed in their error characteristics; a given technique, for instance, could be known to under- rather than overpredict the contribution of a source to an effect. A technique's error characteristics could restrict its use to a specific type of regulatory program. For instance, a technique that systematically overpredicts could be useful in a technology-based program that requires only a conservative screening model; the same technique might not be useful in a program that attempts to base control requirements on a more precise estimate of a source's effects. Similarly, a technique that systematically underpredicts source contributions would be of limited use in a program, such as that prescribed by the Clean Air Act, which takes a preventive approach to environmental problems.

Availability

A source apportionment method should be fully developed, available in the public domain, and ready for regulatory application; otherwise, further research and development should take place before it can be recommended for use in a national visibility protection program.

Some promising source apportionment methods (such as the models for atmospheric formation of size-distributed secondary particles, linked to Mie theory light-scattering calculations) are now being developed but are not ready for widespread use.

Administrative Feasibility

Technical merit alone does not determine the suitability of a source apportionment method. If a particular approach is to be the basis for a national program of visibility protection, it should be structured to fit the administrative requirements of a regulatory program.

Resources

The resources required to apply a particular source apportionment system should be clearly understood. Before a source apportionment method is selected, it should be known how many people, how much time, and how much money are required to start and maintain an assessment of source contributions to visibility impairment in Class I areas. Otherwise, it is unlikely that a regulatory or research program would be established with the amount of support needed to do the work correctly.

Regulatory Compatibility

The source apportionment method should be compatible with the various regulatory frameworks that have been or could be imposed on the national visibility problem.

If a national program were based on the principle of non-deterioration of existing air quality, there might be little need to determine the precise causes of current visibility impairment. A system of source registration and emissions offsets might suffice to meet regulatory needs. Alternatively, a regulatory program might specify national visibility standards and require remedial action to improve visibility in particular Class I areas by a specified amount. In that case, a source apportionment system would have to be able to apportion existing visibility impairment among contributing sources and to forecast whether a changed distribution of emissions would lead to compliance with standards.

Multijurisdictional Implementation

Where several government agencies have jurisdiction over different parts of a regional visibility problem, the source apportionment method should be suitable for use on a common basis by all parties. Responsibilities for visibility protection in Class I areas are now divided among the National Park Service, the Forest Service, the Fish and Wildlife Service, the Environmental Protection Agency, and state agencies. Some simple source apportionment systems, such as plume blight models, might be applicable for use by each of these agencies and could be used nationwide by different agencies acting independently. On the other hand, regional haze analyses that extend over several states and incorporate several Class I areas within a single analytical framework would need large amounts of data and might require a more unified approach to visibility regulation than has been taken to date.

Communication

The source apportionment approach should facilitate open communication among policy makers. One can envision two models of equal technical accuracy, one based on readily understood material balance assumptions, the other consisting of a mathematic simulation that policy makers must accept on faith. The more easily understood model could be preferred. Within the framework of an easily understood model, policy makers could conduct a rapid (if informal) analysis of the effects of alternative policies; rapid analysis and discourse might be impossible with a less understandable model. If policy judgments must be made by officials who do not have technical expertise, then the ease of communicating results to policy makers will be an important consideration in model selection.

Economic Efficiency

The source apportionment method should support an economic analysis directed at finding the least expensive solution to a visibility problem. In addition to being able to identify the source contributions to a regional visibility problem, the method should be capable of being matched to a analysis of the least expensive way to meet a particular visibility improvement goal. Some source apportionment methods, particularly linear methods, are readily linked to cost optimization calculations. Non-linear chemical models can be difficult to use within a system that requires economic optimization.

Flexibility

Source apportionment methods that can be adapted readily to new scientific findings or to the changing nature of a particular visibility problem are preferable to less flexible methods. Conditions outside the range of past experience will probably arise in the future. Some source apportionment systems could be more adaptable than others to new circumstances and new scientific understanding.

Balance

There should be a balance between the resource requirements and the accuracy of a source apportionment system. A source apportionment method might require elaborate field experiments to supply data for simplified calculation schemes whose inherent inaccuracy does not warrant such great expense. One also can envision elaborate calculation schemes whose sophistication exceeds the quality of the available input data. The effort required and the cost required of data collection should strike a reasonable balance with that required for data analysis.

CRITIQUE OF SOURCE IDENTIFICATION
AND APPORTIONMENT METHODS

This section presents the merits and deficiencies of the most widely used source identification and apportionment methods, using the criteria just set forth. Each method has advantages. Some are potentially accurate but have not been thoroughly evaluated or require costly, elaborate data collection. Others are well understood, fast, and inexpensive, but do not accurately address visibility impairment caused by secondary particles.

Two or more methods are often used to obtain a reasonably complete picture of a given visibility problem. The rapid lower order analyses are most effective in the early stages of a source apportionment study because they yield approximate solutions that can guide further studies. The more complex methods then can resolve more difficult technical issues if a more accurate analysis of the visibility problem is needed. The following discussion should therefore be viewed as a guide for successively unraveling various aspects of a visibility problem rather than for selecting a best single method of analysis.

Source Identification Methods

In some cases, sources of visibility impairment in Class I areas can be identified directly by simple empirical methods. Source identification is most often possible for plumes (rather than for regional haze) caused by individual nearby sources. Identification methods include visual or photographic methods, emission inventories, and simple forms of receptor modeling.

Visual and Photographic Systems

The most direct and convincing means of identifying a source that impairs visibility is to visually track a plume from its source. Photographic evidence, in the form of videotape or slides, is essential for verifying these observations. Still photographs can be used to estimate or document visual range and variations in visibility over time. Photography is relatively inexpensive for documenting periods of reduced visibility and concurrent weather conditions (fog and layering of plumes, for example). Slides can be interpreted as described in Chapter 4.

Videotape images generally have poorer resolution than do still photographs, but they can illustrate transport and changes in visibility. Videotape can help researchers to identify wind flows associated with low-visibility episodes and can aid in identifying periods when in-cloud chemistry can affect sulfate formation. Given the relatively low cost of time-lapse film or videotape, this technique should be considered for studies of the dynamics of visibility.

In many areas of the West, fire lookout personnel provide valuable information on visibility in Class I areas, on the sources that can impair visibility, and on the frequency and duration of impairment. Although fire lookouts often photograph Class I area visibility conditions each day for state and federal land management agencies, few agencies require written records of the observations of the sources of visibility impairment. In Oregon, observations and photographs taken by fire lookouts were used to develop prima facie evidence that agricultural burning was severely impairing visibility within the Eagle Cap and Central Oregon Cascade wilderness areas (Oregon Department of Environmental Quality, 1990). Regulations now are being developed to control these sources.

Emission Inventories and Source Activity

Visual and photographic information can be supplemented by emission inventories and source activity data. For example, a fire lookout's observations and photographs often can be confirmed by independent records of source activities (acres of vegetation burned daily, locations and dates of wildfires, and operating schedules of industrial facilities). However, the quality of emission inventories must be assessed carefully, because such inventories often are incomplete, and significant point or area sources can be excluded.

Simple Tracer Applications

When a source emits chemically distinct particles that are suspected to be impairing visibility, it could be possible to obtain evidence of a significant contribution from that source by applying simple receptor modeling methods. For example, at kraft process pulp and paper mills, recovery furnaces can emit large quantities of fine-particle hydrated sodium sulfate. If the emissions from such a furnace are suspected of impairing visibility in a nearby Class I area and if sodium sulfate is a major constituent of the fine-particle mass measured at a receptor site at the area's boundary; then a reasonable argument can be made that the furnace is a significant source of visibility impairment.

Evaluation of Source Identification Methods

Because source identification techniques are intended only to determine the sources of visibility impairment—and not to apportion that impairment quantitatively, it is inappropriate to apply all of the criteria listed earlier to them. It is worth noting, however, that simple qualitative tools are easy for most regulatory agencies to use. Although the information they provide is mostly qualitative, it could be sufficient in some cases to support regulatory action. For example, the direct evidence gained through photographs of plumes impairing visibility in a Class I area can be compelling. Any or all of these methods should be considered as basic elements of a visibility-monitoring program.

Because the methods are relatively simple, their costs generally fall within the resource limits of regulatory agencies. These methods are best suited to air-quality management programs that prohibit nuisances or visible emissions, and they can be used in multijurisdictional programs. Because these methods document visible plumes only, they cannot be used alone for complex source apportionment, nor can they predict the impairment that would be associated with new sources. They might be unable to discriminate the effects of several sources.

Speciated Rollback Models

As defined in Appendix C, speciated rollback models are simple, spatially averaged, conservation-of-mass models disaggregated according to the major chemical components of aerosols (Trijonis et al., 1975,

1988). These models are based on the simultaneous use of several separate rollback calculations, one for each of the chemically distinct major contributors to a regional visibility problem. The assumption is made that the concentration above background of each chemically distinct airborne particle type (e.g., sulfates, nitrates, organic carbon particles, etc.) is proportioned to total regional emissions of a particular primary particle species or gaseous aerosol precursor. Speciated rollback modeling is a relatively simple method of apportioning visibility-impairing pollutants among sources in a region.

Technical Adequacy

Two assumptions must be met for a speciated rollback model to be valid. First, the relative spatial distribution of emissions for each species (or species precursor) must remain the same when the quantity emitted changes. Second, the concentration of each pollutant must be linearly related to that of a controlling precursor; atmospheric transformation and deposition rates must be independent of pollutant concentration. Two other assumptions, generally less problematic, are inherent in the model. It is assumed that the weather patterns will be the same as those represented in the air-quality data base used to construct the model, and it is assumed that the temporal distribution of emissions stays fixed when emission quantities are changed.

Deviations from the assumption of spatially homogeneous emissions changes are likely to occur in areas where air quality is most critical at a single receptor point, where a primary air pollutant is emitted from a source for which stack height is important, where there are local emissions of coarse dust, or where two or more important source types differ radically in spatial distribution and only one of these sources is controlled. Deviations from this assumption are less important if spatially averaged air quality is the important predicted quantity, or if the air-quality problem is widespread, as is the case for regional haze. If, for a particular chemical component one dominant source with a fixed spatial distribution is being controlled, then the assumption of spatial similarity of emissions before and after control usually is met.

The assumption that secondary-particle formation is linear in the emissions of a dominant precursor is justified in some cases but not in others. One factor that affects the rate of secondary-particle generation is the rate of OH radical formation in the atmosphere. The OH radical oxidizes organic compounds, forms HO_2 radicals, and promotes NO conversion to NO_2, nitric acid (HNO_3) formation from NO_2, a fraction of the sulfuric acid (H_2SO_4) formation from SO_2, and O_3 build-up. HO_2 radicals are the major source of hydrogen peroxide (H_2O_2), which is the major cause of production of sulfuric acid [and NH_4HSO_4, $(NH_4)_2SO_4$] through the cloud phase oxidation of SO_2 (and HSO_3^-). Ozone generation in the atmosphere is highly nonlinear and controls many other nonlinear processes involved in secondary-particle formation.

In some cases (near large sources or in coherent plumes) the use of linear rollback models to apportion secondary particles could lead to poor control strategies. For example, much sulfate production probably occurs in clouds as a result of oxidation by H_2O_2. Only a finite amount of H_2O_2 is formed in a given air parcel, depending on the conditions of NO_x, RH, H_2O, O_3, and on the amount of sunlight. If for a particular region, the concentration of SO_2 is twice as great as that of H_2O_2, $[SO_2]$ = 2 x $[H_2O_2]$, then only half of the SO_2 can be oxidized locally. Under such H_2O_2-limited conditions, if SO_2 emissions were cut by 20%, sulfate concentrations would not be reduced by this amount. Where H_2O_2 concentrations exceed SO_2 concentrations, however, sulfate formation would decline linearly with SO_2 emissions, as predicted by rollback calculations. Additional calculations and field investigations are needed to determine whether the linear chemical assumptions inherent in a rollback calculation are met in a particular geographical area.

The recent NAPAP integrated assessment modeling study (South, 1990) suggests that linear rollback works well for sulfates, at least when they are spatially averaged. Using the Regional Acid Deposition Model, (RADM) a mechanistic model, the NAPAP study concluded that a 36% reduction in SO_2 emissions in the eastern United States would produce a 37% reduction in sulfates averaged over the East. Evidently, any nonlinearities in sulfate chemistry were compensated for by spatial distribution effects. Power plants, the controlled source category, are somewhat more efficient at producing sulfate from given SO_2 emissions than are other sources. (R. Dennis, pers. comm., EPA, Research Triangle Park, N.C., 1990).

Optics models based on pollutant extinction (scattering and absorption) efficiencies—extinction per unit mass concentration, obtained from Mie theory, regression analysis, or the scientific literature—would tend to fit most naturally with the speciated rollback model. The use of optics models that invoke extinction efficiency values would preserve a structural consistency in that the approach would be organized by aerosol composition from emission inventory to air-quality model to optics model.

One advantage of speciated rollback models is that their input data requirements usually can be met based on data gathered by most air-pollution-control agencies. The required input data are speciated concentrations of ambient pollutants within the region of interest, speciated data on background pollutant concentrations, and regional emission inventories for each pollutant or controlling precursor. The predicted source separation can be as detailed as that of the emissions inventory.

Rollback models are among the few models that have been subjected to comparison of model predictions with monitored air quality before and after reductions in emissions. Rollback models have been used with mixed success to evaluate the effects of emissions changes (NRC, 1975, 1986; Trijonis, 1979, 1982a; Husar et al., 1979, 1981; Eldred et al., 1981; Husar, 1986, 1990; Sisler and Malm, 1990). Sometimes, poor results from a rollback model can be ascribed to a failure to meet the assumptions of fixed spatial emission patterns and linear relationships between secondary aerosols and precursors. Other failures could result from an insufficient test period (so that meteorological conditions were not equivalent before and after the emissions change), variations in external factors such as background concentrations, or an emission reduction so small that expected changes are statistically insignificant. When the model's assumptions hold within a reasonable approximation, conservation of mass can show that the model is valid.

The speciated rollback model can be applied to any temporal concentration specification, such as annual median, annual average, worst 10th-percentile, worst daily average, or worst one-hour average. The speciated rollback model also can be applied to any region that meets the constraints on the spatial distribution of emissions changes.

Because the model neglects the spatial distribution of emissions, there is a potential for biases if some sources produce lesser air-quality effects because of their locations (for example, sources located downwind of a receptor area). Unless the model is modified to included a transfer coefficient for such sources, it will not correctly credit these sources with the effects of their emissions.

The speciated rollback methodology is fairly simple and does not require a significant model development effort. Some basic research is needed to improve the approach—scientists need a better understanding of organic aerosols and their origins—but such research could be required for any modeling approach.

The speciated rollback model can be analyzed with error propagation techniques and "judgmental" uncertainties to provide error bounds and to reveal the uncertainties most in need of resolution (Trijonis et al., 1988). One advantage of the speciated rollback model is that its errors tend to be confined to subelements of the model. For example, errors in estimating organic or elemental carbon particle emissions have only a secondary, smaller influence on the relative importance of sulfates. Uncertainty in the primary organic aerosol emissions inventory is essentially independent of the uncertainty in the SO_2 emissions inventory.

Administrative Feasibility

Perhaps the greatest advantage of the speciated rollback approach is the elimination of many administrative problems associated with more complex source apportionment techniques. The model is a low-cost approach that does not require highly trained personnel, and it is computationally simple. It requires three basic sets of information: a characterization of the types of particles that impair visibility, a characterization of background pollutant concentrations, and a speciated emissions inventory that delineates the relevant source categories. These three pieces of information are prerequisites for any adequate source apportionment and control study. Because the required data are so basic, the administrative resources devoted to source apportionment are kept within reason. Only a small expansion in federal, state, and local training programs and resources is needed to implement the approach, especially if national guidelines are provided for obtaining appropriate air-quality and emissions data according to specified formats. In many cases, nearly complete data sets already exist.

Not only would minimal resources be required to implement the speciated rollback approach, but also fewer resources might be required to defend any conclusions drawn from its use. Arguments inevitably arise between government regulatory agencies and regulated industries over modeling procedures. These arguments have more fuel when there are many modeling assumptions to be contested. The more numerous the parameters in a particular model, the more likely it is that modeling results obtained by regulated industries will differ from those obtained by control agencies. A simple modeling approach does obviate many arguments and discrepancies. If this modeling approach is selected principally to simplify the defense of the conclusions, then the legal and regulatory framework should specify the simpler approach as the basis for control plans. Any policy decision that suppresses debate over modeling assumptions should be made clearly and openly.

The speciated rollback model is compatible with nearly all regulatory approaches. It is not compatible, however, with a new-source regulation program, such as the current PSD program, in an area where there are not enough existing sources to allow an empirical determination of the baseline relationship between emissions and air quality. Even where there are existing sources, the speciated rollback approach might be inadequate if a new source significantly alters the spatial distribution of emissions.

The simplicity of the speciated rollback model lessens multi-jurisdictional problems because it is easy to maintain consistency among various agencies. It would help to have national guidelines for the identification of controlling precursor species, the collection of aerosol composition data, the compilation of speciated emission inventories, and the estimation of background concentrations.

The simplicity of the speciated rollback model keeps the whole modeling process within an understandable framework. This is important because a fundamental principle of decision making in the face of uncertainty is that the process should be open. The critical technical assumptions made by scientists within the context of a rollback model are clear and understandable, not buried in computational details. It is important for decision makers to grasp the major assumptions and understand how the results might change if alternative assumptions were made. The model allows government officials to judge air-quality modeling results in conjunction with all other aspects of the decision-making process, such as costs, benefits, social values, political tradeoffs, and legal issues.

Economic Efficiency

Economically efficient control strategies can be identified with economic programming models—usually based on linear techniques, but sometimes based on nonlinear programming techniques—that determine least-cost strategies for achieving air-quality goals. The speciated rollback model is certainly compatible for use with these least-cost economic models. In fact, nearly all published least-cost analyses of air quality problems have used rollback or modified rollback as the approach to modeling air quality. Perhaps this is because the model's complexity, uncertainty, and flexibility are complementary to that of economic models.

The speciated rollback model might not be optimal for identifying least-cost strategies for attaining air-quality goals at a single receptor site when those strategies involve a detailed change in the spatial distribution of emissions (in cases when it is obviously most efficient to control sources at specific locations). However, focusing on least-cost strategies for attaining air-quality goals at an individual receptor site neglects the benefits associated with improved air quality at other receptors. The speciated rollback model is more appropriate for analyzing air quality averaged over many receptor sites or for considering a widespread problem such as regional haze.

Flexibility

The speciated rollback model has several advantages in flexibility. Its modest data requirements—just an emissions inventory and speciated air-quality data—make the speciated rollback model easily adaptable to many geographical areas. For areas where such data have been assembled, it can predict changes for new and existing sources. As the information available changes (new emissions data, new air-quality data, revised estimates of background concentrations, or "nonlinearities" in the atmospheric chemistry), the model can be altered easily. The simplicity of the model makes a view of the whole process possible. The importance of new information is readily apparent, and the model does not fall prey to errors that can be hidden in the complicated data sets or algorithms of more complex models.

In other ways, the speciated rollback model is inflexible. It is not well suited for extrapolation to entirely new circumstances, such as the addition of a source in a region where there are no existing emissions, or—more generally—for large changes in the spatial pattern of emissions. Although the model can be modified to incorporate new information about nonlinearities in chemical transformation rates, it does not do this precisely or accurately. Finally, model results cannot be extrapolated to meteorologic conditions other than those on which it is based.

Balance

As noted in the discussion of economic efficiencies, the speciated rollback model appears to be in balance with the degree of sophistication and the uncertainties associated with economic optimization models. When linked with an optics model based on extinction (scattering and absorption) efficiencies, the speciated rollback model also is in balance with the optics model in terms of complexity and data requirements. This entire system of models is computationally simple and has reasonable requirements for emission information, aerometric data, emission-to-air-quality modeling, and optics calculations. The main complexity—keeping track of each pollutant within the emissions inventories, air-quality data bases, and light extinction budgets—is consistent throughout the analysis.

Summary

The speciated rollback model has several strengths: It is straightforward, the data needed to support its use are widely available or easily constructed, and its use is compatible with the personnel resources and budgets, of regulatory agencies. The model and its assumptions are easily communicated to decision makers, and it is readily linked to economic decision-making tools. It makes use of data about the chemical composition of atmospheric aerosols to limit the range of error in its predictions. For example, because predicted source–receptor relationships must match the observed chemical composition of the aerosols and the emissions sources, an error in predicting the effect of fugitive dust controls on fine-particle concentrations will affect only the predictions for airborne dust and not those for the other aerosol components.

The important limitations of the speciated rollback model follow from its assumption that emissions changes within major categories of airborne particles and their precursors are spatially homogeneous, and from the assumption that atmospheric chemical and removal processes are linear in emissions. Furthermore, the model is not suited for extrapolation to source configurations, weather conditions, or atmospheric chemical conditions outside the range of the historical data sets upon which such an empirical model must be built. Care must be taken not to use this model far outside the stated range of conditions to which it applies.

Chemical Mass Balance
Receptor Models

Receptor-oriented models, which are discussed in detail in Appendix C, infer source contributions to visibility impairment at a particular site based on atmospheric aerosol samples from that site. Chemical elements and compounds in ambient samples are often used as tracers for the presence of emissions from particular source types (e.g., knowledge of the lead content of the particles emitted from automobiles burning leaded gasoline historically has been used to compute the amount of motor vehicle exhaust aerosol present in an ambient aerosol sample that contains lead). The committee evaluated two widely used receptor modeling methods. The chemical mass balance (CMB) model is discussed first; regression analysis models are discussed in the next section. Alternative receptor-oriented modeling methods, including models based on factor analysis, are discussed briefly in Appendix C.

Technical Adequacy

The chemical mass balance model is widely used within the regulatory community to identify and quantify the sources of particles that are emitted directly to the atmosphere (primary particles). Although several researchers have used the CMB model to apportion secondary particles in aerosols, the uncertainties associated with these applications need to be better explained before this use will be widely accepted by the scientific community. The following discussion is only about the current ability of the model to apportion primary particles.

The theoretical basis of the CMB model has been carefully examined in validation studies, and the assumptions implicit in application of the CMB model have been identified. The effects of deviations from these assumptions are largely understood. Although the CMB model has been used to apportion light extinction by applying additional assumptions about relative humidity and the physical and optical characteristics of particles, these assumptions, which are needed to link the CMB model to atmospheric optics, require full validation before they are likely to be accepted for regulatory use.

Until recently, a major disadvantage of the CMB model has been the scarcity of the required input data about the chemical composition of emissions. With completion of several large projects to develop profiles of emissions sources (Core, 1989a; Houck et al., 1989) and with updates to the Environmental Protection Agency (EPA) source library, profiles are now available for many source types. Although source profile libraries have grown from just a few profiles in the recent past to hundreds today, there might not yet be enough replicate tests of similar sources to establish the variability of the source profiles or to assess the likely effect of variability in the emissions from the same source over time. Source profiles for a particular industry can change because of changes in technology or raw materials, and new sources will be built. Therefore the process of measuring source profiles must continue. Data on the chemical composition of fine particulate mass at receptors in or near Class I areas are now being routinely collected through the Interagency Monitoring of Protected Visual Environments program, the Northeast States for Coordinated Air Use Management project, and several state visibility monitoring programs, most of which have structured their sampling and analytical protocols specifically for receptor modeling. As can be seen from Table 5-1, some important marker element concentrations often fall below quantification limits. Even with these limitations a great deal of data on aerosol chemistry is becoming available for CMB model applications, although some regions of the country are not yet included in ambient monitoring programs.

Several investigators have completed CMB model validation studies (see Appendix C), that focus on apportionment of primary particles only. Further validation studies of hybrid models designed to apportion secondary particles are needed before such models can be widely used with confidence. Validation studies also are needed for use of the CMB model to apportion light extinction among contributing sources.

TABLE 5-1 Percentage of samples with detectable trace element concentrations in regional fine particle measurement programs

	Grand Canyon National Park (yr)				Shenandoah Valley National Park (yr)			Northeast U.S. (yr)
	1979-86	1985	1987	1988-89	1982-86	1988-89	1980	1989
Season	all	fall	winter	all	all	all	summer	all
Reference[a]	1	2	3	4	2	4	5	6
Samplers[b]	SFU	SCIS	IMPR	IMPR	SFU	IMPR	DICH	IMPR
Time (h)	72	12	12	24	72	24	24	24
Analysis	PIXE	XRF	PIXE	PIXE	PIXE	PIXE	INAA (XRF)	PIXE
Element								
Na	44	*	*	50	58	59	100	83
Mg	34	*	*	14	14	12	*	21
Al	73	100	*	93	88	69	97	60
Si	93	100	97	100	98	99	(97)	98
P	7	100	*	0	8	0	*	0

TABLE 5-1   (continued)

	1979-86	1985	1987	1988-89	1982-86	1988-89	1980	1989
Cl	4	3	*	3	0	0	100	0
K	92	100	87	100	100	100	97	100
Ca	91	100	100	100	91	100	97	98
Sc	*	*	*	*	*	*	91	*
Ti	50	98	33	97	58	91	*	89
V	12	96	*	17	24	16	97	66
Cr	9	84	*	21	9	26	91	19
Mn	23	99	13	53	48	58	94	60
Fe	91	100	99	100	100	100	94	100
Co	*	*	*	*	*	*	97	*
Ni	9	68	49	7	10	16	*	80
Cu	35	83	43	90	47	91	73	93
Zn	43	98	93	95	96	100	97	100
Ga	*	30	*	*	*	*	*	*

TABLE 5-1 (continued)

Element	1979-86	1985	1987	1988-89	1982-86	1988-89	1980	1989
As	*	22	50	20	*	27	91	33
Se	*	76	34	25	*	88	97	55
Br	23	98	62	96	45	97	(100)	99
Rb	*	4	3	32	*	8	*	11
Sr	*	35	7	35	*	7	*	6
Zr	*	1	12	13	*	2	*	21
Pd	*	12	*	*	*	*	*	*
Cd	*	9	*	*	*	*	73	*
Sb	*	37	*	*	*	*	97	*
I	*	*	*	*	*	*	64	*
Cs	*	*	*	*	*	*	64	*
La	*	16	*	*	*	*	82	*
Ce	*	*	*	*	*	*	97	*
Sm	*	*	*	*	*	*	76	*

TABLE 5-1 Continued.

Element	1979-86	1985	1987	1988-89	1982-86	1988-89	1980	1989
W	*	*	*	*	*	*	58	*
Pb	37	88	29	88	83	95	(97)	99
Th	*	*	*	*	*	*	52	*

*=Not reported.

[a](1) NPS network, described by Eldred et al. (1987). Detection frequencies calculated by B. Schichtel, Washington University, from data communicated by T.A. Cahill, University of California at Davis (1990). (2) SCENES 1985 Summer Intensive at Grand Canyon, Bryce Canyon, and Glen Canyon. Network described by Mueller et al. (1986); detection frequencies calculated by D. Pankratz, AeroVironment Inc., and communicated by C. E. McDade, ENSR Inc. (1990). (3) WHITEX network (NPS, 1989, table 3C.8). (4) IMPROVE network, described by Eldred et al. (1990). Detection frequencies communicated by J. Sisler, Colorado State University (1991). (5) Research study, Tuncel et al. (1985). (6) NESCAUM Network, described by Poirot et al. (1990). Detection frequencies communicated by R. Poirot (1990).

[b]SFU: stacked filter unit;

SCIS: size classifying isokinetic sequential aerosol sampler;

IMPR: IMPROVE sampler;

DICH: dichotomous sampler.

The CMB model typically can provide useful source separation for five or six source types. Separation of the effects of two sources is difficult if the source emission profiles are similar. For example, smoke from prescribed burning of forest, from wildfires, and from residential wood combustion cannot be differentiated by the model based on a bulk accounting of the concentrations of chemical elements. The sources must be determined on the basis of additional information, such as emission inventories or plume transport calculations. Source contribution estimates from CMB modeling must often be used along with other techniques, such as dispersion modeling, to provide enough detail for the development of a control strategy.

Model response to a full range of temporal variability is possible if ambient sampling times are short enough. One major advantage of the CMB method is that actual samples collected during episodes of air pollution or impaired visibility can be used to identify source contributions, and the apportionment can be conducted for periods as brief as a few hours, thereby allowing the analyst to examine temporal variability in estimated source contributions.

The chemical mass balance model works well even when it is used for areas with varied topography, whereas mechanistic models for aerosol transport can be difficult to apply in complex terrain. The CMB model uses chemical element tracers rather than meteorologic flow fields to determine source–receptor relationships. However, because the CMB model is founded on analysis of measured historical conditions, it cannot be used to predict the effects of different stack heights or new sources.

Systematic error analysis procedures have been developed for the CMB model, and the results have been published in the model validation studies (Appendix C). The CMB model is fully developed and readily available for regulatory use in the analysis of the source origin of primary aerosol particles. Similar models for the apportionment of secondary particles and light extinction, however, are yet to be developed. Particular attention should be given to systematic model validation, documentation of assumptions, and analysis of model sensitivity to deviations from underlying assumptions. Since many regional haze problems are thought to be dominated by secondary airborne particle formation, this is a serious limitation.

Administrative Feasibility

The resources required to apply the CMB model are well known from experience gained during the past few years of regulatory application. They are compatible with the personnel, skills, and budgets of many regulatory agencies. Expansion of CMB models to apportion secondary particles could be constrained by resource limitations if tracer injection were a required element of the modeling protocol.

Because the CMB model is already commonly applied to controlling particulate matter, it could easily be used within an analogous program to improve visibility. CMB modeling is already a formal part of EPA guidelines for the development of regulations, so further extension of its applications primarily should involve training, software development, and updates to the source profile library.

Multijurisdictional implementation of the CMB model already has occurred, following the release of the computer software by EPA and the Desert Research Institute. Training programs and software updates will be needed as its applications expand. Because the CMB model is based on the simple idea that ambient samples represent a linear combination of source contributions and that the ratios of chemical elements can be used as tracers for those sources, the model is easily understood by decision makers, and its outputs are easily manipulated during discussions of alternative control policies.

Economic Efficiency

Given the CMB models' assumption of a linear emissions-to-air-quality relationship, it is straightforward to link the model to linear programming calculations designed to find the least expensive combination of emission controls needed (Harley et al., 1989). The ease with which the CMB model can be used to find cost-effective solutions to regional air-quality problems is a further advantage of this method.

Flexibility

The CMB model cannot respond flexibly to cases that involve new sources or secondary-particle production. Its use is limited to apportionment of primary emissions from existing sources, but within that constraint, it has found success over a wide range of applications.

Balance

The resources required to apply the CMB model and the usefulness of the method are balanced attractively when the model is used to apportion primary emissions based on ambient and source aerosol chemical composition data. However, if artificial tracer injection were to become a required element of a secondary-particle-modeling protocol, the cost of applying the model would in many cases exceed the budgets available.

Summary

The CMB model has many attractive features. It is readily available, and there is an extensive history of model validation and error propagation studies. It is accepted for primary-particle source apportionment by the regulatory and scientific communities. Its use is compatible with control agency budgets and personnel resources. It is easily used within a broader program designed to identify the least expensive set of emission controls needed to improve primary particulate air quality.

The model has several important limitations, however. It is not fully developed for the partitioning of secondary particles among contributing sources. This can be a significant problem because regional haze often consists largely of sulfate particles. Connections made between CMB air-quality models and atmospheric optical models require further testing. In addition, experience shows that CMB analysis can typically separate no more than five or six source types from one another. CMB model application is therefore limited in airsheds where there are many contributing source types. Given these advantages and limitations, it is likely that the CMB model will find is greatest value within visibility studies as part of a hybrid combination of several models used simultaneously, where the CMB model is applied to the primary particle portion of the problem (see later discussion of hybrid models).

Models Based on Regression Analysis

Receptor models based on regression analysis use empirical relationships between source strengths and ambient concentrations to apportion airborne particulate matter among various kinds of sources. Tracer species that are emitted uniquely from particular source types are often used to derive these relationships. Appendix C describes regression analysis in detail. Here we evaluate its use as a tool in support of regulatory decisions.

Technical Adequacy

Regression analysis in general takes a theoretical model relating source emissions to ambient concentrations and determines parameter values by optimizing the fit to observational data. In principle, the regression model can incorporate data explicitly accounting for most major variables in the relationship between source emissions and ambient conditions, leaving only a few elementary parameters to be inferred. The data necessary for such a comprehensive model would include accurate indicators for the presence of emissions from all significant contributors, along with the ages of their emissions and one or more indexes of atmospheric reactivity.

Often, not all of the observations necessary for regression or other statistical procedures are available. Unlike the CMB models, regression models cannot be used to evaluate an episode that is described by a single instance of sampling. It is typically unknown or unclear how much different sources contribute to the ambient mix, and atmospheric reactivity often is undocumented. As a result, regression analyses often are based on rudimentary models in which important unrelated variables are lumped into the conditions being estimated. The interpretation of the estimates then is sensitive to untested assumptions about the statistical behavior of the implicit variables.

If these deficiencies have been overcome in a particular case and if the source contributions to concentrations of the atmospheric particles have been estimated reliably, then those results can be linked to atmospheric optical properties at the same level of detail as is possible for rollback and CMB models. Many extinction budget studies used to relate rollback and CMB model results to visual range are themselves based on a regression analysis of the relationships between visual range and the properties of airborne particles. At some point it could be possible to merge these regression-based air-quality and optical model calculations. Such an approach has yet to produce successful analyses routinely. Any failure to determine an empirical relationship between source emissions and visual range in a particular case could be the result of misspecification of the model rather than the absence of a cause-and-effect relationship.

The data bases available to support regression analysis are similar to those available to support CMB models. Tables 5-1 and 5-2 show that important elements in the chemical signatures for such important source categories as coal combustion have not been measured with adequate sensitivity in ambient air by many existing monitoring networks in Class I areas. This problem, which affects both regression and CMB models, can be remedied through redesign of future field experiments. More fundamentally, it is not clear that reliable endemic signatures even exist for some source types. For example, Figure 5-1 shows that copper smelters emit selenium and arsenic and thus may be difficult to distinguish from coal combustion in the Southwest by either regression models or CMB models. Selected sources can be inoculated with artificial tracers, but this technology is experimental and carries its own difficulties.

The apportionment of urban primary aerosol particles by regression analysis has been tested on simulated data and has survived various cross-checks with emissions data (Cass and McRae, 1983). Even in this restricted setting, however, regression analysis has not been subjected to the formal trials necessary to verify its performance in operational use by regulatory agencies. The apportionment of secondary aerosol particles has not yet been tested rigorously. The needs for major improvements in the characterization and parameterization of atmospheric chemical reactivity within the structure of regression-based air-quality models are similar to those previously discussed for CMB models. If regression models are to be applied to visibility-relevant studies designed to apportion the sources of secondary airborne particles, improvements in models must be accompanied by thorough and convincing field verification studies.

The source separation achievable with regression analysis depends on the availability of reliable signatures for all significant sources, which in

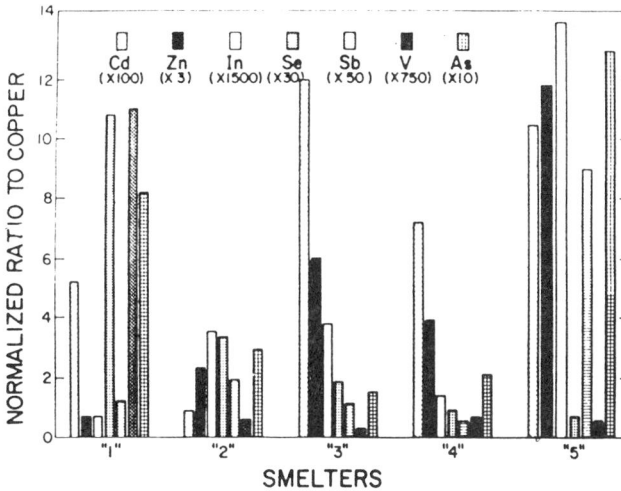

FIGURE 5-1 Chemical signatures of five different copper smelters in southeastern Arizona. The signatures are presented as ratios, relative to copper, of seven elements measured in smelter plumes. The ratio of each element to copper is multiplied by a normalizing factor, as indicated, to facilitate intercomparison. Source: Small et al., 1981.

turn depends on the chemical differentiation of the emissions. In common with CMB and other approaches based on the deconvolution of ambient composition, regression analysis is unimpeded by complex terrain or diverse source configurations. Unlike CMB analysis, the time resolution available from conventional regression models is limited by the fact that the regression relationship is averaged over all the observations used in its derivation.

Standard procedures for determining the error bounds that surround coefficient estimates produced by regression analysis are well understood. These error estimates, however, depend on a close correspondence between the standard assumptions of regression models and the situation being modeled. If the model structure provides a poor approximation of reality, then the error estimates themselves could be wrong. The physical correctness of a regression model, as opposed to its descriptive performance, is difficult to evaluate, because the model is explicitly fit to the data under consideration. Statistical uncertainty and

TABLE 5-2   Commonly Used Endemic Source Tracers[a]

Tracer substance	Source	Representative references	
		Source characterization	Apportionment use
V, Ni	Oil combustion	Miller et al. (1972), Sheffield and Gordon (1986)	White and Roberts (1977), Kleinman et al. (1980), Cass and McRae (1983), Dzubay et al. (1988)
As, Se	Coal combustion	Tuncel et al. (1985), Sheffield and Gordon (1986)	Kowalczyck et al. (1978), Thurston and Spengler (1985), Dzubay et al. (1988), NPS (1989)
Nonsoil K, C^{14}/C^{12}, CH_3Cl	Wood combustion, other contemporary carbon	Dasch (1982), Edgerton et al. (1986)	Lewis et al. (1986), Wolff et al. (1981), Stevens et al. (1984), Lewis et al. (1988), Khalil et al. (1983)

interpretational error may be estimated, but these estimates are sensitive to statistical assumptions and unobserved quantities. Appendix C, demonstrates what results when emissions from an unrecognized or poorly characterized source are proportional to the ambient correlation of those emissions with other emissions, a statistic that is by its nature unknown.

Although the bias is hard to quantify, regression models tend to overestimate the importance of sources that have reliable and easily measurable chemical signatures, whether endemic or injected. This follows from the fact that emissions from unrelated sources tend to correlate in the ambient air, reflecting the common influence of large-scale meteorological factors.

TABLE 5-2 (continued)

Tracer substance	Source	Representative references	
		Source characterization	Apportionment use
Br, Pb	Motor vehicles	Huntzicker et al. (1975), Pierson and Brachaczek (1983)	Friedlander (1973), Kowalczyck et al. (1978), Kleinman et al. (1980), Cass and McRae (1983), Dzubay et al. (1988)
Al, Si, Mn, Fe	Soil-derived material	Miller et al. (1972), Cahill et al. (1981), Batterman et al. (1988)	Friedlander (1973), Kowalczyck et al. (1978), Kleinman et al. (1980), Duce et al. (1980), Cass and McRae (1983), Lewis et al. (1986), White and Macias (1990)

[a]This listing is illustrative rather than comprehensive. Compilations of chemical profiles for many source types are available in Core et al. (1984) and Hopke (1985). Chow and Watson (1989) give a comprehensive listing of other compilations.

Administrative Feasibility

The data base needed for regression analysis requires a considerable investment of resources in field measurements of source and ambient aerosol chemical composition. CMB models also face this problem. As explained earlier when discussing CMB models, that investment in data acquisition already is being made in some parts of the United States. Unlike CMB models, regression models require that these enhanced measurements be repeated over many samples in order to provide an adequate basis for statistical inference via regression analysis. Fortunately, many routine air monitoring networks make successive observations over long periods of time, such that regression analyses can be applied. However, if artificial tracers are required to provide source-specific signatures, they must be injected throughout the sampling period. Indications of the cost for a relatively simple study of a single point source are provided by the $2 million estimated to have been spent on the WHITEX study of the Navajo Generating Station (NPS, 1989) and the $2.5 million initially appropriated for a similar study of the Mohave Power Project.

Resource pressures can adversely affect the equity of an experimental design. Regression analysis tends to overestimate the importance of sources with reliable and easily measurable chemical signatures. If resources, in the form of injected tracers or enhanced sample analyses, can provide good chemical signatures for only a subset of the contributing sources, then the selection of these targets forces a bias in the experimental outcome.

The potential biases inherent in the use of regression models are somewhat offset by the openness of regulatory processes in which such models are used. Regression results can be straightforwardly and succinctly documented, so that they can be understood and replicated, or reworked and reinterpreted, by all interested parties.

Economic Efficiency

Like other models that generate linear source–receptor relationships, regression models yield source apportionment results that can be combined with linear programming economic optimization methods to identify the least expensive way to improve regional air quality.

Flexibility

Like any receptor-oriented method, regression analysis is a tool for studying the existing atmosphere and cannot be directly applied to unbuilt sources. On the other hand, fugitive and intermittent sources, which typically are difficult to model prospectively, pose no particular problems. An evolving understanding of emissions and atmospheric processes can be incorporated into the analysis through improvements in the regression model.

Balance

Provided that the necessary input data are available, the further data analysis required in order to apply a regression model to that data set generally is not so cumbersome as to distort the allocation of resources within a larger regulatory program.

Models for Transport Only and Transport with Linear Chemistry

Two other sets of models for source identification and apportionment are those that assess only the transport of pollutants without regard to the chemical or physical processes that affect the pollutants and those that assess transport coupled with a simple linear chemical transformation process. This family of models includes techniques that range from air parcel trajectory analysis, through models based on prediction of pollutant concentrations via solution of integral or differential equations that govern atmospheric transport and dilution. Appendix C describes these models in detail.

Technical Adequacy

The value of backward trajectory analysis depends on whether the meteorologic data used to produce the analysis are representative and whether the interpolation schemes are accurate. It is scientifically reasonable to interpolate wind information in time and space, but the use of trajectories requires careful interpretation, because the technique has the potential for large uncertainties that can arise if there are errors in observation or interpolation, sparse observations, or a misunderstanding of the vertical structure of the air parcels being transported.

Trajectory analysis alone is insufficient to quantify source contributions, but it can show a qualitative relationship between sources and pollutant concentrations. Moreover, trajectory analysis and other wind flow analyses can help corroborate other source apportionment methods.

The calculation of atmospheric transport requires wind fields that accurately resolve variations in space and time. Wind fields are generated from interpolation of observed or simulated winds. Estimates of transport are therefore fundamentally limited by the spatial and temporal resolution of the input wind fields. Nationwide observation of winds nominally occurs in the United States once every 12 hours with a horizontal resolution of 400-500 km. This spacing in observations was established by the National Weather Service for use in forecasting and is not necessarily sufficient for estimating pollutant transport. To obtain wind data that are more densely packed in time and space, it is usually necessary to survey and accumulate wind observations taken by research groups, state and local agencies, and industrial companies.

There are errors in measurement and interpolation. For relatively quiescent wind conditions, Kahl and Samson (1986) showed that the median error for trajectories drawn from National Weather Service data in the eastern United States after 72 hours of simulation was about 350 km. For convective situations, Kahl and Samson (1988) showed errors of 500 km after 72 hours. Mean absolute errors in estimation of wind velocity components (wind velocity can be divided into separate east–west and north–south components) were found to range from 3.3 to 6.5 m/s. Papers by Kuo et al. (1985), Haagensen et al. (1987), and Draxler (1987) report average trajectory uncertainties of 350 to 425 km after 72 hours. Thus the effects of specific sources on specific receptors in the eastern United States are not clear. The transport of air from large source areas to large receptor areas (for example, entire states) can be treated with reasonable reliability.

Trajectory techniques use regionally representative winds and hence are better for identifying sources hours upwind of the receptor than sources nearby. Small variations in wind measurements can cause trajectories to miss nearby sources, whereas compensating errors over time and the growth of the envelope of potential contributors diminish the sensitivity of trajectories at longer travel times.

Trajectory calculations require a priori assumptions about how vertical atmospheric layers act to transport the pollutant to the receptor. These assumptions can seldom be confirmed because vertical profiles of the transported species are rarely available. The two most common assumptions about the vertical distribution of pollutants are either that the material is well mixed through some predefined layer or that its movement can be represented by the movement of air at a particular level (where the level is defined in terms of height above the ground, a pressure surface, or an isentrope). Winds are averaged over the layer, if applicable, and then interpolated in space and time.

The assumption that pollutants are well mixed could be adequate in the warmer months in the eastern United States where visibility-reducing material is often mixed by convective eddies during episodes of regional haze. However, this assumption might not be appropriate for the complex terrain of the western United States or for plumes from single sources.

Trajectory analyses suffer from their simplicity, results are presented (and too often interpreted) as a single line of air transport from a source or to a receptor. In reality, the transport should be represented by three-dimensional probability fields, the shape and magnitude of which are influenced by the exact elevation of the plume centerline, the amount of vertical mixing, and the vertical and horizontal gradients of wind velocity. Wind velocity is the combination of wind direction and speed. Thus gradients in wind velocity can occur when the wind changes direction when there is no change in speed.

The complete vertical profile of meteorologic factors that affect visibility-reducing gases and particles should be known in space and time. In the future, remote-sensing techniques could make such definition possible, but in the absence of this information it is necessary to estimate the uncertainty in transport paths resulting from the uncertainty in the vertical distribution of meteorologic parameters.

The uncertainty in trajectory paths is believed to increase with travel time and to increase faster in regions of complex topography. The uncertainty can be evaluated by analyzing the divergence of trajectories for different sublayers of the layer in which the pollutants are thought to be transported. Samson (1980) estimated the rate of growth of uncertainty in trajectories drawn from National Weather Service data in the eastern United States to increase linearly at a rate of 5.4 km/hr.

One danger in the use of ensemble trajectory analysis (ETA) is the misinterpretation of results, particularly for areas far from the receptor (in the case of backward trajectories). Consistent trajectories from a particular direction, for example, can indicate a contributing source area or can simply represent the air flow direction on days conducive to pollutant buildup. Also, it is impossible to identify the location or magnitude of the contributing sources along the path of highest likelihood. Finally, ETA techniques are sensitive to individual events for source regions at the fringe of the nominal travel time of the trajectories. That is, one or two trajectories associated with high-arriving pollutant concentrations that traverse a source region through which few trajectories pass will produce apparently strong correlations. For this reason it is necessary to set criteria for a minimum number of trajectories. Below that number, the results are questionable. Using estimated two-dimensional probability fields, Keeler and Samson (1989) disallowed ETA for those areas in which the probability of transport over the sampling period dropped to less than 1×10^{-8} km^{-2}.

Individual trajectory accuracy can be improved by increasing the spatial or temporal resolution of observations or incorporating hydrodynamic models to make the interpolations more realistic.

Estimates of source–receptor relationships based on transport models that calculate pollutant concentrations rather than just air parcel trajectories have some advantages over trajectory analysis alone. The use of either Lagrangian particle trajectory models or Eulerian grid transport models allows modeled wind flow to vary with height. Because vertical wind velocity shear is an important component of horizontal dispersion, this should result in a more realistic evaluation of pollutant transport.

Lagrangian particle trajectory models could be adapted for use with optical models because they simulate the distribution of particles downwind of sources. This makes them useful for evaluating plume blight. Eulerian grid models and Lagrangian particle models for transport alone are about equally useful when applied to regional haze problems where individual plumes have merged to form a widespread region of low visibility.

Lagrangian transport models that include linear chemistry are relatively inexpensive tools for evaluating potential source effects. The approximation that sulfate formation rates are linear in SO_2 concentrations is thought to be reasonable when oxidants (OH in gas-phase oxidation, H_2O_2 and O_3 in aqueous-phase oxidation) are plentiful. Concentrations of these oxidants have been measured only sparingly in the remote Southwest. One must confirm that these conditions hold before using linear chemistry approximations in regions where few measurements have been made.

Administrative Feasibility

Most air pollution control agencies employ persons who have used simple Gaussian plume air-quality models. These personnel can easily learn to use simple trajectory models as well. Trajectory analyses based on interpolation of wind observations usually require the use of an engineering workstation, and large-scale problems can require the use of a mainframe computer. The analyses can be performed by a trained computer technician and can be applied to ETA by air-quality technicians. Conclusions drawn from individual trajectories should be reviewed by a trained meteorologist who can identify inconsistencies in estimated trajectory paths.

Advanced hydrodynamic modeling and the associated transport estimation require personnel who can initialize and apply the necessary algorithms. These models could be beyond the capabilities of some regulatory agencies. Such analyses yield useful estimates of the three-dimensional structure of plume travel and are best applied to individual episodes of unusually heavy pollution. Such analysis are necessarily more expensive than simple two-dimensional trajectory analyses.

Transport models that incorporate linear chemistry require considerably fewer resources than do the fully explicit coupled-transport and chemical models described later. Nonetheless, the choice of linear proportionality constants and the application of the codes require an understanding of the chemical and physical processes that govern the transport, transformation, and deposition of the pollutants. Linear transformation models unaccompanied by special measurements of all model parameters are of questionable value in the estimation of haze sources, because uncertainties in these parameters limit confidence in model simulations.

Transport modeling calculations can be used within each of the regulatory frameworks discussed earlier. Transport models that predict pollutant concentrations could be useful in predicting the effects of new sources. Trajectory models that show the paths of air parcels from source to receptor lend themselves to graphic displays and video animations that are easy to communicate to decision makers. Gaussian plume models are widely used by regulatory agencies, and some standardization has occurred that facilitates communication between different jurisdictions that use the same modeling tools. The more advanced trajectory and hydrodynamic modeling methods could require extensive model validation studies before they are understood well enough for regulatory use.

Economic Efficiency

The results of simple trajectory streaklines alone are not conducive to coupling with linear programming economic optimization methods. However, ensemble trajectory analysis that yields estimates of linear source–receptor relationships, as well as Eulerian or Lagrangian transport models that compute pollutant concentrations and can incorporate linear chemistry, can be combined with such economic optimization models (see Harley et al., 1989).

Flexibility

Trajectory modeling as described here is a receptor-oriented tool used to diagnose the likely sources of observed pollutant concentrations. As such it cannot be applied to unbuilt sources except to demonstrate the likelihood of transport from the new source to a receptor. Transport models that compute pollutant concentrations (with or without the addition of a linear transformation and deposition module) can be used to estimate the expected change in concentrations with a changing source configuration, assuming fixed meteorologic conditions and a linear response in transformation rates.

Balance

Because the personnel required to apply simple trajectory models and Gaussian plume pollutant concentration models are available to most air pollution control agencies, the use of these models in the context of a visibility protection program is not likely to create an imbalance between personnel needs and available staff. The more advanced trajectory and hydrodynamic modeling methods can require both a team of highly skilled specialists and a concurrent field experiment program to acquire data for model application and validation. Care must be taken when using the advanced transport models to match the needs for field experiments with the input data requirements of such models. The temptation to run improved theoretical models using inadequate input data should be resisted.

Mechanistic Models for Transport and Explicit Chemistry

Comprehensive mechanistic air-quality models are three-dimensional representations of physical and chemical atmospheric processes. They simulate the interactions of many chemicals and analyze processes on the time scale of air pollution episodes. The most detailed models now under development explicitly treat airborne particle mechanics in addition to gas-phase transport, removal, and chemical processes. These models are extensions of regional models originally developed to predict gaseous pollutant concentrations. Consequently, much of this critique is an assessment of the existing regional gas-phase models on which the aerosol models are based. The critique focuses on current approaches to aerosol modeling rather than on fully operational, three-dimensional Eulerian regional aerosol and visibility models, because such models are not yet available. See Appendix C for further discussion of mechanistic models.

Models that are based on physical and chemical atmospheric processes can be used to determine source–receptor relationships and to assess the merits of various pollution control strategies. Statistical relationships or other empirical models based primarily on aerosol and optical measurements might be inadequate for this task, especially when complex chemical processes such as secondary-particle formation are involved. Mechanistic modeling requires substantially more resources than do other source apportionment techniques since large mainframe computers usually are needed to manage the model and extensive, high-quality data are needed for model execution and evaluation. These factors, along with other issues associated with technical and administrative feasibility, economic efficiency, flexibility, and balance, need to be considered when determining the range of applicability of the mechanistic modeling approach in source apportionment.

Mechanistic models can be envisioned that provide definitive source apportionment for visibility impairment in complex airsheds with multiple sources, but these models are not yet available. The most serious barrier to their development is an inadequate understanding of the phenomena that determine atmospheric particle size distributions. Theoretical descriptions of aerosol processes that occur in controlled laboratory experiments are well advanced, but the relative importance of competing processes as they occur in the atmosphere is not fully understood. The atmospheric data sets needed for the development and testing of advanced aerosol models are not readily available. Such information can be obtained only through experimental studies of atmospheric aerosol phenomena.

Technical Adequacy

Mechanistic models are theoretically grounded in fundamental physical and chemical principles. Mathematic descriptions of transport, chemical reactions, removal mechanisms, and airborne particle dynamics are derived from basic principles. Models that explicitly include aerosol dynamics would provide the most detailed analysis of particle size and composition distribution. Because particle size distribution is an important determinant of visibility impairment, such models could be extremely useful. However, these models are hard to develop because of mathematic difficulties and difficulties associated with imperfect understanding of atmospheric aerosol physics and chemistry. Complete models for predicting particle size distributions and chemical composition are not available for regulatory use.

There are simpler models that can predict from first principles the total concentrations of some secondary species (sulfates and nitrates) as well as primary particles, but they cannot predict particle size distributions. Those models could provide detailed information about the source contributors to atmospheric pollution. Translation of their findings into source effects on visibility requires simplifying assumptions about particle size distributions, in particular that the secondary particles are in the fine-particle size regime (i.e., diameters less than 2.5 μm) with a size distribution like that observed during field experiments. Although the use of assumed size distributions is attractive for practical reasons, there are serious doubts about the accuracy of visibility model calculations based on typical size distributions rather than distributions specific to the time and place being studied. Larson and Cass (1989), for example, examine the use of size distribution data taken at Pasadena, California, to "shape" the submicron size distribution of bulk fine-particle sulfates, nitrates, and carbon concentration data taken elsewhere in Southern California prior to entering the data into Mie theory light-scattering calculations. Their results show that imposing typical size distributions on bulk chemical composition data (an analogue for the bulk chemical composition predictions of today's secondary-particle models) produce light extinction predictions that are in many cases far inferior to the results obtained if the size distribution and chemical composition data are taken concurrently. If models that predict bulk fine-particle chemical properties (but not size distributions) are to be used as part of a visibility study, and if the accuracy generally expected of mechanistic models is maintained, then the translation of pollutant concentration predictions into effects on visibility will probably need to be supported by size distribution measurements made at the time and place studied.

Whether the model includes aerosol dynamics explicitly and calculates the composition-dependent particle size distribution or assumes that secondary species exist as fine particles with an empirically determined size distribution, the model output can readily be made compatible with optics models. The main outputs from the regional models under development are gas and particle concentrations. Particle size information either will be explicitly calculated or will be assumed to be described by representative size distributions. In addition, the particle and gas concentrations will be available over the gridded modeling domain for prescribed time periods with hourly resolution or better. This detail provides information that is sufficient for detailed optics model calculations for any time of the day and for any sight path in the modeling domain.

Input data requirements for mechanistic models with high spatial and temporal resolution include data about weather patterns, pollutant emissions, and gas and particle concentrations. Meteorologic data for input into the chemical transport model can be obtained most effectively from a combination of observed data and a dynamic model (Seaman and Stauffer, 1989). Emissions and ambient data for particles and precursor gases are harder to acquire than are meteorologic data. Although sulfur and NO_x emissions and ambient concentrations are widely available for the United States, data on ammonia, organic particle precursors, and primary fine-particle species and size distributions are much less comprehensive or reliable (Placet et al., 1990).

Acquiring accurate input data for primary particle size distributions is technically difficult and expensive for regional models, and the lack of such information frequently limits the accuracy of model predictions. It is especially difficult to obtain accurate information on particles from fugitive sources, such as windblown dust and smoke from wild fires or controlled burns. In practice, size distributions of primary-particle emissions are often based on crude estimates. The accuracy of mechanistic visibility models can be improved significantly by ensuring that primary-particle emissions are carefully characterized.

Evaluation of complex mechanistic models is a multifaceted exercise. Individual processes can be evaluated independently using laboratory or well-defined field data. The entire model is best evaluated by using carefully designed, high-quality field observations of gas-phase precursors and particle size distributions along with the emissions and meteorologic information for the temporal and spatial scales of interest.

Unfortunately, much current work on mechanistic model development is not tied closely to field verification programs. Many atmospheric aerosol phenomena are poorly understood and cannot be accurately predicted with mechanistic models. Examples include the formation of fine particles by nucleation; the physical and chemical properties of organic aerosols, including their affinity for water and their distributions between the gas and particle phases; the effect of heterogeneous (incloud and noncloud) chemical transformations on the size distribution of secondary sulfates and the degree to which various chemical species are internally and externally mixed; and the associated effect of mixing characteristics on particle dynamics. The effort required to understand such phenomena undoubtedly exceeds the effort required to develop the models.

Mechanistic visibility models can, in principle, achieve useful and precise source separation provided that detailed emissions inventories are available for the important pollutants. The models can be designed with enough flexibility to accommodate the maximum feasible information on sources. The high costs of acquiring model input data that provide precise source and chemical discrimination place important limits on the use of such models.

Comprehensive mechanistic models generally are used to study individual air pollution episodes of a few days' duration. For applications where information for longer time is needed, such as annual or seasonal averages, the mechanistic models could be executed for the entire period of interest. The computer time required to perform adequate analysis for periods even as long as a season could be prohibitively expensive. If this is the case, model runs that represent carefully selected episodes during the period of interest could be executed and the results averaged to provide the desired long-term values. Procedures for selecting and weighting representative episodes need to be developed for visibility modeling.

Mechanistic models can respond to the full range of geographic settings likely to be encountered, subject to the availability of suitable input data. Mechanistic transport and chemistry models can operate on domains comparable to or smaller than the operational domains of the models that provide meteorologic input data for the mechanistic models. Meteorologic models that successfully operate over all of North America have been demonstrated. Consequently, mechanistic models have been and can continue to be successfully applied to whole regions of the United States. Successful application of these models to domains with complex terrain depends on the adequacy of the data provided by the meteorologic models. For example, detailed mechanistic modeling studies of visibility in the Grand Canyon would require meteorological data that describe air flow channeled by the canyon walls as well as air flow over the rim of the canyon. These sorts of detailed small-scale phenomena are not addressed explicitly in models with large (>20 km $\times$ 20 km) grids.

The three-dimensional structure of mechanistic models allows them to respond to the full range of source configurations. In particular, the effects of ground level versus elevated sources can be addressed. Generally, comprehensive mechanistic models can be designed to accommodate the details of plume rise and mixing of sources. In addition, grid sizes can be chosen to provide adequate resolution. This detail provides a means for evaluating a variety of sources.

Within the Eulerian framework, individual plumes generally are averaged over a grid cell. Therefore, Eulerian grid-based mechanistic models might not be suitable for studies of single-source plume-dominated problems. The influence of individual plumes can be examined through special treatment of subgrid-scale processes or, for the case of sulfate, by "tagging" the plume (Chang et al., 1990a).

Systematic error analysis in the context of mechanistic modeling must account for errors associated with mathematic approximations made to simplify computations, with the assumed size distributions and mass emissions rates of primary particles, and with parameterizations used to describe the effects of atmospheric chemical and physical processes on particle size distributions. Approaches to error analysis that have been proposed for gas-phase models can, in principle, be extended to evaluate visibility models. However, reliable estimates of uncertainty will require a more thorough understanding of atmospheric aerosol processes as well as comparisons between atmospheric observations and model simulations. It is not known how accurately atmospheric optical properties can be calculated directly from source emissions data using fully mechanistic models.

Mechanistic modeling is in principle a highly equitable source apportionment method, provided that emissions inventories are detailed enough. All emission sources are treated by the same set of computational rules, which are derived from fundamental physical principles.

The emissions from some source classes, such as fugitive dust sources, are difficult to specify in the context of the gridded emissions inventories used by mechanistic models. Any such errors in emissions input data will be reflected as a bias in the model results. Errors also will occur as a result of uncertainties about atmospheric aerosol processes. In principle, the limitations introduced by potential or known errors could be identified by sensitivity analysis.

Administrative Feasibility

The administrative feasibility of using mechanistic visibility models for regulatory purposes will be governed by the resources that these models require. Mechanistic models require substantially more computer and personnel resources than do other source apportionment methods.

Resources are needed to obtain data on emissions, meteorology, chemical concentration, and atmospheric optics for model initialization, execution, and evaluation. The overall study design and analysis require additional resources. These requirements cannot be quantified precisely, because they vary with the scope of the application, the availability of data, the treatment of particle dynamics, and the availability of knowledgeable personnel.

Full-scale mechanistic models must therefore be used judiciously. Because of resource limitations, it is advisable to consider developing a range of analytical tools, including the complete mechanistic model and other simpler model versions. The comprehensive model could establish the range of applicability of the simpler models; the less resource-intensive models could then be used to examine the effects on visibility of different policy options.

Mechanistic models can be used in all of the regulatory frameworks discussed earlier. They are particularly well-suited for predicting the effects of new or modified sources on visibility because the models are not based on existing patterns of emissions. These models also are valuable for determining source–receptor relationships when secondary-particle formation is important.

Mechanistic models are well suited for assessing a wide variety of emissions control strategies. Because emissions must be clearly described as input data to the model, it is generally straightforward to simulate the effect on emissions and air quality of specific source controls. Also, because the models can provide high spatial resolution over large domains, they can be used directly to evaluate multijurisdictional visibility problems.

Economic Efficiency

Mechanistic models can be used, in principle, to support regional economic analyses of the least expensive regional air- quality controls. Because of the detailed source description built into these models, nearly any control policy can be simulated. However, because these models are highly nonlinear chemically, their use in the search for economically attractive combinations of emissions controls is a tedious and expensive process that involves systematic perturbation of the model inputs. Linear programming techniques used with simpler air-quality models generally cannot be used with the nonlinear mechanistic models.

Flexibility

Mechanistic modeling is particularly flexible in that new sources can be incorporated easily into the modeling structure. Furthermore, models can be structured to permit the incorporation of new information on the behavior of airborne particles. The information provided by these models can be used to calculate a range of visibility indexes, including aerosol optical properties such as light extinction, color, texture, and contrast.

Balance

Mechanistic models demand a balance between field data acquisition and computational analysis. Problems can arise because both of these activities require a high level of effort. To analyze one regional visibility problem, several years' effort by a team of experts could be required for field data acquisition, followed by several years' effort by a team of computational experts to digest those data through the modeling system. If either aspect of the analysis is cut short because of time or other resource constraints, the analysis probably will not work. If the proper balance between field data collection and modeling support cannot be assured, then agencies should take a less intensive approach that can be completed with the time and budget allowed. Technical input to the regulatory process often is needed in less time than that required for mechanistic model development and application. Any approach taken should ensure that the time and resource demands of mechanistic models do not distort individual regulatory proceedings. The construction and testing of such models could be pursued as a long-term goal, so that verified base case analyses of regional visibility problems could be developed before any particular emissions control policies had to be tested. These completed case studies would be available for use in the rapid evaluation of emergent policy issues.

Summary

The major advantage of mechanistic models is that they simulate chemical and physical processes in three dimensions and therefore are suitable for predicting the effects of new and modified sources. These models also can be used to evaluate a range of visibility indexes include extinction, sight path, and scene-specific information. Because of this versatility, mechanistic models provide analyses that can be more readily linked to human judgments of visibility changes than do many simpler models. Coordination of model development with atmospheric studies could lead to experimental designs that answer important questions about model development and may also ensure that models are closely tied to observations.

The accuracy of current models is limited by incomplete understanding of atmospheric aerosol phenomena, especially particle size distribution. Consequently, mechanistic modeling for source apportionment of visibility impairment is still in its infancy. Models will need to evolve substantially to be valid over the range of conditions found in the atmosphere. It also is essential to verify model components and model simulations by comparison with field observations. The tendency to develop aerosol models in isolation from field measurements should be avoided.

Resource requirements will limit the direct use of complex mechanistic visibility models in some applications. The role of mechanistic models needs to be considered carefully for each application. One approach would be to use less advanced models to answer simple questions quickly and reserve those questions that require high accuracy for study by more complex mechanistic models. Such options need to be explored when determining the role of mechanistic models in a national program for visibility protection.

Hybrid Models

Hybrid models, which are described in Appendix C, are formed by combining two or more techniques. Models relating source emissions to ambient visibility are not rigid structures, but flexible assemblages of modules that are often interchangeable. The alternative modeling approaches discussed in Appendix C represent only an illustrative few of those that can be assembled from available modules. This section sketches some considerations involved in the construction of hybrid models, and illustrates some of the possibilities.

A source-receptor model for visibility must account for three factors that together determine the relationship: transport, transformation, and optical impact. These factors are listed in causal order: each factor affects, but is not significantly affected by, the factor after it. Modules for an aerosol's optical impact were described in Chapter 4, so we focus here on transport and transformation.

Transport governs the dilution of all aerosol species, and completely determines the source-receptor relationship for long-lived primary species. Windfield-driven ("source-oriented") transport modules estimate dilution from meteorological conditions, the bases for the estimates ranging from sophisticated hydrodynamic theory to simple empirical parameterizations. Mass balance ("receptor-oriented") models derive dilution directly from observed ambient concentrations of conserved substances.

Transport also governs the atmospheric age of emissions, along with exposure to solar radiation, liquid water, and other factors that modulate gas-to-particle conversion. While the age of an air parcel can be estimated through chemical mass balance techniques in certain circumstances, that information is generally available only from windfield modules.

Atmospheric transformations eliminate significant fractions of some species, convert others from the gas to the particle phase, and distribute condensed material among particles of differing size and composition. Many of the conversion and loss mechanisms are approximately linear in emissions, in the sense that the fraction of emissions converted or lost in unit time is nearly independent of concentration. The conversion of SO_2 to $SO_4^=$ in clear air is often represented as an example of such a process.

In the most general case, transformations involve some irreversible nonlinear processes. The details of particle size distribution and morphology are likely to reflect such influences. The contents of an air parcel then depend on its dilution history, and transport and transformation are inextricably coupled. The ambient impact of multiple sources in this situation can be accurately captured only by a fully mechanistic Eulerian model.

Numerous self-contained modules for transport, chemical transformation, pollutant dry deposition, and distribution of pollutants between gas and particle phases are presently available as documented and portable computer software. Various source-receptor models can be assembled from these components, cafeteria fashion, and this is in fact the way that present state-of-the-art single-source models are constructed.

The atmospheric aerosol often consists of a mixture of chemically distinct substances emitted from dissimilar sources that have been mixed into the same air mass. Different contributors to this aerosol are often most easily tracked by different air-quality models. For example, contributions from fugitive soil or road dust sources are easily identified by chemical mass balance methods; transport models that use fluid mechanics have great difficulty representing fugitive dust sources correctly because accurate emissions inventories for these sources are hard to assemble. On the other hand, particle formation can be tracked by certain mechanistic models for atmospheric transport and chemical reaction; while chemical mass balance models that are fully satisfactory for secondary-particle source apportionment have not been developed. A combination of these two modeling approaches is potentially more accurate and efficient than either method would be if used separately. Alternatively, mechanistic modules can be incorporated into rollback and receptor models. An example of the latter would be the unitization of an equilibration module within a speciated rollback model to relate ammonium nitrate concentrations to NO_x and ammonia sources.

Because each candidate method likely to be considered as part of a hybrid has been critiqued separately, it is not necessary to repeat that discussion here. When evaluating the merits of a particular hybrid model, the user should list the elements of a complete description of the airborne particles (sulfates, nitrates, organic carbon particles, and soil dust, among others, as well as size distribution and other physical properties relevant to subsequent optical calculations). Then it should be determined which method is proposed as the basis for describing each subset of aerosol properties. A check should be made to ensure that all relevant aerosol properties are described by some portion of the model. The user can then refer to the critiques of individual methods to assess the strengths and weaknesses of each part of the model.

SINGLE-SOURCE MODELING PROBLEMS

The main thrust of the committee's advice has been directed at the description and evaluation of models relevant to the analysis of regional haze problems. We believe that the nation's most widespread and important visibility problems are in fact regional haze problems and that tools should be selected that address the real problem. However, as a practical matter, we realize that some important decisions will be made in the near future under present regulatory paradigms that focus on analysis of proposed new sources one source at a time. For that reason, we feel that it is useful to provide advice on model selection in those cases that focus attention on a single-source.

First, we will set the stage for this discussion by detailing some of the important distinctions between the single-source and regional haze perspective on a visibility problem. Then we will provide a brief critique of available models and modeling approaches relevant to single source problems.

Widespread Haze and Plume Blight

There are two distinct ways that pollutants can impair visibility:

• When an observer is inside an extensive polluted airmass, this airmass will degrade the appearance of the surrounding scenery. The airmass itself will appear to the observer as a diffuse haze obscuring distant objects.

• When an observer is outside a well-defined plume of polluted air, the plume will appear as an identifiable object, defacing the sky or background scenery. The visual effect of such plumes is referred to as plume visibility or plume blight.

Widespread haze and plume blight are idealized categories which do not exhaust the range of possible forms of visibility impairment. They can be viewed as extreme cases between which there is a continuum of possibilities for visibility impairment. However, such an all-encompassing approach does not offer much promise for the development of operational tools. At a practical level, the concepts of haze and plume blight serve as adequate frameworks for most analyses.

The most pervasive and pressing threat to visibility in Class I areas is regional haze arising from the combined emissions of many sources. As discussed in the committee's interim report (NRC, 1990), there are also concerns in some near-pristine regions over the potential for widespread haze from large individual sources. The committee has therefore focused on models used to determine the sources of multi- and single-source haze rather than plume blight. Under current guidelines, however, the main tools for judging the effects of proposed new sources are plume blight models, which explicitly assume that the observer is located outside the plume (EPA, 1980a, 1988a). Federal land managers typically base their permitting decisions on the potential visibility of a proposed source's emissions to observers in relatively clean air, rather than the contribution of these emissions to widespread haze. Since plume blight models are used widely and can yield useful information about the visibility effects of some sources, the committee has evaluated these models as well. However, the committee does not support the predominant use of plume blight models to determine the effects of individual sources on visibility. As discussed elsewhere in the report (see Chapter 7), if significant progress is to be made in preventing and remedying visibility impairment in Class I areas, source contributions not just to plume blight but also to regional haze must be considered.

Widespread haze and plume blight typically occur on different spatial scales, are often attributable to different pollutants, and differ markedly in their sensitivity to atmospheric and observation conditions (see Chapter 4). Hazes affecting national parks and wilderness areas are usually meso- to synoptic-scale phenomena, with extents of 100 km and up. The plumes responsible for plume blight usually occur closer to a source (EPA, 1988a), although they may be seen at great distances through clean air. Haze consists primarily of particles that scatter and absorb light, whereas plume blight is often due largely to light absorption by nitrogen dioxide (NO_2) gas (White et al., 1985). Haze can form under a variety of atmospheric conditions. Plume blight, in contrast, is sensitive to stability and other airmass characteristics (EPA, 1988a). Emissions that form a dramatic elevated plume in stable early-morning air can disappear from view by late morning, when they have been entrained and diluted by the deepening mixing layer. Even optically thick plumes can be difficult to see if the background air is itself hazy. Widespread haze and NO_2-dominated plume blight are fairly insensitive to illumination and are moderately sensitive to sun-target-observer geometry. Plume blight due to light scattering by particles is extremely sensitive to both factors (White and Patterson, 1981; White et al., 1986). The passage of a cloud across the sun can instantly change the color of a particle plume from bright white to dark brown or cause it to disappear. Changes in viewing angle can have similar effects.

The differences between widespread haze and plume blight lead to differences in the models designed to determine their effects. Since plume blight occurs near sources, plume blight models treat advection and dispersion much more simply. The standard EPA models rely on standard Gaussian plume parameterizations developed to predict ground-level effects (EPA, 1970). Some of these models have been described and evaluated by Latimer and Samuelsen (1978), Eltgroth and Hobbs (1979), Bergstrom et al. (1981), and White et al. (1985). Because plume blight occurs near sources and is often of concern in clean, dry air, plume blight models also treat atmospheric chemistry fairly simply. The EPA screening model (EPA, 1988a), for example, neglects the conversion of SO_2 to sulfate. Even more sophisticated models (e.g., EPA, 1980a) incorporate only a rudimentary mechanism for the gas-phase oxidation of SO_2 to form sulfate particles, along with minimal O_3-NO_x chemistry for the production and loss of NO_2. Plume blight models do, however, treat primary particle emissions in considerable detail (White and Patterson, 1981; White et al., 1986).

The visual effects of approximately uniform haze are adequately characterized for many purposes by a single number, the light scattering or total extinction coefficient. The effects of plume blight, however, depend on the scattering coefficient, absorption coefficient, and scattering phase function (giving the angular distribution of scattered light) of both plume and background, along with the width, distance, and elevation of the plume (Latimer and Samuelsen, 1978; White and Patterson, 1981; White et al., 1986). Plume blight models therefore require detailed calculations of radiative transfer.

Measurement strategies for characterizing widespread haze and plume blight also differ. The composition of a plume from a tall stack can be difficult to ascertain without airborne measurements (Richards et al., 1981). In-stack measurements can be unreliable indicators of plume particle characteristics, and ground-level measurements of the entrained plume can be dominated by background. Haze, on the other hand, is routinely sampled by fixed surface observatories (Chapter 4). Plume blight is by definition remotely sensed; it is most naturally documented by teleradiometer or camera (Seigneur et al., 1984). While haze can also be remotely sensed, it is more reliably documented by ambient light extinction measurements.

As noted above, current regulatory programs rely heavily on plume blight models for judging the visibility effects of proposed new sources. However, a proposed source's potential for plume blight does not necessarily correlate well with its potential to form widespread single-source haze or with its contribution to regional haze. The question of correlations has not received much study, but certain qualitative observations can be made. In support of a correlation, a source's potential for both plume blight and haze increases with its emission rate. Moreover, incremental increases in both plume blight and haze are most noticeable in otherwise clean air. However, haze is due predominantly to secondary particles, while plume blight is mostly due to NO_2 and primary particles. Source potentials for forming haze and plume blight are thus not necessarily comparable across source types emitting different pollutants. Furthermore, haze and plume blight are enhanced by different atmospheric conditions, so their incidence frequencies may not be comparable. Finally, plume blight cannot occur where local topography restricts sightpaths, while haze can occur anywhere.

Critique of Single-Source Plume Blight Models

Models for the visual appearance of coherent plumes from single sources fall into two categories. The simpler plume blight models recommended for use by EPA are modified Gaussian plume dispersion models developed for time-averaged estimates of plume concentrations and visual appearance. More advanced single-source models incorporate explicit chemical reactions and aerosol processes within the dispersing plume. These will be referred to as reactive plume models. See Appendix C for more information on both kinds of models.

Technical Adequacy

The simple plume blight models and more complex reactive plume models both are derived from solutions to the atmospheric diffusion equation and have a well-documented theoretical basis. Both types of models have been adapted for use with atmospheric optical models. However, the simpler models have several limitations based in their representation of atmospheric mixing, their abbreviated chemistry, and their geometric assumption that the plume must be seen from the outside. One weakness of single-source models in general arises from the sensitivity of the horizontal integral of light extinction to vertical dispersion. Given the instantaneous nature of plume perception, an accurate depiction of many plumes requires capturing their often looping or meandering appearance. Gaussian plume (averaged) models have a symmetric geometry that cannot capture such instantaneous and irregular plume structure. Even for the more advanced models that can depict meandering plumes, data on the temporal and spatial patterns of turbulence at the elevation of plumes are difficult to obtain.

The simpler plume blight models are most appropriate for use in cases (often near the source) where light extinction within the plume is due to primary particles and NO_2. In cases where light extinction is dominated by secondary particles formed within the plume, a reactive plume model that can track particle formation should be used. In cases where the observer is located within the plume (within a plume traveling down a canyon, for example), a reactive plume model should be used in which the observer can be inside the plume.

The input data required by single-source models are difficult to obtain. Particle size distributions and water content measured at the high temperatures inside stacks before the pollutants are released to the atmosphere can be quite different from the size distribution and water content of particles beyond the tip of the stack where the plume cools rapidly. One would like to use the composition of the plume a few tenths of a kilometer downwind as the initial condition for a simple plume model. Such measurements are difficult to make; they can require the use of aircraft for sampling. One alternative is to sample from the stack, using a dilution sampler that cools and dilutes the effluent prior to measurement, thereby mimicking the cooling that occurs in the early stages of plume insertion into the atmosphere.

The EPA-recommended plume blight models and similar models proposed by others have been evaluated against data collected for this purpose at a variety of large point sources (White et al., 1985, 1986). Plumes were viewed against the sky in all cases. The accuracy with which plumes' essentially instantaneous appearance could be predicted was limited by the statistical treatment of fluctuating dispersion characteristics in all models. When observed rather than predicted dispersion parameters were used as inputs to the plume chemistry and optics modules, the EPA and Environmental Research and Technology models satisfactorily reproduced the observed appearance of NO_2-dominated plumes. The EPA and other models have been much less successful in reproducing the appearance of plumes that have significant particle loadings.

Reactive plume models have undergone limited evaluation against field data, largely because data sets sophisticated enough to support these models are hard to obtain. Most data sets do not include size-resolved plume particle composition; background particle concentrations; or background concentrations of ammonia, hydrogen peroxide, or reactive hydrocarbons. The most comprehensive evaluation study to date was carried out by Hudischewskyj and Seigneur (1989).

Single-source plume models obviously do not have to separate the effects of many contributing sources. The temporal resolution of single-source models is usually a few minutes or longer; actual plumes can vary in appearance in just seconds. The simpler, Gaussian plume-based visibility models cause difficulty if they are used for rough terrain, because they cannot represent a plume that impinges directly on such elevated features. Even when the plume does not impinge directly on such features, it can be difficult to see a plume when there is elevated terrain in the background. Two versions of EPA's plume visibility model are in general circulation, PLUVUE and PLUVUE II. PLUVUE II has been known for some time to have coding errors that render it essentially unusable in situations where the plume is viewed against terrain rather than sky. PLUVUE has recently been found to have an error that can cause it, too, to generate faulty output for terrain backgrounds. Both models are now being corrected by EPA (D. Latimer, pers. comm., Latimer and Associates, Denver, Colo.).

Administrative Feasibility

The personnel resources required for the use of simple Gaussian plume blight models are readily available to most air-pollution control agencies. Indeed, most agencies already perform calculations using these models as part of the new-source review process. The more complex reactive plume models must be applied by PhD-level personnel.

The simple Gaussian models are compatible with current regulatory programs for the prevention of significant deterioration of air quality in relatively clean areas; indeed, they are about the only tools used routinely for visibility analysis by regulatory agencies. The reactive models are not yet widely used within the regulatory community, but efforts in that direction should be encouraged because these models could address a wider range of conditions than can the simple plume blight models. Simulated photographs that show how plumes would look when viewed against the background sky can help communicate the results of plume models (Williams et al., 1980).

Flexibility

The Gaussian single-source plume blight models that have been adopted for regulatory use are fairly inflexible. To the regulator, this could appear advantageous in that all model users will have to make similar simplifying assumptions to apply those models to a particular problem. The disadvantage is that the assumptions (that straight plumes are formed, dominated by NO_2 and primary particles) often do not correspond to the situation being modeled. The reactive plume models, while more difficult to apply, are better able to represent actual conditions. They also provide a more flexible framework for incorporating advances in the scientific understanding of plume structure and aerosol processes.

Balance

The balance between data collection and data analysis is not likely to be distorted when using simple Gaussian plume blight models, as the data and personnel resources needed for their use are modest. Programs that involve reactive plume models must be carefully structured. Funds must not run short before the experiments needed to acquire input data for a reactive plume model are completed, or the models will have to be run using assumed rather than measured emissions and ambient data.

Bridging the Gap Between Near-Source
Models and the Regional Scale

If a proposed single new source is made large enough, it may have significant effects on regional haze. A succession of large sources analyzed and built one at a time, also eventually can in the aggregate affect regional haze. For that reason, even in single source siting decisions, it may be necessary to consider effects on visibility at spatial scales greater than those of plume blight models or reactive plume models. The clear answer to how to do this is that the proposed new source or modified existing source should be introduced into an appropriate multiple source regional-scale model chosen from those described in Appendix C and in the regional haze section of this chapter. This requires that data bases on regional air quality and visibility and on the other sources already present in the region be maintained. An interagency working group on air quality modeling recently has been formed to assess the feasibility of multiple-source regional air quality modeling conducted by public agency staffs in support of single new source siting decisions (P. Hanrahan, pers. comm., Oregon Department of Environmental Quality, 1992). Their preliminary conclusion is that this is highly desirable and technically feasible. A multiple source Lagrangian puff model with linear chemistry probably will be selected for use initially, followed by a search for a model with a more explicit description of atmospheric chemistry and aerosol processes. The model will be driven by the U.S. Environmental Protection Agency's national emissions data base, presumably as modified by any more exact data available locally. Factors that will determine the rate of progress toward achieving an operational model at the regional scale in the west include issues of coordination between the various agencies, and access to slightly larger computers than are presently owned by most such agencies. A training program will be needed to diffuse the operational skills needed to support such models throughout the affected agencies. The most important missing piece of this system, within a regulatory context, is the present lack of clear criteria for judging how much of an increment to a regional haze problem due to a new single source is "too much."

SELECTION OF MODELS TO ADDRESS OTHER
AIR-QUALITY PROBLEMS

Visibility degradation is just one of many important air-quality problems, some of which have been or are being considered as objects of federal, state, or local legislation and regulation. Control measures that focus on one issue, such as ecosystem damage caused by acid rain or the health effects of fine particles, can alleviate other air-quality problems, such as visibility degradation in Class I areas. This complicates the task of the regulator, who might be accustomed to dealing with individual issues separately.

The source apportionment models discussed in this report can provide the technical basis for determining the effects of proposed controls on the attainment of several air-quality goals. In analyzing several problems simultaneously, one must choose a model that can be applied to all the problems at hand. In many cases complex mechanistic models will describe a broad range of physical and chemical processes associated with the major gaseous and particulate pollutants. As they analyze visibility impairment, these models can simultaneously determine the concentrations, fluxes, and effects of primary emissions as well as secondary oxidants, acids, and aerosols.

Simpler approaches could be applied to analyses of limited scope. For example, simpler rollback models can be used to make a preliminary examination of the effect of sulfur reduction mandated by the 1990 Clean Air Act Amendments on visibility in Class I areas in the eastern United States (Trijonis et al., 1990). Such an analysis can provide useful information about visibility and acid deposition, because sulfate is a dominant component of eastern haze, and sulfuric acid is a major component of acid rain in the East. If NO_x or volatile organic compound controls are to be evaluated, then more complex models would be preferred because the interactions associated with these chemicals are more complex.

The possible visibility effects of PM_{10} controls in nonattainment areas also illustrate the complications that arise when multiple regulatory goals and multiple sources are taken into account. The question in this case is whether controls adopted for countywide PM_{10} abatement are adequate to produce acceptable visibility in nearby Class I areas. If a small PM_{10} nonattainment area is the only major source of airborne particles in the region, then there might be a simple way to assess the effects of local PM_{10} controls on broader scale visibility. However, if the local PM_{10} nonattainment area is only a small part of a large multiple-source region, then more complex modeling than that needed for a study of PM_{10} control will be required to sort out the various source contributions to the regional visibility problem.

An assessment of the potential benefits on visibility of existing and expected legislation and regulation aimed at alleviating other air-quality problems should be an integral part of the decision-making process. If visibility is considered in the selection of air-quality models that will be used to analyze those other problems, then the effects on visibility of policies directed at other air-quality problems are more likely to be taken into account.

SUMMARY

The committee evaluated several alternative methods that could be used alone or in combination to analyze the effects of individual emissions sources or source classes on atmospheric visibility. Empirical methods range from qualitative tools, such as photography, through models based on material balances, such as speciated rollback and chemical mass balance models, to techniques based on statistical inference. We also have described models that are derived from the basic equations that govern atmospheric transport and that sometimes include chemical reaction and aerosol processes.

Many of the more empirical models already have been developed almost to their fullest potential, and therefore a fairly clear picture of how these models could fit into a comprehensive source apportionment program is emerging. Photographic methods and other simple source identification systems provide an inexpensive way to qualitatively implicate single emissions sources that create visible plumes in or near Class I areas. Photographs, videotape, and film can provide sufficient evidence for regulatory action in simple cases, but they are not likely to be useful for source apportionment of widespread regional hazes.

Speciated rollback models linked to light extinction budget calculations represent perhaps the only complete system of analysis that can be used for regional haze source apportionment throughout the United

States based on data available today. The feasibility of such analyses is demonstrated for several regional cases in Chapter 6: A speciated roll-back model is used to develop a preliminary description of the likely origin of visibility impairment in the eastern, southwestern, and north-western regions of the United States. The speciated rollback model is best suited to analyses of regional haze and is not suited to projecting the effects of changes in single members of a group of similar sources or the effects of proposed new sources in areas where the air is now clean. New sources are outside the realm of the past experience upon which rollback models are built.

Receptor-modeling methods are valid, well accepted, and widely available for source apportionment of primary particles. Regression analysis and chemical mass balance techniques each have appropriate uses; regression analysis requires less prior knowledge of emissions characteristics, but it is more susceptible to model specification errors and consequent biases. However, neither model is yet developed to the point of acceptance for the purpose of apportioning secondary particles (sulfates, for example) among sources. It is not known whether further research will lead to a fully successful receptor-oriented model for sec-ondary-particle formation; certainly many attempts are being made to expand the applicable range of receptor models. Receptor models cur-rently must be used in conjunction with other models for secondary-particle formation. This simultaneous use of more than one model is useful in many cases, and therefore one can expect chemical mass bal-ance receptor models to be used as part of comprehensive systems of source apportionment (either as a primary tool for source apportionment or to check the consistency of findings obtained by other methods).

Mechanistic source apportionment models for use in the analysis of regional haze problems are also in considerable demand. In principle, such models could contain a physically based cause-and-effect descrip-tion of atmospheric processes that could provide detailed and accurate predictions. In practice, mechanistic models exist in pieces that have not been fully assembled and tested. Their completion and testing should be pursued as a high priority.

Some mechanistic models are available that present a partial picture of the effects of source emissions on pollutant concentrations. They can trace transport paths from source to receptor and in some cases can predict concentrations of secondary particles, such as sulfates and ni-trates. These partial mechanistic models probably will find use in regional visibility analysis, but must be combined with other empirical methods (such as the imposition of measured particle size distributions) to form a complete source apportionment system.   Whether these semiempirical, semimechanistic models can produce predictions that are more accurate than those of a fully empirical speciated rollback model is an interesting research question that should be investigated.

There is a clear need for a comprehensive and versatile single-source model for use in predicting the effects of large new sources.   An advanced single-source model should not be restricted to Gaussian plume model formulations.   Further research will be needed to create such a model, because current plume blight models do not include the atmospheric aerosol processes that lead to light scattering and absorption by particles.   Efforts are underway by public agencies in the western states to develop the operational capability to evaluate the effect of proposed large single sources on regional scale visibility.   The procedure is to maintain a baseline multiple source regional model into which data on a single new source can be inserted.   This development effort should be encouraged.

In summary, we will consider methods for source apportionment that are either available or could be put together from available components. Following our emphasis on a nested approach in which models of increasing difficulty and accuracy are chosen, the most attractive systems are judged to be as follows:

Regional haze assessment:
• Speciated rollback models;
• Hybrid combinations of chemical mass balance receptor models with secondary-particle models;
• Mechanistic transport and secondary-particle formation models used with measured particle size distribution data to facilitate light-scattering calculations.

Analysis of existing single sources close to the source:
• Photographic and other source identification methods (in simple cases only);
• Hybrid combinations of chemical mass balance or tracer techniques with secondary-particle formation models that include explicit transport calculations and an adequate treatment of background pollutants;
• The most advanced reactive plume models available, hybridized

with measured data on particle properties in such plumes and accompanied by an adequate treatment of background pollutants.

Analysis of new single sources close to the source:
• The most advanced reactive plume models available, hybridized with measured data on particle properties in the plumes of similar sources and accompanied by an adequate treatment of background pollutants.

Analysis of single sources at the regional scale:
• Insertion of the single source in question into an appropriately chosen multiple-source description of the regional haze problem.

The hybrid models mentioned above are available to the extent that the necessary pieces of the modeling systems exist. Any novel combination of existing models should be carefully evaluated.

For the reasons explained in the introduction to this chapter, most source apportionment studies would benefit from the use of several candidate models, and hence groups of models rather than single models are noted above. We emphasize that the skill and knowledge of the personnel executing a modeling study are often more important in determining the quality of the study than is the choice of the modeling method.

We recommended research to achieve several goals. First, fully developed mechanistic models for the chemical composition, size distribution, and optical properties of atmospheric particles and gases should be created and tested. Two types of mechanistic models are needed: an advanced reactive plume aerosol process model for analysis of single-source problems close to the source and a grid-based multiple-source regional model for analysis of regional haze. In pursuit of those objectives, a program of careful field experiments and data analysis must be designed and conducted to support the use of aerosol process models (for example, to collect data on the degree of uptake of water by airborne particles) and to better characterize emission sources (to measure the chemical composition and size distribution of primary particles at their source in addition to the gaseous precursor emissions). Finally, experimental programs must be designed and conducted to test the performance of completed models of all kinds against field observations on emissions and air quality.

6

Emission Controls
and Visibility

The Committee on Haze in National Parks and Wilderness Areas is charged with developing working principles for assessing the relative importance of anthropogenic sources of emissions that contribute to haze in Class I areas and for assessing various alternative measures for source control. For assessing controls, the committee has focused on principles that can be derived from scientific knowledge of visibility impairment.

Any effective strategy to accomplish the congressionally established goal of remedying and preventing anthropogenic visibility impairment in Class I areas (Clean Air Act, §169A) requires limiting the emissions of pollutants that reduce visibility. Visibility problems in Class I areas are mostly the result of regional haze, rather than the effect of emissions from one or a few individual sources at specific sites. Therefore, a strategy that relies only on influencing the location of new sources, although perhaps useful in some situations, would not by itself prove effective. Moreover, such a strategy would not, of course, remedy the visibility impairment caused by existing sources.

Because most impairment of visibility is regional, an effective program to improve visibility in Class I areas must operate over large geographic areas. Not only would such a program benefit Class I areas, but it would improve visibility outside of these areas. Efforts to reduce haze in Shenandoah National Park or Great Smoky Mountains National Park would improve visibility in large parts of the East; the same is true for Class I areas in the West. Class I areas cannot be regarded as potential islands of clean air in a polluted sea.

The first step in designing a visibility strategy is to characterize the particles and gases that impair visibility and, to the extent possible, to apportion the impairment among contributing sources. The source apportionment methods (see Chapter 5 and Appendix C) should be appropriate to the temporal and geographic scale of the visibility problem. A visibility strategy could include controls on several different pollutants and source types. Although, sulfur dioxide, the precursor to sulfates, is often the most important single cause of impairment, control of fine and coarse particulate matter, oxides of nitrogen, ammonia (which is possibly a limiting factor in nitrate formation), and volatile organic compounds (VOCs) also must be considered.

In this chapter the committee provides an example, using a speciated rollback model, to apportion anthropogenic light extinction among source types in the eastern, southwestern, and northwestern United States. This exercise is not intended as an operational evaluation of regional haze in those regions. Instead, it is presented to illustrate some of the issues that arise in any apportionment of visibility impairment.

The next step in designing a strategy is to determine whether control measures exist or can be developed to reduce the emissions that impair visibility. Appendix D describes control options for the principal source categories identified in the example. (There could, of course, be situations where other sources are also of concern.) Many of these control techniques are effective and commercially available; others are still under development.

Next, it is necessary to assess the effectiveness of alternative control measures. The committee used the speciated rollback modeling exercise to estimate the visibility improvements that would result from application of commercially available pollution control methods and technologies to major contributing sources in the three regions modeled. The committee's analysis shows that the application of these controls could noticeably improve visibility in the areas modeled.

This exercise is not intended to signify the committee's recommendation for adopting any specific control strategy. Such a decision involves issues outside the bounds of science and the committee's expertise. Rather, the exercise allows an estimate to be made of the extent to which visibility could be improved with commercially available technology. Others must decide whether the public welfare requires the improvement and whether a program to do so would best use a technology-based approach or some alternative, such as an air-quality management approach or market-based mechanisms (see Chapter 3). Similarly, the committee does not recommend the use of a specific methodology to generate alternative control strategies for analysis. Appendix E illustrates the use of linear programming and cost-effectiveness analysis to do so; however, there are a variety of other less formal techniques.

The committee is aware of the limitations of its identification of control measures. First, there can be differences of opinion about what control technologies are proven and commercially available. Such judgments are, like choices about how to shape a control strategy, ultimately based on economic and policy considerations as well as on technology. Second, the committee has not assessed the effects of such strategies as land-use planning and increased energy efficiency that rely on changes in behavior rather than on the application of control techniques. This is because of our lack of knowledge about the possible effects of such strategies; it is not the result of any disposition against them.

The committee did not estimate the cost of the modeled control program; such an estimate is beyond the committee's expertise. Moreover, a cost estimate would not be very informative in isolation from cost estimates for other control strategies. Approaches that rely on behavioral changes or that provide incentives for increasing efficiency and reducing the cost of controls might be more cost-effective than the approach modeled.

Because of data limitations, the modeling exercise does not cover the entire United States; it excludes Alaska, Hawaii, and those portions of the West outside the Pacific Coast, Idaho, Nevada, and the Four Corners states. The excluded area contains important Class I areas such as Yellowstone National Park.

APPORTIONMENT OF REGIONAL HAZE
USING A SPECIATED ROLLBACK MODEL

The committee used a speciated rollback model to apportion anthropogenic light extinction among source types in three large regions of the United States—the East (states east of the Mississippi River), the Southwest (California, Nevada, Arizona, New Mexico, Utah, and Colorado), and the Pacific Northwest (Oregon, Washington, and Idaho). This modeling exercise is presented to illustrate issues that arise in any apportionment, and is not intended as an endorsement of speciated rollback modeling in favor of other approaches.

Appendix C describes the limitations of speciated rollback modeling. The model presented here treats haze in an aggregated, average sense for the entire modeled region. It does not indicate which source areas within the region contribute the most pollution to a specific receptor area, such as a national park. Moreover, it does not distinguish an episodic source, whose impact is pronounced on a few days and nonexistent on most, from a continuous source whose constant effect is imperceptible. More-advanced models (see Chapter 5 and Appendix C) could provide more detailed information about source–receptor relationships.

To address problems with regional haze, speciated rollback models must be extended beyond their typical use as a tool for assigning pollutant concentrations to sources. The extended rollback models developed here pertain to anthropogenic *light extinction*, not just aerosol concentrations. The apportionment is achieved in two steps: First, visibility impairment is allocated among aerosol types by means of light extinction budget calculations (see Chapter 4), and then each aerosol type is related to emissions through the linear rollback model.

This analysis is concerned only with anthropogenic visibility impairment. This means that the results depend on the apportionment of material between natural and anthropogenic sources, which is itself an estimate. The relative contributions of natural and anthropogenic sources to the organic and dust portions of the airborne particles are very poorly characterized; natural sources are arbitrarily assumed here to account for one-third and one-half of the respective totals. The footnotes to Table 6-1 summarize the natural-anthropogenic partitioning of the other aerosol types; the consistent use of fractions is intended to highlight the approximate nature of the values. Based on these assumptions, the fractions of average light extinction that results from anthropogenic sources are about seven-eighths in the East, five-eighths in the Pacific Northwest, and three-eighths in the cleanest areas of the Southwest.

Table 6-1 presents extinction budgets that relate the anthropogenic fraction of light extinction to aerosol constituents in the three regions. The numbers are based on the overall extinction budget by aerosol constituents and the portion of each constituent that results from anthropogenic sources. As described in the footnotes to Table 6-1 and the references cited therein, deriving these two sets of information is a non-trivial task involving the assemblage and cross comparison of numerous data sets.

In each region, all anthropogenic pollutants are assumed to be emitted within that region. Emissions from major sources across the border in

TABLE 6-1 Visibility Model Results: Anthropogenic Light Extinction
    Budgets[a]

	East[b]	Southwest[c]	Northwest[d]
Sulfates	65	39	33
Organics	14	18	28
Elemental carbon	11	14	15
Suspended dust	2	15	7
Nitrates	5	9	13
Nitrogen dioxide	3	5	4

[a]Percentage contribution by specific pollutant to anthropogenic light extinction in three regions of the United States.

[b]Based on Table 9, Table 18, Figure 45, Appendix A, and Appendix E of NAPAP Visibility SOS/T Report (Trijonis et al., 1990). It is assumed that sulfates (3% natural) account for 60% of non-Rayleigh extinction, organics (33% natural) account for 18%, elemental carbon (3% natural) accounts for 10%, suspended dust (50% natural) accounts for 4%, nitrates (10% natural) account for 5%, and nitrogen dioxide (10% natural) accounts for 3%.

[c]Based on Table 9, Table 18, Figure 45, Appendix A, and Appendix E of the NAPAP Visibility SOS/T Report (Trijonis et al., 1990). It is assumed that sulfates (10% natural) account for 33% of non-Rayleigh extinction, organics (33% natural) account for 20%, elemental carbon (10% natural) accounts for 12%, suspended dust (50% natural) accounts for 23%, nitrates (10% natural) account for 8%, and nitrogen dioxide (10% natural) accounts for 4%.

[d]Extinction efficiencies (relative to organics) are chosen as 1.5 for sulfates, 2.5 for elemental carbon, 0.3 for fine crustal materials, and 1.5 for nitrates (Trijonis et al., 1988, 1990). Coarse dust extinction is assumed to be three times fine dust extinction (Trijonis et al., 1988, 1990). Natural aerosol particle fractions are assumed to be one-tenth for sulfates, one-third for organics, one-tenth for elemental carbon, one-half for crustal materials, and one-tenth for nitrates. These assumptions are applied using the fine mass concentrations in Trijonis et al., (1990). The percentage contribution for nitrogen dioxide is assumed to be 4%.

northern Mexico and southern Canada are neglected, as is the anthropogenic background that is found in the most remote regions of the globe. Because the analysis neglects the anthropogenic portion of the extra-regional background, it will tend to overestimate the effects of changing regional emissions. This problem would become more severe if the analysis were conducted for smaller study regions. Table 6-2 lists, for each region, the percentage contribution from each major source type to emissions of pollutants that contribute to haze formation. Primary particles and gaseous precursors of secondary particles are included.

Tables 6-3 through 6-5 apportion light extinction using the speciated rollback model. In keeping with the overall level of the analysis, the following simplistic precursor relationships are assumed:

- Sulfate concentration is proportional to emissions of sulfur oxides (SO_x).
- Half of the organic airborne particle concentration is proportional to emissions of primary organic particulate matter (PM); the other half is proportional to emissions of VOC.
- The elemental carbon concentration is proportional to emissions of elemental carbon.
- The suspended dust concentration is proportional to emissions of suspended dust.
- Half of the nitrate concentration is proportional to emissions of ammonia (NH_3); the other half is proportional to emissions of oxides of nitrogen (NO_x).
- The nitrogen dioxide (NO_2) concentration is proportional to emissions of NO_x.

The relationship for nitrate reflects an arbitrary assumption about the distribution of circumstances in which NO_x or NH_3 is the limiting reactant in forming ammonium nitrate particles, and the relationship for organics reflects an assumed partitioning between primary and secondary material. These two relationships are clearly more uncertain than the others.

Tables 6-3 through 6-5 show that in all three regions, electric utilities, gasoline-fueled vehicles, and diesel-fueled mobile sources are either the top three anthropogenic sources of light extinction or are three of the top four sources. According to this modeling approach, in the East, electric utility SO_x alone are responsible for slightly more than one-half of anthropogenic light extinction. This is because sulfates are the predominant component of anthropogenic haze in the East and electric utilities are the predominant emitter of SO_2 in the East. For the Southwest and Northwest, however, no single source category is dominant.

To illustrate the use of the tables, we can estimate the effect on visibility of the SO_2 emission reductions in the East called for by the 1990 Clean Air Act amendments. According to NAPAP (NAPAP, 1991a,b), the amendments require about a 36% reduction in SO_2 emissions in the East (equivalent to a 46% reduction in electric utility SO_2 emissions; see Table 6-2). From Table 6-3, we deduce that a 46% reduction in SO_2 from electric utilities will produce a 46% $\times$ 51% = 23% reduction in anthropogenic light extinction in the East. Because about seven-eighths of total light extinction in the East is anthropogenic, the total improvement in light extinction would be 7/8 $\times$ 23% = 20%. This result is in agreement with the NAPAP visibility assessment that used light extinction budgets similar to those used here but that was based on a deterministic model for sulfate transport and chemical reaction in the atmosphere, the Regional Acid Deposition Model (RADM).

POTENTIAL VISIBILITY IMPROVEMENTS FROM EMISSION CONTROLS

This section assesses the potential for visibility improvements from application of commercially available technology to control emissions. The assessment is for illustration and should not be construed as advocating a specific control strategy. The method could, however, be applied to estimate the effectiveness of emissions reductions being considered by policy makers, especially to compare the effects of alternative control strategies.

Appendix D describes control methods for major sources of visibility impairment in the regions modeled: electric utilities, industrial coal combustion, the petroleum and chemical industries and industrial oil combustion, nonferrous smelters, diesel-fueled and gasoline-fueled motor vehicles, fugitive dust, feedlots and livestock waste management, residential wood burning, forest management burning, and organic solvent evaporation. These methods are summarized in Table 6-6.

216

TABLE 6-2 Percentage Contribution of Source Categories to Emissions in the East, Southwest, and Northwest[a,b,c]

	SO$_x$	Organic Particles	VOCs	Elemental Carbon[c]	Suspended Dust	NH$_3$	NO$_x$
East							
Electric utilities (nearly all from coal-fired power plants)	78	Neg	Neg	Neg	Neg	Neg	39
Diesel-fueled mobile sources (trucks, buses, train, jets, ships, heavy equipment)	1½	Neg	Neg	47	Neg[d]	Neg	16
Gasoline vehicles	1	34	31	29	Neg[d]	Neg	26
Petroleum and chemical industries and industrial oil combustion	4½	Neg	11	Neg	Neg	Neg	Neg
Industrial coal combustion	7	Neg	Neg	Neg	Neg	Neg	Neg
Residential wood burning	Neg	20	13	15	Neg	Neg	Neg
Fugitive dust (presumably predominantly from on-road and off-road traffic)	Neg	Neg	Neg	Neg	100	Neg	Neg
Feedlots and livestock waste management	Neg	Neg	Neg	Neg	Neg	66	Neg
Miscellaneous	8	46	45	9	Neg	34	19

217

TABLE 6-2 (continued)

	SO_x	Organic Particles	VOCs	Elemental Carbon[c]	Suspended Dust	NH_3	NO_x
Southwest							
Electric utilities (nearly all from coal-fired power plants)	33	Neg	Neg	Neg	Neg	Neg	19
Diesel-fueled mobile sources (trucks, buses, trains, jets, ships, heavy equipment)	12	5	Neg	52	Neg[d]	Neg	23
Gasoline vehicles	5	38	42	31	Neg[d]	Neg	32
Petroleum and chemical industries and industrial oil combustion	22	Neg	12	Neg	Neg	Neg	Neg
Copper smelters	19	Neg	Neg	Neg	Neg	Neg	Neg
Fugitive dust (presumably predominantly from on-road and off-road traffic)	Neg	Neg	Neg	Neg	100	Neg	Neg
Residential wood burning	Neg	8	5	6	Neg	Neg	Neg
Feedlots and livestock waste management	Neg	Neg	Neg	Neg	Neg	75	Neg
Miscellaneous	9	49	41	11	Neg	25	26

TABLE 6-2 (continued)

	SOx	Organic Particles	VOCs	Elemental Carbon[c]	Suspended Dust	NH3	NOx
Northwest							
Electric utilities (nearly all from coal-fired power plants)	30	Neg	Neg	Neg	Neg	Neg	8
Diesel-fueled mobile sources (trucks, buses, trains, jets, ships, heavy equipment)	12	Neg	Neg	37	Neg[4]	Neg	29
Gasoline vehicles	4	15	31	16	Neg[4]	Neg	36
Petroleum and chemical industries and industrial oil combustion	19	Neg	10	Neg	Neg	Neg	Neg
Residential wood burning	Neg	22	25	22	Neg	Neg	Neg
Forest management burning	Neg	45	13	20	Neg	Neg	Neg
Fugitive dust (presumably predominantly from on-road and off-road traffic)	Neg	Neg	Neg	Neg	100	Neg	Neg
Feedlots and livestock waste management	Neg	Neg	Neg	Neg	Neg	81	Neg
Primary metallurgical process	8	Neg	15	Neg	Neg	Neg	Neg
Organic solvent evaporation	Neg	Neg	15	Neg	Neg	Neg	Neg
Miscellaneous	27	18	6	5	Neg	19	27

TABLE 6-2   (continued)

SO_x, oxides of sulfur; VOCs, volatile organic compounds; NH_3, ammonia; NO_x, oxides of nitrogen; Neg, negligible.

[a]Based on the 1985 NAPAP Inventory (Zimmerman et al., 1988a; Battye, pers. comm., E.H. Pechan and Associates, Durham, N.C., 1990), with the exception that copper smelter emissions have been updated to 1988 (E. Trexler, pers. comm., DOE, Washington, D.C., 1990) because of the large recent changes in emissions from that source category.

[b]Source categories are included for a region only if they are estimated to contribute at least 2% of total anthropogenic haze. Individual emissions are noted as Negligible (Neg) if the anthropogenic haze contribution from that source emission is less than 0.5%.

[c]The elemental carbon/organic carbon fractions of $PM_{2.5}$ emissions are assumed to be 70/20 for diesel-fueled mobile sources, 10/40 for gasoline vehicles, 10/45 for residential wood burning, and 6/60 for forest management burning (Gray, 1986; Trijonis et al., 1988; Core, pers. comm., Oregon Department of Environmental Quality, Portland, Oregon, 1990). It is assumed that all other urban sources account for the same amount of organic PM emissions as the total of gasoline and diesel mobile sources but only for 10% as much elemental carbon (Gray, 1986; Trijonis et al., 1988).

[d]Suspended dust from vehicular traffic is treated as a separate category.

TABLE 6-3 Percentage Apportionment of Anthropogenic Light Extinction in the Eastern United States by Speciated Rollback Model

Source	Total Source Contribution	Sulfates ∝SOx	Organics 1/2∝org PM 1/2∝VOC	Elemental Carbon ∝EC PM	Suspended Dust ∝dust PM	Nitrates 1/2∝NOx 1/2∝NH₃	NO₂ ∝NOx
Species contribution		65	14	11	2	5	3
Electric utilities (nearly all from coal-fired power plants)	54	52	Neg	Neg	Neg	1	2
Gasoline vehicles	11	1	5	3	Neg[a]	1	1
Diesel-fueled mobile sources (trucks, buses, trains, jets, ships, heavy equipment)	6	1	Neg	5	Neg[a]	Neg	Neg
Industrial coal combustion	5	5	Neg	Neg	Neg	Neg	Neg
Petroleum and chemical industries and industrial oil combustion	4	3	1	Neg	Neg	Neg	Neg
Residential wood burning	4	Neg	2	2	Neg	Neg	Neg
Fugitive dust (presumably predominantly from on-road and off-road traffic)	2	Neg	Neg	Neg	2	Neg	Neg
Feedlots and livestock waste management	2	Neg	Neg	Neg	Neg	2	Neg
Miscellaneous	12	4	6	1	Neg	1	Neg

[a]Suspended dust from vehicular traffic is treated as a separate source category. Neg = Negligible

TABLE 6-4 Percentage Apportionment of Anthropogenic Light Extinction in the Southwestern United States by Speciated Rollback Model

	Total Source Contributions	Sulfates $\propto SO_x$	Organics $1/2\propto$org PM $1/2\propto$VOC	Elemental Carbon $\propto$EC PM	Suspended Dust $\propto$dust PM	Nitrates $1/2\propto NO_x$ $1/2\propto NH_3$	NO_2 $\propto NO_x$
Species contributions	16	39	18	14	15	9	5
Gasoline vehicles	15	2	7	4	Neg[a]	1	2
Electric utilities (nearly all from coal-fired power plants)	15	13	Neg	Neg	Neg	1	1
Fugitive dust (presumably predominantly from on-road and off-road traffic	15	Neg	Neg	Neg	15	Neg	Neg
Diesel-fueled mobile sources (trucks, buses, trains, jets, ships, heavy equipment)	14	5	Neg	7	Neg[a]	1	1
Petroleum and chemical industries and industrial oil combustion	10	9	1	Neg	Neg	1	Neg
Copper smelters	7	7	Neg	Neg	Neg	Neg	Neg
Feedlots and livestock waste management	3	Neg	Neg	Neg	Neg	3	Neg
Residential wood burning	2	Neg	1	1	Neg	Neg	Neg
Miscellaneous	18	3	9	2	Neg	3	1

[a] Suspended dust from vehicular traffic is treated as a separate source category. Neg = Negligible.

TABLE 6-5 Percentage Apportionment of Anthropogenic Light Extinction in the Northwestern United States by Speciated Rollback Model

	Total Source Contribution	Sulfates ∝SO$_x$	Organics 1/2∝org PM 1/2∝VOC	Elemental Carbon ∝EC PM	Suspended Dust ∝dust PM	Nitrates 1/2∝NO$_x$ 1/2∝NH$_3$	NO$_2$ ∝NO$_x$
Species contributions		33	28	15	7	13	4
Diesel-fueled mobile sources (trucks, buses, trains, jets, ships, and equipment)	13	4	Neg	6	Neg[a]	2	1
Gasoline vehicles	12	1	6	2	Neg[a]	2	1
Electric utilities (nearly all from coal-fired power plants)	11	10	Neg	Neg	Neg	1	Neg
Forest management burning	11	Neg	8	3	Neg	Neg	Neg
Residential wood burning	10	Neg	7	3	Neg	Neg	Neg
Petroleum and chemical industries and industrial oil combustion	7	6	1	Neg	Neg	Neg	Neg
Fugitive dust (presumably predominantly from on-road and off-road traffic	7	Neg	Neg	Neg	7	Neg	Neg
Feedlots and livestock waste management	7	Neg	Neg	Neg	Neg	5	Neg
Primary metallurgical processes	3	3	Neg	Neg	Neg	Neg	Neg
Organic solvent evaporation	2	Neg	2	Neg	Neg	Neg	Neg
Miscellaneous	19	9	4	1	Neg	3	2

[a]Suspended dust from vehicular traffic is treated as a seperate source of energy

Neg = Negligible.

TABLE 6-6 Control Methods for Visibility-Impairing Emissions Sources[a]

Source category	Emissions	Control methods	Comments
Electric utilities	SO_x	Wet flue gas desulfurization (FGD, also known as scrubbing)	Predominant U.S. scrubbing technology; can reduce SO_2 emissions by 90–95%
		Spray dry FGD	Mostly used for low-sulfur coal; 80–90% SO_2 removal efficiency could be possible for high-sulfur coal; lower capital cost than wet FGD
		Switching to low-sulfur coal or natural gas	Raises fuel costs; low-sulfur coal supplies uncertain
		Furnace sorbent injection	Under demonstration; moderate removal efficiency, low capital cost
		Duct sorbent modification	Under development; moderate removal efficiency, low capital cost

TABLE 6-6 (continued)

Source category	Emissions	Control methods	Comments
	NO_x	Combustion modification	Available for new boilers; under demonstration for retrofit; 40–60% NO_x removal
		Selective catalytic reduction	Post-combustion method; more expensive than combustion modification; alone, it can reduce NO_x emissions by 80%
		Selective noncatalytic reduction	Under development; could be less expensive than selective catalytic reduction
	SO_x, NO_x	Environmental dispatching	Involves sending out electricity from lowest-emitting plants first; raises fuel cost
		Clean coal technologies: atmospheric fluidized bed combustion, pressurized fluidized bed combustion, integrated gasification combined	Under demonstration; could be cost-effective for both SO_2 and NO_x removal
Industrial coal combustion	SO_x	Wet FGD	Predominant scrubbing method; can reduce SO_2 emissions by 90–95%
		Spray dry FGD	Can reduce SO_2 emissions by 80–90%
	NO_x	Combustion modification	Can reduce NO_x emissions by about 50%

		Flue gas treatment	Can reduce NO_x emissions by 40% under optimum conditions (or by 80% using catalytic conversion); retrofit might not be feasible
	SO_x, NO_x	Atmospheric fluidized bed combustion	Available for industrial use; comparable in cost to wet scrubbing
Petroleum and chemical industries and industrial oil combustion	SO_x	Feedstock desulfurization; wet scrubbing; Claus tailgas treatment	Can reduce SO_x emissions during refinery processes by 80–90%; Can reduce SO_2 emissions from hydrogen sulfide-to-sulfur conversion by 99%
	VOCs	Better selection and maintenance of seals and packing materials; improved tank-filling procedures; vapor recovery systems	Can reduce VOC emissions from petroleum refineries, petrochemical plants, and petroleum transport and marketing by 90% or more

TABLE 6-6 (continued)

Source category	Emissions	Control methods	Comments
Copper smelters	SO_x	Use of sulfuric acid plants to recover SO_2	Can reduce SO_2 emissions by 97% or more
Diesel-fueled mobile sources	SO_x	Use of low-sulfur fuel	Low-sulfur requirements to take effect in 1993
	Elemental carbon	Advanced direct-injection engines; improved electronic control; reduced oil consumption	Sufficient for attainment of 1991 particulate emission standards
		Trap oxidizers	Under development; might be needed for attainment of stricter standards that take effect in 1994
	Elemental carbon, NO_x	Turbocharging and intercooling	Now used by almost all diesel engines
		Catalytic converters	Mature technology for gasoline vehicles, but new for diesel vehicles; requires use of low-sulfur fuel; might be needed to meet 1994 emission standards
		Enhanced inspection and maintenance (I/M)	Could especially reduce emissions from older vehicles
		Alternative fuels (methanol, natural gas)	Demonstration fleets of alternatively fueled heavy-duty vehicles now exist

Source	Pollutant	Control measure	Comments
Gasoline vehicles	SO_x	Use of low-sulfur fuel	
	VOCs	Vapor recovery systems	Already required on gas station pumps in some urban areas
		Reformulated gasoline	Usable in existing vehicles; costs several cents per gallon more than regular gasoline
	VOCs, NO_x	Three-way catalytic converter	Principal tailpipe technology for VOC, NO_x, and carbon monoxide reduction; improved catalyst coatings could further reduce emissions
		Improved electronic control; better air–fuel management; engine improvements	Should enable manufacturers to meet 1994 VOC emission standards
		Enhanced inspection and maintenance	I/M requirements strengthened by 1990 Clean Air Act amendments
		Alternative fuels (e.g., methanol, natural gas, electricity)	Alternatively fueled light-duty vehicles are under development; NO_x reductions from fuels other than electricity could be limited

TABLE 6-6 (continued)

Source category	Emissions	Control methods	Comments
Fugitive dust	Dust particles	Improved agricultural land management; roadway surface improvements; road sweeping and flushing; improved material storage and transfer	Technologies are widely available and applied; control of natural dust emissions is often impracticable.
Feedlots and livestock waste management	NH_3	Application of chemical additives to animal waste	Can reduce NH_3 emissions from feedlots by about 50%
Residential wood burning	Organic particles, elemental carbon	Improved woodstoves	Emissions from best existing stove technology stoves are about 80% lower than from conventional woodstoves
Forest management burning	Organic particles, elemental carbon	Reduction of number of acres burned; increased use of residues; burning under increased fuel moisture; helitorch ignition; rapid mop-up	Together, these practices can reduce emissions by about 50%
Organic solvent evaporation	VOCs	Reformulation of solvents; use of nonsolvent-based alternatives	Some alternatives are available; others are under development

[a]See Appendix D for additional information.

Table 6-7 shows the emission reductions assumed for this assessment. The reductions are based on application of commercially available controls alone—that is, controls of proven effectiveness that can be applied or purchased today. For example, SO_x reductions from electric utilities are based on application of wet and spray-dry flue-gas desulfurization, but not on the sorbent injection and clean coal technologies still being demonstrated. Reductions in motor vehicle emissions are based on engine improvements, better electronic control, and enhanced inspection and maintenance, but not on the use of alternative fuels or of unproven technologies such as trap oxidizers. Reductions greater than those assumed in Table 6-7 would of course be possible as advanced technologies become available.

Emission reductions are specified as ¼, ½, or ¾ as a reflection of their approximate nature. For a sensitivity analysis, these reductions are replaced by 20%, 40%, and 60% for a low estimate and by 30%, 60%, and 90% for a high estimate. All emission reductions are relative to the 1985 NAPAP inventory used as the basis for the speciated rollback modeling analysis.

The approximate haze improvements associated with these reductions are shown in Tables 6-8, 6-9, and 6-10. The reductions of Table 6-7 have been multiplied by the source allocations of Tables 6-3 through 6-5 to yield the anthropogenic light extinction improvements of Tables 6-8 through 6-10. For example, assuming electric utility SO_x accounts for 52% of anthropogenic light extinction in the East (Table 6-3), then a 75% reduction in electric utility SO_x (Table 6-7) would yield a 39% reduction in eastern anthropogenic light extinction (Table 6-8). The 1990 Clean Air Act amendments will require a reduction of approximately 50% in electric utility SO_2 emissions nationwide by the year 2000 (most of the reduction is to take place in the East). The notes to Tables 6-8 through 6-10 state the total percentage decreases in anthropogenic light extinction for the entire control strategy, i.e., the sum of all the entries in each table.

Table 6-11 summarizes the visibility improvements that would result from the example emission reductions. For the East, anthropogenic light extinction would be reduced by an estimated 59%. Because natural visibility impairment is small compared to anthropogenic impairment in the East, there would be nearly an equivalent reduction in total light extinction, and median visual range would more than double. As indi-

TABLE 6-7 Estimated Commercially Available Percentage Reductions of Visibility-Impairing Emissions

Source category	Emission	Reduction[a]
Electric utilities	SO_x	¾
	NO_x	½
Industrial coal combustion	SO_x	¾
	NO_x	½
Diesel-fueled motor vehicles	SO_x	¾
	NO_x	½
	Elemental carbon	¾
Gasoline vehicles	SO_x	¾
	Organic particles + VOC	½
	NO_x	½
Petroleum and chemical industries and industrial oil combustion	SO_x	¾
	VOC	½
Fugitive dust	Dust particles	¼
Feedlots and livestock waste management	NH_3	½
Residential wood burning	Organic particles, VOCs	¼
	Elemental carbon	¼
Forest management burning	Organic particles, VOCs	½
	Elemental carbon	½
Organic solvent evaporation	VOC	½
Copper smelters		b
	SO_x	

[a]These are illustrative reductions, in most cases from individual units, that are technically feasible using commercially available technologies and methods. They provide the bases for assigning 25%, 50% or 75% reduction in the base scenario of the illustrative assessments shown in Tables 6-8 to 6-11 (and for selecting 20%/40%/60% and 30%/60%/90% for the low and high reduction scenario of the sensitivity analyses).

[b]Copper smelters, which remain a significant source of SO_x in the southwest, were not analyzed for potential SO_x reductions because they are already controlled, and the prospects for further reductions are uncertain.

TABLE 6-8 Speciated Rollback Model Percentage Reductions[a] in Eastern U.S. Anthropogenic Light Extinction for Emission Reductions of Table 6-7. Total Reduction in Eastern U.S. Anthropogenic Light Extinction: 59% ± 12%[b]

Source	Sulfates ∝ SO_x	Organics ½∝org PM ½∝VOC	Elemental Carbon ∝ELC PM	Suspended Dust ∝Dust PM	Nitrates ½∝ NO_x ½∝NH_3	NO_2 ∝NO_x
Electric utilities	39				½	1
Gasoline-fueled motor vehicles	1	2½			½	½
Diesel-fueled motor vehicles	1		4			
Industrial coal combustion	4					
Petroleum and chemical industries and industrial oil combustion	2	½				
Residential wood burning		½	½			
Fugitive dust				½		
Feedlots and livestock waste management					1	

[a]The numbers in the table indicate percentage reductions in anthropogenic light extinction achieved by controls applied to each source category and emissions type. Emission sources are presented in the order of their relative contribution to anthropogenic light extinction.

[b]The spread of the results is obtained by replacement of 25%/50%/75% emission reductions with 20%/40%/60% and 30%/60%/90% reductions.

TABLE 6-9 Model Percentage Reductions[a] in Southwest U.S. Anthropogenic Light Extinction for Emission Reductions of Table 6-7. Total Reduction in Southwest U.S. Anthropogenic Light Extinction: $40\% \pm 8\%$[b]

Source	Sulfates $\propto SO_x$	Organics ½ $\propto$ org PM / ½ $\propto$ VOC	Elemental Carbon $\propto$ ELC PM	Suspended Dust $\propto$ Dust PM	Nitrates ½ $\propto NO_x$	Nitrates ½ $\propto NH_3$	NO_2 $\propto NO_x$
Gasoline-fueled motor vehicles	1½	3½			½		1
Electric utilities	9½				½		½
Fugitive dust				4			
Diesel-fueled motor vehicles	4		5		½		½
Petroleum and chemical industries and industrial oil combustion	7	½					
Feedlots and livestock waste management						1½	

[a]The numbers in the table indicate percentage reductions in anthropogenic light extinction achieved by emission reductions applied to each source category and emissions type. Sources are presented in order of their relative contribution to anthropogenic light extinction.

[b]The spread of the results is obtained by replacement of 25%/50%/75% emission reductions with 20%/40%/60% and 30%/60%/90% reductions.

TABLE 6-10 Model Percentage Reductions[a] in Northwest U.S. Anthropogenic Light Extinction for Emission Reductions of Table 6-7. Total Reduction in Northwest U.S. Anthropogenic Light Extinction: $41\% \pm 8\%$[b]

Source	Sulfates $\propto SO_x$	Organics ½ $\propto$ org PM ½ $\propto$ VOC	Elemental Carbon $\propto$ ELC PM	Suspended Dust $\propto$ Dust PM	Nitrates ½ $\propto NO_x$ ½ $\propto NH_3$	$NO_2 \propto NO_x$
Diesel-fueled motor vehicles	3		4½		1	½
Gasoline-fueled motor vehicles	1	3			1	½
Electric utilities	7½				½	
Forest management burning		4	1½			
Residential wood burning		2	1			
Petroleum and chemical industries and industrial oil combustion	4½	½		2		
Fugitive dust						
Feedlots and livestock waste management					2½	
Organic solvent evaporation		1				

[a]The numbers in the table indicate percentage reductions in anthropogenic light extinction achieved by emission reductions applied to each source category and emissions type. Sources are presented in order of their relative contribution to anthropogenic light extinction.
[b]The spread of the results is obtained by replacement of 25%/50%/75% emission reductions with 20%/40%/60% and 30%/60%/90% reductions.

TABLE 6-11   Summary of Visibility Improvements from the Example Control Program

	East	Southwest	Northwest
Percentage reduction in anthropogenic light extinction[a]	59	40	41
Percentage reduction in total average light extinction[b]	52	15	26
Percentage increase in median visual range	108 ($\sim$25→50 km)[c]	18 ($\sim$150→180 km)	35 ($\sim$70→95 km)

[a]From Tables 6-8 through 6-10.
[b]Assuming natural contributions to total light extinction are 1/8 in the East, 5/8 in the Southwest, and 3/8 in the Northwest. Current median visual ranges are from Figure 2-1.
[c]Median visual range before and after control.

cated in Table 6-8, the majority of the improvement would come from controls on electric utility SO_x.

In the Southwest and Northwest, anthropogenic light extinction would be reduced by about 40%. Accounting for the larger relative contributions from natural sources in those less polluted regions, the reductions in total average light extinction are estimated to be only 15% in the Southwest and 26% in the Northwest. On worst-case pollution days, when natural contributions are relatively less important, light extinction would be reduced correspondingly more. As shown in Tables 6-9 and 6-10, the major source categories that contribute to the improvements would be motor vehicles (gasoline and diesel), electric utilities, industrial petroleum and chemical sources, and (in the Northwest) forest management burning.

It is of interest to compare the visibility improvements from the example emissions reductions to the improvements necessary to accomplish the national goal of no anthropogenic impairment of visibility from widespread haze. Achieving the latter goal under average conditions (not for worst-case pollution days) would require anthropogenic light extinction reductions of approximately 98.5% in the East, 83% in the Southwest, and 94% in the Northwest. As noted above, anthropogenic light extinction constitutes about seven-eighths of the average total light extinction in the East, five-eighths in the Northwest, and three-eighths in the cleanest areas of the Southwest. It is assumed that "no anthropogenic impairment" on the average means that average anthropogenic light extinction is less than 10% of average natural light extinction. Choosing 15% rather than 10% for this threshold would change the required anthropogenic light extinction reductions from 98.5% to 98% in the East, 83% to 75% in the Southwest, and 94% to 91% in the Northwest. The potential reductions in light extinction from the example emission reductions—59 $\pm$ 12% in the East, 40% $\pm$ 8% in the Southwest, and 41% $\pm$ 8% in the Northwest—fall far short of this.

The above analysis supports the following conclusions regarding the visibility improvement that would result from scenarios based upon the application of commercially available controls:

• In the East, anthropogenic and total light extinction would be reduced by more than one-half. In the Southwest and Northwest, reductions in anthropogenic light extinction would be less than one-half, with total (anthropogenic plus natural) reductions of about one-fourth.

• The key to improving visibility substantially in the East is to control SO_x emissions from electric utilities.

• Improving visibility substantially in the Southwest and Northwest will require the control of many source categories—especially electric utilities, diesel-fueled and gasoline-fueled motor vehicles, petroleum and chemical industrial sources, forest management burning, and fugitive dust.

• The visibility improvements in all regions would fall far short of the national goal of no anthropogenic impairment from regional haze.

The above analysis is only an example calculation—not an operational formulation and evaluation of a national haze control program. Nevertheless, the qualitative aspects of these conclusions do not appear sensitive to the details of the calculation. This insensitivity to specific numerical assumptions is supported by the compartmentalized nature of errors in the speciated rollback model (see Appendix C) and by the rudimentary sensitivity analysis in Tables 6-8 through 6-10.

RELATIONSHIP BETWEEN VISIBILITY AND OTHER AIR-QUALITY PROBLEMS

Visibility is just one of many air-quality problems. The pollutants that impair visibility contribute to other environmental problems, some of which have been or are being considered as objects of federal, state, or local legislation or regulation. For example, controls aimed at reducing acid rain or lowering ambient concentrations of ozone and PM_{10} could improve visibility in Class I areas; conversely, controls aimed at improving visibility could alleviate other air-quality problems. Policy makers should weigh these linkages in the design and assessment of possible control strategies.

The source attribution models discussed in Chapter 5 and Appendix C can provide the technical basis for examining the effects of proposed controls on multiple air-quality problems. Complex mechanistic models that explicitly consider a broad range of chemical interactions are best suited for this analysis. Simpler approaches could be applied to analyses of limited scope. For example, a rollback approach can be used to estimate the effect of SO_2 reductions that stem from the 1990 Clean Air Act amendments on visibility in eastern Class I areas (Trijonis et al., 1990). Such an analysis can provide useful information, because sulfates are a dominant component of eastern regional haze, and sulfuric acid is a major component of eastern acid rain. If the effects of NO_x or VOC controls were examined, the more complex models would be necessary because the chemical interactions of these pollutants are more complex.

The committee recognizes that these approaches must be balanced against other considerations. Although it could be theoretically attractive to calculate beforehand all relevant advantages and disadvantages of proposed controls on emissions, this might not be possible or even desirable in the process of formulating public policy. Moreover, many social and economic factors must be taken into account in deciding how to protect and improve visibility in Class I areas.

CONCLUSIONS

An effective program to improve visibility in Class I areas must reduce emissions of pollutants that impair visibility. Because most visibility impairment in Class I areas is caused by regional haze, an effective program must operate over large geographic areas.

The committee used a speciated rollback model to apportion anthropogenic light extinction in the East, Southwest, and Northwest. Its analysis indicates that in the East, SO_2 emissions from electric utilities cause more than one-half of all anthropogenic light extinction. In the Southwest and Northwest, no single source category is dominant; the major sources include gasoline- and diesel-fueled motor vehicles, electric utilities, industrial petroleum and chemical sources, and (in the Northwest) forest management burning.

Control technologies are available to reduce emissions from all major sources of haze. However, some of these technologies are expensive or of limited effectiveness. Continued support for research and development by government and industry is needed for efforts to improve the cost-effectiveness of existing emissions control technologies and to develop new technologies, especially low-emission technologies for fossil-fuel-based electricity generation (such as coal gasification and fuel cells), more efficient energy use technologies, and renewable sources of energy.

The committee estimated the visibility improvements that would result from the application of all commercially available controls in the three regions in the model. This analysis should not be construed as prescribing a specific control strategy. Estimated reductions in anthropogenic light extinction were about 60% in the East, and 40% in the Northwest and Southwest. The corresponding reductions in total light extinction were about 50% in the East, 25% in the Northwest, and 15% in the Southwest. The results fall far short of the national goal of no man-made impairment of visibility. The analysis indicates that the key to improving visibility in the East is to control SO_2 emissions from power plants. In the West, substantial improvements in visibility require that many source categories be controlled.

In designing and assessing strategies to improve visibility in Class I areas, it is important to consider linkages with other air-quality problems, such as acid rain, PM_{10}, and lower-atmosphere ozone. Many social and economic considerations must also be taken into account in designing control strategies.

7

Conclusions and Recommendations

The preceding chapters described the scientific and legal framework of efforts to protect and improve visibility, as well as methods for attributing visibility impairment to sources and for assessing alternative control measures. This chapter discusses the implications of current knowledge for future regulatory and research efforts. The committee did not presuppose a particular form for a visibility program because the design of a program involves many policy issues outside the bounds of science and the committee's expertise. However, present scientific knowledge about visibility impairment in Class I areas has several implications for policy makers.

GENERAL CONCLUSIONS

Any effective strategy to accomplish the statutory goal of remedying and preventing anthropogenic visibility impairment in Class I areas must limit emissions of pollutants that can cause regional haze.

Incontrovertible scientific evidence links emissions of air pollutants to the formation of haze that limits visibility and degrades the visual environment. Almost all the effects of air pollution on visibility are caused by airborne particles. In most cases, visibility degradation is caused by five kinds of particulate substances (and associated particulate water): sulfates (SO_4^{2-}), nitrates (NO_3^-), organic matter, elemental carbon, and

239

soil dust. Although some of these particles are emitted directly into the atmosphere, others, such as SO_4^{2-}, are the result of the transformation in the atmosphere of gaseous emissions such as sulfur dioxide (SO_2). Because of their physical properties, airborne particles and their gaseous precursors can exist in the atmosphere for several days; during this time, winds can carry these materials great distances (e.g., typically hundreds of miles). This leads to the formation of a widespread regional haze.

Visibility problems in Class I areas are predominantly the result of regional haze from many sources, rather than individual plumes caused by a few sources at specific sites. Therefore, a strategy that relies only on influencing the location of new sources, although perhaps useful in some situations, would not be effective in general. Moreover, such a strategy would not remedy the visibility impairment caused by existing sources until these sources are replaced.

Progress towards the national goal of remedying and preventing anthropogenic visibility impairment in Class I areas will require regional programs that operate over large geographic areas.

Because most visibility impairment in Class I areas results from the transport by winds of emissions and secondary airborne particles over great distances, focusing only on sources immediately adjacent to Class I areas—as under the current program—is unlikely to improve visibility effectively.

A program that focuses solely on determining the contribution of individual emission sources to visibility impairment is doomed to failure. Instead, strategies should be adopted that consider many sources simultaneously on a regional basis.

Nonetheless, the assessment of the contribution of individual sources to haze will remain important in some situations. For instance, a regional emissions management approach to haze could be combined with a strategy to assess whether locating a new source at a particular location would have especially deleterious effects on visibility. The committee has set out working principles in Chapter 5 and Appendix C for attributing visibility impairment to single sources.

The committee doubts, however, that such attributions could be the basis for a workable visibility protection program. It would be extremely time-consuming and expensive to try to determine the percent contribution of individual sources to haze one source at a time. For instance, the efforts to trace the contribution of the Navajo Generating Station to haze in the Grand Canyon National Park took several years and cost millions of dollars without leading to quantitatively definitive conclusions. Moreover, there are (and probably will continue to be) considerable uncertainties in ascertaining precise relationships between individual sources and the spatial pattern of regional haze.

Visibility policy and control strategies might need to be different in the West than in the East.

Haze in the East and in the West differ in important ways. Haze in the East is six times more intense than in the West because of the much higher level of pollution in the East. Were all anthropogenic pollution to disappear, visibility would remain greater (by about 50 %) in the West. In relatively clean areas, small increases in pollutant concentrations can markedly degrade visibility; increases of the same magnitude are less noticeable in more polluted areas. Hence, visibility in Class I areas in the West is particularly vulnerable to increased levels of pollution. Moreover, the West contains most of the nation's large national parks and wilderness areas, which can be fully appreciated only when visibility is excellent. The East, however, contains a large population to enjoy the benefits of any improvement in visibility in that region.

In the East, SO_2 emissions from coal-fired power plants account for about one-half of all anthropogenic light extinction. Reductions of these emissions are expected to occur in the next two decades as a result of the 1990 Clean Air Act amendments' acid rain control program. In the West, no single source category dominates; therefore, an effective control strategy would have to cover many source types, such as electric utilities, gasoline- and diesel-fueled vehicles, petroleum and chemical industrial sources, forest management burning, and fugitive dust.

Efforts to improve visibility in Class I areas also would benefit visibility outside those areas.

Because most visibility impairment is regional in scale, the same haze that degrades visibility within or looking out from a national park also degrades visibility outside it. Class I areas cannot be regarded as potential islands of clean air in a polluted sea.

Reducing emissions for visibility improvements could help alleviate other air-quality problems, just as other types of air-quality improvements could help visibility.

The substances that contribute to regional haze also contribute to a variety of other undesirable effects on human health and the environment. For example, SO_2 is a precursor of sulfuric acid in acid rain; oxides of nitrogen (NO_x) and volatile organic compounds (VOCs) are precursors of lower-atmosphere ozone; and fine atmospheric particles are a respiratory hazard. Such particles can influence climate by interacting with incoming solar radiation and by modifying cloud formation. Policy makers should consider the linkages between visibility and other air-quality problems when designing and assessing control strategies.

Achieving the national visibility goal will require a substantial, long-term program.

As shown in Chapter 6, the application of all commercially available control technology would reduce, but not eliminate, visibility impairment in mandatory Class I areas. Policy makers might develop a comprehensive national visibility improvement strategy as the basis for further regulatory action. Policy makers also might establish milestones against which progress toward the visibility goal could be measured.

Various indices can be used to characterize visibility and measure progress. Of these, the extinction coefficient (which includes scattering and absorption as components) is the most fundamental and can be linked most closely to the chemical composition and physical properties of the atmosphere. In addition, the extinction coefficient can be quantitatively related to other indexes, including human judgment of visibility.

Current scientific knowledge is adequate and control technologies are available for taking regulatory actions to improve and protect visibility. However, continued national progress toward this goal will require a greater commitment toward atmospheric research, monitoring, and emissions control research and development.

The slowness of progress to date is due largely to a lack of a commitment to an adequate government effort to protect and improve visibility and to sponsor the research and monitoring needed to better characterize the nature and origin of haze in various areas. As discussed in Chapter 3, the federal government has accorded the national visibility goal less priority than other clean-air objectives. Even to the extent that Congress has acted, the Environmental Protection Agency (EPA), the Department of Interior (DOI), and the Department of Agriculture have been slow to carry out their regulatory responsibilities or to seek resources for research.

This lack of commitment can be seen by reviewing this committee's recommendations, as well as those drafted in 1985 by EPA's Interagency Visibility Task Force. The task force (now defunct) consisted of representatives from EPA, DOI, the Departments of Agriculture and Defense and was formed to develop long-term strategies for remedying regional haze and to recommend a long-range program to restore visibility in the national parks. The task force's research recommendations include many of those highlighted by this committee, such as the study of human perception, airborne particle characteristics, and atmospheric optics. Although several task force recommendations were implemented (e.g., establishment of an eastern visibility monitoring network and expansion of monitoring into more wilderness areas managed by the Forest Service), many were not. The gaps in scientific understanding of visibility impairment are a symptom of the absence of a strong national commitment to enforcing the visibility protection provisions of the Clean Air Act.

SUMMARY OF CONCLUSIONS

To accomplish statutory goals on visibility, emissions of pollutants that cause visibility impairment must be limited.

Visibility programs must take large geographic areas into consideration because the visibility problem is regional.

Many sources must be considered simultaneously for visibility improvement.

Visibility policy and control strategies may need to be different in the East and West.

Improving visibility in Class I areas will also improve visibility in other areas.

Visibility improvements will help alleviate other air quality problems.

A long-term program is required to achieve the national visibility goal.

Current scientific knowledge is adequate and controls are available for improving and protecting visibility, however, continued national progress requires a greater commitment toward atmospheric research, monitoring, and emission control research and development.

RECOMMENDED RESEARCH STRATEGIES

The committee was charged with recommending strategies for alleviating critical scientific and technical gaps in the areas of visibility and aerosol monitoring, source apportionment, and emissions control technology. The committee considered what measures might be taken to understand better sources of haze, possible means of reducing emissions from those sources, and alternative ways of preventing future visibility impairment in Class I areas. The remainder of this chapter summarizes the committee's recommendations to fill technological and data gaps in research and program implementation.

The committee emphasizes that the existence of technological and data gaps does not imply that it would be premature to take further regulatory action, if otherwise warranted, to protect and enhance visibility in the nation's Class I areas. Regulatory action in the environmental field typically is taken without complete scientific knowledge; this is inherent in a statutory scheme like the Clean Air Act that is based on a philosophy of prevention. A decision as to whether or how to take regulatory action is therefore not wholly one of science but also involves policy considerations, such as the relative risks of overprotection and underprotection, which were outside the committee's purview.

Although the causes of visibility impairment are reasonably well understood, additional research is still necessary in some areas.

These areas include research on atmospheric transport and transformations of visibility-impairing pollutants, the development of models integrating these complex processes to relate source emissions to effects, and improved instrumentation for routine monitoring and for obtaining data that can be used to evaluate model performance. Such research would make possible a more effective and efficient visibility improvement program. Monitoring and research must be closely coordinated so that progress in research can be implemented in the monitoring program. However, better models are not enough. Any model, even the simplest or most refined, depends on good empirical data on the airborne particles that cause haze and on their sources. Greater resources are needed to develop these data.

Steps must be taken to ensure that visibility-related research is of high quality.

Resources are limited; therefore, the committee believes that precautions should be taken to ensure that the visibility protection activities in the federal land management agencies, EPA, the Department of Energy, and state air agencies are of the highest possible quality. In addition, a greater effort is needed for formal publication of scientific work in independent, peer-reviewed literature.

EPA should build upon and expand its efforts to track the success of the PSD program. In particular, information is needed about the control of new sources with the potential to damage visibility in Class I areas and about the effects on such areas of the new emission trading programs under Title IV of the Clean Air Act.

EPA, the states, and the public need to be able to track the implementation of programs for controlling emissions of new, modified, and existing sources of visibility impairment. For instance, the present PSD program requires that new or modified major emitting facilities in clean air areas install the best available control technology and comply with a system of increments intended to limit pollution degradation in national parks and wilderness areas. In the early 1980s, EPA sponsored a series of studies that furnished much valuable information about the extent to which the PSD program resulted in additional control of new sources. This data base, however, has not been updated since 1984. In addition, the new permitting program established by Title V of the Clean Air Act creates a promising data base for tracking emission changes near national parks, especially those resulting from Title IV's acid rain control program.

EPA's visibility screening model needs to be revised to consider a proposed source's contribution to regional haze.

The new source review provisions of the PSD program require the states and federal land managers to complete an analysis of Class I area visibility impairment of the proposed source. Unfortunately, the EPA's

VISCREEN model used in this analysis is inadequate, because it ignores SO_2-to-SO_4^{2-} conversion and considers the visibility impairment caused by the NO_2 and primary particulate plume only as viewed from clean air. EPA's guidance on the use of the model is also outdated. In addition, VISCREEN is not a cautious model for screening haze effects and therefore is not consistent with EPA's PSD impact assessment policies.

Research on relating human judgments of visibility to objective measures, such as light extinction, should continue, with the results used to inform decision makers and the public about the perceptibility of forecasted visibility changes.

Development of control strategies will require communication among the scientific community, decision makers, and the public. During this process, it would be useful to relate light extinction to human perception. Considerable progress has been made by researchers in developing techniques that produce photographic simulations of visibility changes. Over the past decade, progress also has been made in relating human judgments of visibility and various perceptual cues, such as contrast and visual range, to visibility changes measured by light extinction.

An independent science advisory panel with EPA sponsorship should be established to help guide the research elements of the national visibility program.

Such a board could address the need for wider participation among the scientific community in visibility issues.

SUMMARY OF RECOMMENDED RESEARCH STRATEGIES

Additional research is necessary in several areas of visibility impairment.

Visibility-related research must be of the highest quality.

The success and future of the PSD program must be determined.

The contribution of individual sources to regional haze must be included in the EPA visibility screening model.

Research relating human judgments of visibility to objective measures should be continued.

EPA should form an independent science advisory panel for the national visibility program.

RECOMMENDED MONITORING STRATEGIES

A consensus should be developed on the specific instrumentation to be used for monitoring light extinction. Standards should be established for the performance characteristics of the instrumentation.

Progress toward the visibility goal should be measured in terms of the extinction coefficient, and extinction measurements should be routine and systematic. Extinction either can be measured directly using transmissometry or inferred from measurements of absorption and scattering. Instruments for making such measurements have been developed and can yield reasonably accurate results. Such instrumentation can be cost-effective, reliable, and relatively easy to use. Desirable refinements in optical instrumentation are summarized below.

The monitoring program should be able to relate visibility degradation to the sources of the impairment on a scale commensurate with regional haze events and the distribution of major emissions sources.

The monitoring program should be designed to characterize the existing patterns of haze, trends in those patterns, and the pollutants that contribute to visibility degradation on a regional scale.

National visibility monitoring networks, notably the Interagency Monitoring of Protected Visual Environments (IMPROVE) network, have been established to assess visibility impairment in the nation's Class I areas, to identify sources of haze, and to track trends in impairment conditions. The committee identified several scientific, technical, and data gaps in existing networks that should be addressed.

If national visibility monitoring networks are to accomplish their goals, a long-term commitment to establishing and financially supporting monitoring programs is essential.

This commitment is critical in view of the adoption of Title IV of the Clean Air Act Amendments of 1990 (acid deposition control), which is expected to result in an annual 10-million-ton reduction in SO_2 emissions from 1980 levels. This amounts to an estimated 36% reduction in SO_2 emissions in the eastern states, with parallel reductions in SO_4^{2-} concentrations. These reductions are expected to result in perceptible improvements in visibility.

The SO_4^{2-} reductions achieved through Title IV offer an exceptional opportunity to track visibility improvements associated with reductions in SO_2 emissions from coal-fired power plants. To document adequately the visibility benefits from Title IV, monitoring networks in the eastern states need to be expanded and long-range funding ensured. Otherwise, EPA will lack information needed to measure whether reasonable progress is being made toward the national visibility goal and will be unable to meet its responsibility under Section 169A(a)(4) of the act to ensure such progress.

Future measurement programs should devote increased attention to quality assurance and control.

Because of the importance of the IMPROVE network to the regulatory process, strengthening IMPROVE's quality assurance and control program should be a high priority. Most of this effort should go into quality-control procedures integrated into routine monitoring and should be designed to characterize and document the performance of IMPROVE's measurements.

Any quality-assurance program should include, at the management level, periodic systems and performance audits by independent outsiders. External audits are not a substitute for internal quality control but should be designed to provide timely information on potential problems and on departures from quality-control specifications.

Greater attention should be given to the implications of planned changes in airport visibility monitoring for research on visibility impairment.

The transition from human to automated airport visibility monitoring that is planned by the National Weather Service, the Federal Aviation Administration, and the Department of Defense has unfortunate implications for monitoring haze. The Automated Surface Observing System (ASOS) will provide little or no information on the magnitude and extent of haze. In particular, this system cannot quantify haze levels corre-

CONCLUSIONS AND RECOMMENDATIONS 251

sponding to visual ranges exceeding 10 miles; prevailing visual range is typically much greater than this.

The historical record of ambient haze levels is based largely on airport visibility data from human observers. There is no other source of such information. If airport visual range measurements cease, no information will be available for assessing temporal and spatial trends. It is especially important that such trends be documented during the coming decade, so that the effect of acid rain controls on visibility can be determined.

Airports should be equipped with integrating nephelometers that are sensitive enough to measure the range of haze levels encountered in the atmosphere.

Automated and human observer measurements should overlap for approximately 1 year to ensure consistency between the measurements. In this way, the collection of valuable data on visibility trends could continue.

SUMMARY OF MONITORING RECOMMENDATIONS

Agreement must be reached on the instrument to be used for monitoring light extinction.

Monitoring programs must relate regional visibility degradation to the sources of impairment.

A long term commitment to a national visibility monitoring program is required.

Quality assurance and control must be given increased attention.

The implications of changes in monitoring airport visibility on visibility research must be determined.

Airports should be equipped with integrating nephelometers.

RECOMMENDED MEASUREMENT METHODS

Aerosol Measurements

Current measurement techniques permit useful estimates of the average contributions of major aerosol constituents to atmospheric visibility impairment. However, several measurement methods must be developed or improved to refine present understanding of visibility impairment.

An accurate method for measuring particulate organic and elemental carbon, particularly at the low concentrations found in and near national parks, needs to be developed.

Although organic and elemental carbon often are major constituents of atmospheric particles, atmospheric measurements of particulate carbon using different sampling and analytical methods can disagree by as much as a factor of 5. Discrepancies tend to be larger in national parks and wilderness areas with relatively low particle concentrations. This uncertainty limits quantitative understanding of the contribution of organic and elemental carbon to visibility impairment. Methods that can characterize the organic aerosol composition in more detail also are needed. Such methods could provide important insights into the origins and properties of organic particles.

A method for the routine measurement of water content of airborne particles needs to be developed.

Water is another important particulate constituent that is difficult to measure. Hygroscopic species such as SO_4^{2-}, NO_3^-, and possibly certain organic acids absorb water and significantly reduce visibility, especially at high relative humidities. At present, there is no routine method of measuring particulate water content.

Greater attention should be given to measuring airborne particle size distributions.

Significant progress has been made during the past decade in measuring aerosol composition as a function of particle size. This work has shown that variations in size distributions can lead to differences of a factor of 2 in scattering and absorption efficiencies. Unfortunately, measurements of aerosol chemical composition typically are based on filter samples that provide inadequate information on size-dependent variations in extinction efficiencies. Fine and coarse particle filter samples should be supplemented with measurements of size-resolved chemical composition to address the importance of particle size in atmospheric optical phenomena.

Additional research on inlet sampling efficiencies should be conducted especially for submicron-size particle sampling from aircraft.

While it has long been recognized that inlet sampling losses for coarse particles can be significant, recent work suggests that loses of submicron-size visibility-impairing particles also can be important during sampling from aircraft. Aircraft sampling is often required to characterize the extent of regional haze, and to study phenomena including secondary airborne particle production. It is important to have sampling inlets that are known to be accurate for visibility-impairing particles.

Commercially available instruments are needed for continuous measurements of particulate sulfates, organic and elemental carbon, nitrates, and elemental composition.

Such instruments would be valuable for visibility research and monitoring as well as for work on other atmospheric problems involving airborne particles. Information on temporal variations in particulate composition can provide further clues about the sources, transport, and atmospheric processing of airborne particles. Some progress has been made toward the continuous measurement of particulate sulfur and carbon loadings, and elemental composition can be measured with a time resolution of several hours. However, the instruments used for such measurements usually are custom built and not readily available.

The development of self-powered instrumentation for visibility monitoring and research should be accelerated.

Visibility monitoring is often carried out in remote regions where electric power is not available, which limits the range of measurements that can be conducted. In recent years, significant progress has been made in the development of solar- or battery-powered instrumentation for visibility-related measurements, but major limitations remain, particularly with regard to measurements of gas and particle composition.

Optical Measurements

High-sensitivity integrating nephelometry should be used for routine visibility monitoring.

Factors considered in making this recommendation include reasonable cost, potential wide dynamic range in sensitivity, and ease of installation, operation, and calibration. In addition, nephelometer data can be compared with airborne particle measurements at the same point to determine the relative contribution of different pollutants to haze.

The fact that nephelometers measure only scattering and not total extinction (scattering plus absorption) is both a strength and weakness of this method. By supplementing nephelometer measurements with independent data for absorption coefficients, the separate contributions of scattering and absorption to extinction can be determined. This can be useful for evaluating the relative contributions of various types of particulate matter (e.g., soot versus SO_4^{2-}, NO_3^-, and organics) to haze. The widespread use of integrating nephelometers has been hindered by the lack of a readily available, easily serviced, and electronically up-to-date instrument. The committee recommends the development of an instrument with adequate sensitivity for good and poor visibility conditions. Issues that need to be considered in the design of such a nephelometer include inlet losses of coarse particles and decreases in aerosol relative humidity associated with heating by the instrument.

Efforts should be made to develop instrumentation for continuous measurements of particle absorption coefficients.

This instrumentation should be sensitive to the range of absorption coefficients obtained from atmospheric observations. Although instrumentation for routine, continuous, real-time monitoring of scattering and extinction coefficients is available, similar instrumentation for absorption coefficients is not. Instead, particle absorption coefficients usually are determined from filter samples. Photoacoustic spectrometers can provide real-time absorption data, but sensitivities are limited under good visibility conditions, and those one-of-a-kind instruments are expensive to build and difficult to operate. It would be desirable to have an instrument for routine, continuous measurements of absorption coefficients to characterize atmospheric optical properties more completely.

SUMMARY OF MEASUREMENT METHODS
RECOMMENDATIONS

Aerosol Measurements

Accurate methods for measuring particulate organic and elemental carbon must be developed.

A method for the routine measurement of aerosol water content must be developed.

Increased emphasis should be given to measuring aerosol particle size distributions.

Research should be conducted on inlet sampling efficiencies, especially for submicron-size particle sampling from aircraft.

Continuous measurements are required for aerosol sulfate, organic and elemental carbon, nitrate and trace elements.

Development of self-powered instrumentation for visibility measurements must be accelerated.

Optical Measurements

The integrating nephelometer should be used for routine visibility monitoring.

Instrumentation for continuous particle absorption measurements must be developed.

RECOMMENDED SOURCE APPORTIONMENT
MODELING RESEARCH

In Chapter 5 and Appendix C, the committee reviewed methods for identifying and quantifying source contributions to visibility impairment. These methods include simple source identification techniques, speciated rollback models, receptor models, and mechanistic models. The committee found that there is no single best source apportionment technique for a given visibility problem and recommended a nested progression from simple through more advanced and precise techniques. This section discusses research and data collection needed for the development and validation of source apportionment methods.

Emission Inventories

Better data on source emissions are needed.

Source emissions data are essential to most haze apportionment models. Speciated rollback models require data on aggregate emissions of SO_x, NO_x, VOCs, NH_3, and primary airborne particles; chemical mass balance (CMB) receptor models require additional information on endemic tracers such as key trace metals found in source exhaust; and mechanistic models require further data on the particle size distributions in the primary source exhaust.

Unified procedures are needed for testing individual sources that will simultaneously meet the input data requirements of the modeling systems recommended in Chapter 5. More source tests should include collection of the aerosol chemical data required by speciated rollback and CMB models and the size distribution data for fine airborne particles required by mechanistic visibility models (in addition to the usual particle mass and gaseous pollutant emission data). Efforts should be directed at identifying and measuring additional endemic tracers that might be useful in detecting the contribution of one or more key sources at a particular ambient sampler. Such tracers would be identified using molecular or isotopic characteristics rather than using those of elements.

Cheaper and more reliable vehicle exhaust emission measurement techniques are needed.

Source emissions data must be integrated accurately into overall emission inventories.

Despite major advances made through the National Acid Precipitation Assessment Program (NAPAP), the integration of improved single-source emissions data into national or regional inventories is not yet satisfactory. The committee believes that further thought and planning need to be directed at this issue.

As discussed in Chapter 4, the most important anthropogenic emissions from a visibility standpoint are SO_2, primary organic particles, gaseous VOCs, primary elemental carbon particles, soil dust materials, ammonia, and nitrogen oxides. The national inventories for gaseous sulfur and nitrogen oxides are of good quality. The inventories for ammonia and soil dust are of lesser quality and need further attention. Inventories for ammonia and soil dust are less critical than those for some other pollutants because they usually are only minor contributors to regional haze.

From a visibility standpoint, the national inventories needing the most improvement are those for primary elemental and organic carbon particles and for VOCs that lead to secondary organic particle formation.

Inventories for elemental carbon and primary organic particles essentially are nonexistent except for Los Angeles.

Model Validation and Field Studies

Model validation studies employing existing data sets should be supported. Mechanistic models as well as hybrid receptor-oriented methods should be included in future model validation studies.

The most convincing tests of air-quality model performance are those directly based on actual measurements. Comprehensive field studies are an expensive but indispensable step toward the collection of atmospheric data sets against which new and existing models can be tested. Several measurement programs and mechanistic-model development programs provide opportunities to test model performance at little additional expense. Possible candidates are the 1990 winter study of the Grand Canyon sponsored by the Salt River Project and project MOHAVE, which included comprehensive atmospheric chemistry, aerosol optics, and wind measurements. The redundancies and cross checks available in the resulting data should facilitate the comparison of differing source apportionment approaches. More potential test data will come from the application of the NAPAP's Regional Acidic Deposition Model to the comprehensive data base from the Denver Brown Cloud study.

Receptor Models

This section focuses on research needs specific to haze in national parks and wilderness areas. The essential inputs to any receptor-oriented haze apportionment model are a collection of source signatures and estimates of chemical transformation mechanisms and reaction rates. The study of atmospheric reactions is motivated by a broad array of scientific and policy issues, ranging from the global cycling of sulfur, nitrogen, and carbon to the photochemistry of urban smog. This large complex of issues can be expected to drive the evolution of reaction models, which receptor methods could exploit with little additional work. This discussion therefore focuses on strategies for increasing the availability and quality of source signatures.

The chemical mass balance model is widely accepted in the management and study of local particulate air quality. Regression analysis, although not ready for operational use, is an accepted research tool. Both approaches have been tested successfully on simulated data sets for primary urban airborne particles. However, those endorsements are largely irrelevant to most haze applications, which involve mostly secondary particles at regional scales. Hybrid models have been proposed by several investigators but have not been validated. Such models should be tested on simulated haze data sets incorporating realistic operational factors, such as measurement uncertainty, variable emissions, and covarying source strengths. The results from model comparison studies would greatly increase present understanding of modeling errors, sensitivities to uncertainties, and the relative merits of differing receptor modeling approaches.

Substantial source testing as well as measurements of ambient concentrations resulting from particulate and gaseous emissions are needed to improve emissions profiles for sources of haze.

The development of source signatures has been driven largely by the apportionment of total suspended particles and PM_{10} in urban, suburban, and industrial areas. Much of the success demonstrated by chemical-signature methods has been in identifying coarse primary dusts. These dusts, although major contributors to total particle mass, are minor contributors to light extinction. Moreover, some of the most reliable tracers, such as lead for motor vehicles, are disappearing from emissions as the result of control measures. A major data gap exists in present knowledge of source emissions profiles that would be useful in distinguishing types of sources, including domestic wood burning, agricultural and prescribed burns and wildfires, gasoline- and diesel-fueled motor vehicles, and coal combustion.

Standard protocols for the release and sampling of tracers should be developed, along with formal field studies to verify these protocols.

The identification of haze sources in remote locations may require the use of injected tracers which can be measured at very low ambient concentrations. Artificial tracers have been widely used to trace airflows in meteorological studies. Their use in quantitative haze apportionment studies has received less attention.

Resources should be devoted to the development of novel, inexpensive, and relatively short-lived tracers for use in multiple-tracer studies.

In many studies, distinct artificial tracers are needed for several different sources. The suite of tracers now used in regional applications is very limited, comprising only heavy methane and certain perfluorocarbons. Those tracers have tropospheric lifetimes of decades; thus, large-scale operational use would degrade their value by increasing background levels. Useful tracers need not have such long lifetimes; a lifetime of weeks is adequate for haze studies.

Mechanistic Models

Research efforts should continue toward the completion of advanced mechanistic models for the calculation of source effects on visibility. The two highest priorities are for: (1) an advanced reactive plume aerosol processes model for use on single-source problems, and (2) a grid-based multiple source regional model for use in analysis of regional haze problems.

Recent advances in mechanistic models (see Chapter 5) might greatly improve simulations of atmospheric physical and chemical processes that affect secondary particle formation, particle size distributions, dispersion, and deposition over long distances. As a result, future models may provide far better estimates of the impairment caused by multiple sources in complex airsheds than do models now available.

Research is needed to improve the understanding of phenomena that determine atmospheric particle size distributions.

This is the most serious barrier to the development of advanced mechanistic models. For example, better understanding is needed of particle emissions and of atmospheric aerosol processing phenomena including nucleation, gas-to-particle conversion, and cloud processing to complete the development of reliable, advanced models. Such information can only be obtained through field measurements, but little work of this kind has been done during the past decade. Field studies need to be designed carefully to advance the understanding of aerosol processes that affect size distributions.

SUMMARY OF SOURCE APPORTIONMENT
MODELING RECOMMENDATIONS

Emission Inventories
Better data on source emissions are needed.

Source emissions data must be accurately integrated into overall emission inventories.

Inventories for elemental and organic carbon particles and VOCs that lead to particles require improvement.

Model Validation and Field Studies
Existing data sets must be used in model validation studies.

Mechanistic models and hybrid receptor-oriented methods should be subjected to further validation studies.

Receptor Models
Source testing and ambient concentration measurements are required for improved emissions profiles.

Inexpensive and short-lived tracers should be developed.

Standard protocols and field validation studies are required for the release and sampling of tracers.

Mechanistic Models
An advanced reactive plume aerosol processes model is required for single-source problems.

A grid-based multiple source regional model is required for the analysis of regional haze.

A comprehensive understanding of the factors that determine particle size distributions is required.

RECOMMENDED CONTROL
TECHNOLOGY RESEARCH

Better economic modeling techniques are needed to assess the most cost-effective methods to reduce emissions of relevant pollutants from their different sources.

A variety of control technologies for point and area sources will have to be applied to achieve the emission reductions required to restore clear air to the nation's Class I areas. The emissions reduction capabilities and the costs of commercially available control technologies for stationary and mobile point sources are well documented. However, it is difficult to translate unit costs for a specific technology into aggregate costs for overall emissions reductions for an urban area or region.

Cheaper and more effective pollution-control technologies should be developed.

Technological advances that reduce the cost and improve the effectiveness of controls could not only improve visibility but also alleviate other environmental problems such as acid rain and tropospheric ozone. More cost-effective controls for power plants and motor vehicles could be especially beneficial. Areas where further research and development are desirable include:

• Combined NO_x and SO_2 control technologies for power plants and industrial boilers, at a cost less than the aggregate cost of flue gas desulfurization for SO_2 control and selective catalytic reduction (SCR) for NO_x control;
• Low-cost, low-temperature SCR for NO_x control;
• Better particulate control for diesel vehicle engines;
• More efficient batteries for electric vehicles and innovative methods for increasing vehicle range before recharging. (Electricity is potentially a very clean motor vehicle energy source relative to most alternatives even when the environmental effects of its production and delivery are taken into account.)

Important research efforts already are under way, for example, in the Clean Coal Technology Demonstration Program administered by the Department of Energy.

For longer-term visibility improvements, an integrated assessment of industrial, economic, and social changes that would minimize emissions of haze-causing pollutants should be carried out.

Promising prospects for emission reduction include the following: replacement or repowering of existing plants by high-efficiency and low-emission technologies, such as coal gasification and fuel cells, or by nonfossil generation systems; increased emphasis on efficient energy use in all sectors; industrial process modification to minimize aggregate emissions; and development of low-emission transportation systems.

SUMMARY OF RECOMMENDED
CONTROL TECHNOLOGY RESEARCH

Better economic models are needed to assess the most cost-effective methods to reduce pollutant emissions.

Cheaper and more effective control technologies should be developed.

An integrated assessment of industrial, economic, and social changes that would minimize emissions of haze-causing pollutants should be carried out.

SUMMARY

Present scientific knowledge has important implications for the design of programs to protect and improve visibility. What is needed, overall, is the recognition that any effective visibility protection program must be aimed at preventing and reducing regional haze. An effective program must, therefore, control a broad array of sources over a large geographic area. Such a program would mark a considerable break from the present approach of focusing on visible plumes from nearby sources and of attempting to determine the effects of individual sources on visibility impairment.

Although visibility impairment is as well understood as any other air-pollution effect, gaps in knowledge remain. Filling these gaps will require an increased national commitment to visibility protection research. We believe that the time has come for Congress, EPA, and the states to decide whether to make that commitment.

References

ACGIH (American Conference of Governmental Industrial Hygienists). 1989. Air Sampling Instruments for Evaluation of Atmospheric Contaminants, 7th ed., S.V. Hering, ed. American Conference of Governmental Industrial Hygienists, Cincinnati, Ohio.

ACSPCT (Arizona Copper Smelter Pollution Control Technology). 1977. First Annual Report on Arizona Copper Smelter Pollution Control Technology, Arizona Department of Health Services, Phoenix, Ariz. April.

Adams, K.M. 1988. Real-time, in situ measurements of atmospheric optical absorption in the visible via photoacoustic spectroscopy. I. Evaluation of photoacoustic cells. Appl. Opt. 27:4052–4056.

Adams, K.M., L.I. Davis, Jr., S.M. Japar, and D.R. Finley. 1990. Real-time, in situ measurements of atmospheric optical absorption in the visible via photoacoustic spectroscopy. IV. Visibility degradation and aerosol optical properties in Los Angeles. Atmos. Environ. 24A:605–610.

Aden, G.D., and P.R. Buseck. 1983. A minicomputer procedure for quantitative EDS (energy-dispersive spectrometer) analyses of small particles. Pp. 195–201 in Microbeam Analysis-1983, R. Gooley, ed. San Francisco, Calif.: San Francisco Press, Inc.

Air Resources Specialists, Inc. 1988. Visibility Monitoring and Data Analysis Report for Spring 1986 Through Winter 1988 Monitoring Seasons, Vol. 2, Appendix A, Seasonal Summary Reports. NPS Contract #CX-0001-07-0010. Prepared for the National Park Service, Air Quality Division, Fort Collins, Colo. June.

Aitchison J., and J.A.C. Brown. 1957. The Lognormal Distribution: With Special References to its Uses in Economics. Cambridge: Cambridge University Press.

Allard, D., and I. Tombach. 1981. The effects of non-standard conditions on visibility measurement. Atmos. Environ. 15:1847–1857.

Appel, B.R., T. Tokiwa, and M. Haik. 1981. Sampling of nitrates in ambient air. Atmos. Environ. 15:283–290.

Andreae, M.O. 1986. The ocean as a source of atmospheric sulfur compounds. Pp. 331–362 in The Role of Air-Sea Exchange in Geochemical Cycling, P. Buat-Menard, ed. Dordrecht, the Netherlands: D. Reidel Publishing Co.

Ashbaugh, L.L. 1983. A statistical trajectory technique for determining air pollution source regions. J. Air Pollut. Control Assoc. 33: 1096–1098.

Ashbaugh, L.L., W.C. Malm, and W.Z. Sadeh. 1985. A residence time probability analysis of sulfur concentrations at Grand Canyon National Park. Atmos. Environ. 19:1263–1270.

Batterman S.A., Dzubay T.G., and R.E. Baumgardner. 1988. Development of crustal profiles for receptor modeling. Atmos. Environ. 22:1821–1828.

Baumgardner, D., B. Huebert, and C. Wilson. 1992. Meeting Review: Airborne Aerosol Inlet Workshop, 27-28 Feb. and 1 March, 1991. NCAR/TN-362+1A. National Center for Atmospheric Research, Boulder, Colo.

Beck, R.W., and Associates. 1986. PANORAMAS/Pacific Northwest Regional Aerosol Mass Apportionment Study, Vol. 1, Final Report. A Joint Study of Regional Haze within the States of Oregon, Washington, and Idaho. Prepared for the PANORAMAS Steering Committee of the States of Oregon, Washington, and Idaho. Portland, Oreg.

Belsley, D.A., E. Kuh, and R.E. Welsch. 1980. Regression Diagnostics: Identifying Influential Data and Sources of Collinearity. New York: John Wiley and Sons.

Bergstrom, R.W., C. Seigneur, B.L. Babson, H.-Y. Holman, and M.A. Wojcik. 1981. Comparison of the observed and predicted visual effects caused by power plant plumes. Atmos. Environ. 15:2135–2150.

Berner, A., C.H. Luerzer, F. Polh, O. Preining, and P. Wagner. 1979.

The size distribution of the urban aerosol in Vienna. Sci. Total Environ. 13:245–261.

Biswas, P., C.L. Jones, and R.C. Flagan. 1987. Distortion of size distributions by condensation and evaporation in aerosol instruments. Aerosol Sci. Technol. 7:231–246.

Bohren, C.F., and D.R. Huffman. 1983. Absorption and Scattering of Light by Small Particles. New York: John Wiley and Sons.

Bond, R.G. 1972. Air pollution. In Handbook of Environmental Control: Air Pollution, R.G. Bond, C.P. Straub, and R. Prober, eds. Cleveland, Ohio: CRC Press.

Bowman, L.D., and R.F. Horak. 1972. Continuous ultraviolet absorption ozone photometer. Anal. Instrum. 10:103–108.

Bradley, J.T. 1989. Visibility Sensor and the Automated Visibility Parameter. Pp. 5–9 in Automated Surface Observing System (ASOS) Progress Report Vol. 5, Number 1. March.

Bradley, J.T., and S.M. Imbembo. 1985. Automated Visibility Measurements for Airports. Paper AIAA-85-0191. Presented at the American Institute of Aeronautic and Astronautics (AIAA) 23rd Aerospace Sciences Meeting, Reno, Nev., Jan. 14-17.

Bradshaw, J.D., M.O. Rodgers, D.D. Davis. 1982. Single photon laser-induced fluorescence detection of NO and SO_2 for atmospheric conditions of composition and pressure. Appl. Opt. 21:2493–2500.

Braman, R.S., T.J. Shelley, and W.A. McClenny. 1982. Tungstic acid preconcentration and determination of gaseous and particulate ammonia and nitric acid in ambient air. Anal. Chem. 54:358–364.

Braman, R.S., M.A. de la Cantera, and Q.X. Han. 1986. Sequential selective hollow tube preconcentration and chemiluminescence analysis system for nitrogen oxide compounds in air. Anal. Chem. 58:1537–1541.

Breger, Stewart, Elliott, and Hawkins. 1991. Providing economic incentives in environmental regulation. Yale J. Regulation 8:463–.

Britt, H.I., and R.H. Luecke. 1973. The estimation of parameters in nonlinear, implicit models. Technometrics 15:233–247.

Brorström, E., P. Grennfelt, and A. Lindskog. 1983. The effect of nitrogen dioxide and ozone on the decomposition of particle-associated polycyclic aromatic hydrocarbons during sampling from the atmosphere. Atmos. Environ. 17:601–605.

Brost, R.A., P.L. Haagenson, and Y.-H. Kuo. 1988. Eulerian simulation of tracer distribution during CAPTEX. J. Appl. Meteor. 27: 579-593.

Cadle, S.H., P.J. Groblicki, and P.A. Mulawa. 1983. Problems in the sampling and analysis of carbon particulate. Atmos. Environ. 17: 593-600.

Cahill, T.A., L.L. Ashbaugh, R.A. Eldred, P.J. Feeney, B.H. Kusko, and R.G. Flocchini. 1981. Comparisons between size-segregated resuspended soil samples and ambient aerosols in the western United States. Pp. 269-285 in Atmospheric Aerosol: Source/Air Quality Relationships, E.S. Macias and P.K. Hopke, eds. ACS Symposium Series 167. American Chemical Society, Washington, D.C.

Cahill, T.A., P.J. Feeney, R.A. Eldred, and W.C. Malm. 1987. Size/time/composition data at Grand Canyon National Park and the role of ultrafine sulfur particles. Pp. 657-667 in Visibility Protection: Research and Policy Aspects, P.S. Bhardwaja, ed. APCA Publication No. TR-10. Air Pollution Control Association, Pittsburgh, Penn.

Cahill, T.A., R.A. Eldred, N. Motallebi, and W. C. Malm. 1989. Indirect measurement of hydrocarbon aerosols across the United States by nonsulfate hydrogen-remaining gravimetric mass correlations. Aerosol Sci. Technol. 10:421-429.

Calvert, J.G., and W.R. Stockwell. 1984. The mechanism and rates of the gas phase oxidation of sulfur dioxide and nitrogen oxides in the atmosphere. Pp. 1-62 in Acid Precipitation: SO_2, NO, NO_2 Oxidation Mechanisms: Atmospheric Considerations., J.G. Calvert, ed. Boston, Mass.: Butterworth Publishers.

Calvert, J.G., A. Lazrus, G.L. Kok, B.G. Heikes, J.G. Walega, J. Lind, and C.A. Cantrell. 1985. Chemical mechanisms of acid generation in the troposphere. Nature 317:27-39.

Campbell, D., S. Copeland, T. Cahill, R. Eldred, C. Cahill, J. Vesenka, and T. VanCuren. 1989. The Coefficient of Optical Absorption from Particles Deposited on Filters: Integrating Plate, Integrating Sphere and Coefficient of Haze Measurements. Paper 89-151.6. Presented at the 82nd Annual Meeting and Exhibition of the Air & Waste Management Association, Anaheim, Calif., June 25-30.

CARB (California Air Resources Board). 1989. Instrumental Measurements of Visibility Reducing Particles: Staff Report and Technical Support Document. State of California Air Resources Board, Sacramento, Calif. January.

Carmichael, G.R., and L.K. Peters. 1987. A Mesoscale Acid Deposition Model: Preliminary Applications and a Guide for User Interface. EPA/600/3-87/027. U.S. Environmental Protection Agency, Research Triangle Park, N.C.

Carmichael, G.R., L.K. Peters, and T. Kitada. 1986. A second generation model for regional-scale transport/chemistry/deposition. Atmos. Environ. 20:173-188.

Carrales, M., Jr., and R.W. Martin. 1975. Sulfur Content of Crude Oils. Information Circular 8676. Bureau of Mines, U.S. Department of Interior. Washington, D.C.: U.S. Government Printing Office.

Cass, G.R. 1978. Methods for Sulfate Air Quality Management with Applications to Los Angeles. Ph.D. Dissertation. Division of Engineering and Applied Science, California Institute of Technology, Pasadena, Calif.

Cass, G.R. 1979. The relationship between sulfate air quality and visibility with examples in Los Angeles, Calif. Atmos. Environ. 13:1069-1084.

Cass, G.R. 1981. Sulfate air quality control strategy design. Atmos. Environ. 15:1227-1249.

Cass, G.R., and G.J. McRae. 1981. Minimizing the cost of air pollution control. Environ. Sci. Technol. 15:748-757.

Cass, G.R., and G.J. McRae. 1983. Source-receptor reconciliation of routine air monitoring data for trace metals: An emission inventory assisted approach. Environ. Sci. Technol. 17:129-139.

Cass, G.R., R.W. Hahn, R.G. Noll, W.P. Rogerson, G. Fox, and A. Paranjape. 1982. Implementing Tradable Permits for Sulfur Oxides Emissions. A Case Study in the South Coast Air Basin. Report 22. Environmental Quality Laboratory, California Institute of Technology, Pasadena, Calif.

CFR Title 40, Part 60.531, Subpart Aaa; Standards for Performance of New Residential Wood Heaters.

Chandrasekhar, S. 1960. Radiative Transfer. New York: Dover.

Chang, J.S., R.A. Brost, I.S.A. Isaksen, S. Madronich, P. Middleton, W.R. Stockwell, and C.J. Walcek. 1987. A three-dimensional Eulerian acid deposition model: Physical concepts and formulation. J. Geophys. Res. 92:14681-14700.

Chang, J.S., P. Middleton, and L. Hash. 1990. Simulations of regional sulfate distributions with RADM. Pp. 726–737 in Visibility and Fine Particles, C.V. Mathai, ed. A&WMA Publication No. TR-17. Air and Waste Management Association, Pittsburgh, Penn.

Charlson, R.J., H. Horvath, and R.F. Pueschel. 1967. The direct measurement of atmospheric light scattering coefficient for studies of visibility and pollution. Atmos. Environ. 1:469–478.

Charlson, R.J., W.M. Porch, A.P. Waggoner, and N.C. Ahlquist. 1974. Background aerosol light scattering characteristics: Nephelometric observations at Mauna Loa Observatory compared with results at other remote locations. Tellus 26:345–359.

Covert, R.J., R.J. Charlson, and N.C. Ahlquist. 1972. A study of the relationship of chemical composition and humidity to light scattering by aerosols. J. Appl. Meteor. 11:968–976.

Charlson, R.J., D.S. Covert, T.V. Larson, and A.P. Waggoner. 1978. Chemical properties of tropospheric sulfur aerosols. Atmos. Environ. 12:39–53.

Charlson, R.J., J. Langner, H. Rodhe, C.B. Leovy, and S.G. Warren. 1991. Perturbation of the northern hemisphere radiative balance by backscattering from anthropogenic sulfate aerosols. Tellus 43AB: 152–163.

Chow, J.C. 1985. Development of a Composite Modeling Approach to Assess Air Pollution Source/Receptor Relationships. Ph.D. Dissertation. Harvard School of Public Health, Harvard University, Cambridge, Mass.

Chow J.C., and J.G. Watson. 1989. Summary of particulate data bases for receptor modeling in the United States. Pp. 108–133 in Receptor Models in Air Resources Management, J.G. Watson, ed. A&WMA Publication No. TR-14. Air and Waste Management Association, Pittsburgh, Penn.

Christian, G.D., and J.E. Reilly, eds. 1986. Instrumental Analysis, 2nd ed. Boston: Allyn and Bacon, Inc.

Cichanowicz, J.E., C. Dene, C. Robie, and J. Jarvis. 1990. Postcombustion NO_x controls: Potential options for present and future generation. P. 14-5-1 in Proceedings: GEN-UPGRADE-90. EPRI Report GS-6986. Electric Power Research Institute, Palo Alto, Calif. September.

Clark, W.C., and R.E. Munn, eds. 1986. Sustainable Development of the Biosphere. Cambridge, England: Cambridge University Press.

Clarke, A.D. 1982. Integrating sandwich: A new method of measurement of the light absorption coefficient for atmospheric particles. Appl. Opt. 21:3011–3020.

Cochran, W.G. 1968. Errors of measurement in statistics. Technometrics 10:637–666.

Commins, B.T., and P.J. Lawther. 1957. Volatility of 3,4-benzpyrene in relation to the collection of smoke samples. Br. J. Cancer 12: 351–354.

Cooper, J.W., and J.G. Watson. 1980. Receptor oriented methods of air particulate source apportionment. J. Air Pollut. Control Assoc. 30:1116–1125.

Core, J.E. 1985. Visibility in Oregon's Wilderness and National Park Lands (1982-1984). Air Quality Division, Oregon Department of Environmental Quality, Portland, Oreg. September.

Core, J.E. 1989a. The Pacific Northwest Source Profile Project: Source profile development protocols for PM$_{10}$ receptor modeling. Pp. 134–144 in Receptor Models in Air Resources Management, J.G. Watson, ed. A&WMA Publication No. TR-14. Air and Waste Management Association, Pittsburgh, Penn.

Core, J.E. 1989b. Air quality and forestry burning: Public policy issues. In The Burning Decision: Regional Perspectives on Slash, D. Hanley, J. Kammenga, and C. Oliver, eds. Institute of Forest Resources, University of Washington, Seattle, Wash.

Core, J.E., J.A. Cooper, P.L. Hanrahan, and W.M. Cox. 1982. Particle Dispersion Model Evaluation: A New Approach Using Receptor Models. Paper No. 82-3.1. In Proceedings of the 75th Annual Meeting and Exhibition of the Air Pollution Control Association, Pittsburgh, Penn.

Core, J.E., J.J. Shah, and J.A. Cooper. 1984. Receptor Model Source Composition Library. EPA 450/4-85-002. U.S. Environmental Protection Agency, Research Triangle Park, N.C. November.

Countess, R.J. 1990. Interlaboratory analyses of carbonaceous aerosol samples. Aerosol Sci. Technol. 12:114–121.

Countess, R.J., G.T. Wolff, and S.H. Cadle. 1980. The Denver winter aerosol: A comprehensive chemical characterization. J. Air Pollut. Control Assoc. 30:1194–1200.

Coutant, R.W., L. Brown, J.C. Chuang, R.M. Riggin, and R.G. Lewis. 1988. Phase distribution and artifact formation in ambient air sampling for polynuclear aromatic hydrocarbons. Atmos. Environ. 22: 403–410.

Covert, D.S., and J. Heintzenberg. 1984. Measurement of the degree of internal/external mixing of hygroscopic compounds and soot in atmospheric aerosols. Sci. Total Environ. 36:347–352.

Covert, D.S., J. Heintzenberg, and H.C. Hansson. 1990. Electro-optical detection of external mixtures in aerosols. Aerosol Sci. Technol. 12:446–456.

Crane, S.D. 1989. Best existing stove technology. In Proceedings of the 28th Annual Meeting of the Pacific Northwest International Section. Air and Waste Management Association, Spokane, Wash. November.

Currie, L.A., R.W. Gerlach, C.W. Lewis, W.D. Balfour, J.A. Cooper, S.L. Dattner, R.T. DeCesar, G.E. Gordon, S.L. Heisler, P.K. Hopke, J.J. Shah, G.D. Thurston, and H.J. Williamson. 1984. Interlaboratory comparison of source apportionment procedures: Results for simulated data sets. Atmos. Environ. 18:1517–1537.

Daisey, J.M., and T.J. Kneip. 1981. Atmospheric particulate organic matter: Multivariate models for identifying sources and estimating their contributions to the ambient aerosol. In Atmospheric Aerosol: Source/Air Quality Relationships, E.S. Macias and P.K. Hopke, eds. ACS Symposium Series 167. American Chemical Society, Washington, D.C.

Dalton, S.M. 1990. Worldwide SO_2 control technology overview. 15-1-1 in Proceedings of GenUpgrade '90, Vol. 3. EPRI Report GS-6986. Electric Power Research Institute, Palo Alto, Calif. September.

Dalton, S.M. 1991. The 1990 U.S. Clean Air Act - The Effect on FGD Market and Design. Paper presented at the 2nd International Conference on Desulfurization, Sheffield, England. March 20-21.

Danielson, J.A., ed. 1973. Air Pollution Engineering Manual, 2nd ed. EPA Publication AP-40. Office of Air Quality Planning and Standards, U.S. Environmental Protection Agency, Research Triangle Park, N.C.

Danilatos, G.D. 1988. Foundations of environmental scanning electron microscopy. Adv. Electron. Electron Phys. 71:109–250.

Danilatos, G.D., and R. Postle. 1982. The environmental scanning electron microscope and its applications. Scanning Electron Microsc.

1:1–16.

Dasch, J.M. 1982. Particulate and gaseous emissions from wood-burning fireplaces. Environ. Sci. Technol. 16:639–645.

Dasgupta, P.K., W.L. McDowell, and G.S. Rhee. 1986. Porous membrane-based diffusion scrubber for the sampling of atmospheric gases. Analyst (London) 111:87–90.

Daum, P.H., T.J. Kelly, J.W. Strapp, W.R. Leaitch, P. Joe, R.S. Schemenauer, G.A. Isaac, K.G. Anlauf, and H.A. Wiebe. 1987. Chemistry of a winter stratus cloud layer: A case study. J. Geophys. Res. 92(D-7):8426–8436.

Davidson, C.I., and Y.-L. Wu. 1989. Dry deposition of particles and vapors. Pp 103-109 in Acidic Precipitation, Vol 3:: Sources, Deposition, and Canopy Interactions, S.E. Lindberg, A.L. Page, and S.A. Norton, Eds. New York: Springer-Verlag.

Demerjian, K.L. 1978. Oxidant modeling status. Pp. 1–3 in Quality Analysis in Transportation Planning. Transportation Research Record Series 670. Washington, D.C.: National Academy Press.

de Vera, E.R., E. Yeung, and M. Imada. 1974. Equivalency determination and calibration procedure for a UV absorption ozone monitor. Pp. 18 in Air Industrial Hygiene Laboratory, Report No. 160. Berkeley: California State Department of Health.

De Wiest, F., and D. Rondia. 1976. On the validity of determinations of benzo(a)pyrene in airborne particles in the summer months. Atmos. Environ. 10:487–489.

Dickerson, R.R., A.C. Delany, and A.F. Wartburg. 1984. Further modifications of a commercial NO_x detector for high sensitivity. Rev. Sci. Instrum. 55:1995–1998.

DOE (U.S. Department of Energy). 1991. Clean Coal Technology Demonstration Program. DOE/FE-0219P. U.S. Department of Energy. Washington, D.C.: U.S. Government Printing Office.

DOI (U.S. Department of the Interior). 1982. Preliminary Certification of No Adverse Impact on Theodore Roosevelt National Park and Lostwood National Wildife Refuge Under Section 165(d)(2)(C)(iii) of the Clean Air Act. 47 Fed. Reg. 30,222 (July 10, 1982).

DOI (U.S. Department of the Interior). 1990. Preliminary Notice of Adverse Impact on Shenandoah National Park Under Section 165(d)(2)(C)(ii) of the Clean Air Act. 55 Fed. Reg. 38,403 (Sept. 18, 1990).

DOI (U.S. Department of the Interior). 1992. Preliminary Notice of Adverse Impact on Great Smoky Mountains National Park Under Section 165(d)(2)(C)(ii) of the Clean Air Act. 57 Fed. Reg. 4,465 (Feb. 5, 1992).

Draper, N.R., and H. Smith. 1981. Applied Regression Analysis, 2nd ed. New York: John Wiley and Sons.

Draxler, R.R. 1987. Sensitivity of a trajectory model to the spatial and temporal resolution of the meteorological data during CAPTEX. J. Appl. Meteor. 26:1577–1588.

Draxler, R.R., and R.D. Taylor. 1982. Horizontal dispersion parameters for long-range transport modeling. J. Appl. Meteor. 21:367–372.

Driedger, A.R., III, D.C. Thornton, M. Lalevic, and A.R. Bandy. 1987. Determination of part-per-trillion levels of atmospheric sulfur dioxide by isotope dilution gas chromatography/mass spectroscopy. Anal. Chem. 59:1196–1200.

Duce, R.A., C.K. Unni, B.J. Ray, J.M. Prospero, and J.T. Merrill. 1980. Long-range atmospheric transport of soil dust from Asia to the tropical North Pacific: Temporal variability. Science 209:1522–1524.

Duntley, S.Q., A.R. Boileau, and R.W. Preisendorfer. 1957. Image transmission by the troposphere I. Optic. Soc. Am. J. 47:499–.

Dzubay, T.H., L.E. Hines, and R.K. Stevens. 1976. Particle bounce errors in cascade impactors. Environ. Sci. Technol. 10:229–234.

Dzubay, T.G., R.K. Stevens, W.D. Balfour, H.J. Williamson, J.A. Cooper, J.E. Core, R.T. De Cesar, E.R. Crutcher, S.L. Dattner, B.L. Davis, S.L. Heisler, J.J. Shah, P.K. Hopke, and D. L. Johnson. 1984. Interlaboratory comparison of receptor model results for Houston aerosol. Atmos. Environ. 18:1555–1566.

Dzubay, T.G., R.K. Stevens, G.E. Gordon, I. Olmez, A.E. Sheffield, and W.J. Courtney. 1988. A composite receptor method applied to Philadelphia aerosol. Environ. Sci. Technol. 22:46–52.

Eatough, D.J., V.F. White, L.D. Hansen, N.L. Eatough, and E.C. Ellis. 1985. Hydration of nitric acid and its collection in the atmosphere by diffusion denuders. Anal. Chem. 57:743–748.

Eatough, N.L., S. McGregor, E.A. Lewis, D.J. Eatough, A.W. Huang, and E.C. Ellis. 1988. Comparison of six denuder methods and a filter pack for the collection of ambient $HNO_3(g)$, $HNO_2(g)$, and

$SO_2(g)$ in the 1985 NSMC study. Atmos. Environ. 22:1601–1608.

Eatough, D.J., N. Aghdale, M. Cottam, T. Gammon, L.D. Hansen, E.A. Lewis, R.J. Farber. 1990. Loss of semi-volatile organic compounds from particles during sampling on filters. Pp. 146–156 in Visibility and Fine Particles, C.V. Mathai, ed. A&WMA Publication No. TR-17. Air and Waste Management Association, Pittsburgh, Penn.

Edgerton, S.A., M.A.K. Khalil, and R.A. Rasmussen. 1986. Source emission characterization of residential wood-burning stoves and fireplaces: Fine particle/methyl chloride ratios for use in chemical mass balance modeling. Environ. Sci. Technol. 20:803–807.

Edwards, A.L. 1984. An Introduction to Linear Regression and Correlation, 2nd ed. New York: W.H. Freeman.

Eldred, R.A. 1988. IMPROVE Sampler Manual, Version 2. Air Quality Group, Crocker Nuclear Laboratory, University of California at Davis, Davis, Calif. January.

Eldred, R.A., L.L. Ashbaugh, T.A. Cahill, R.G. Flochinni, and M.L. Pitchford. 1981. The Effect of the 1908 Smelter Strike on Air Quality in the Southwest. AQG 81-025. Air Quality Group, Crocker Nuclear Laboratory, University of California at Davis, Davis, Calif.

Eldred, R.A., T.A. Cahill, P.J. Feeney, and W.C. Malm. 1987. Regional patterns in particulate matter from the National Park Service network, June 1982 to May 1986. Pp. 386–396 in Visibility Protection: Research and Policy Aspects, P.S. Bhardwaja, ed. APCA Publication TR-10. Air Pollution Control Association, Pittsburgh, Penn.

Eldred, R.A., T.A. Cahill, L.K. Wilkinson, and P.J. Feeney. 1989. Particulate Characterization at Remote Sites Across the U.S.: First Year Results of the NPS/IMPROVE Network. Paper No. 89-151.3. Presented at the 82nd Annual Meeting and Exhibition of the Air and Waste Management Association, Anaheim, Calif., June 25-30.

Eldred, R.A., T.A. Cahill, L.K. Wilkinson, and P.J. Feeney, J.C. Chow, and W.C. Malm. 1990. Measurement of fine particles and their chemical components in the IMPROVE/NPS networks. Pp. 187–196 in Visibility and Fine Particles, C.V. Mathai, ed. A&WMA Publication No. TR-17. Air and Waste Management Association, Pittsburgh, Penn.

Eltgroth, M.W., and P.V. Hobbs. 1979. Evolution of particles in the plumes of coal-fired power plants-II. A numerical model and comparison with field measurements. Atmos. Environ. 13:953-976.

Ely, D.T., J.T. Leary, D.M. Ross, and T.R. Stewart. 1991. The establishment of the Denver visibility standard. In Proceedings of Air and Waste Management Association Annual Meeting, Vancouver, B.C., June 16-21.

Ensor, D.S., and A.P. Waggoner. 1970. Angular truncation error in the integrating nephelometer. Atmos. Environ. 4:481-487.

EPA (U.S. Environmental Protection Agency). 1970. Workbook of Atmospheric Dispersion Estimates. Office of Air Programs Publication AP-26. Office of Air Programs, U.S. Environmental Protection Agency, Research Triangle Park, N.C.

EPA (U.S. Environmental Protection Agency). 1973. 38 Fed. Reg. 25,678 (Sept. 14, 1973).

EPA (U.S. Environmental Protection Agency). 1979. Protecting Visibility: An EPA Report to Congress. EPA 450/5-79-008. Office of Air Quality Planning and Standards, U.S. Environmental Protection Agency, Research Triangle Park, N.C. October.

EPA (Environmental Protection Agency). 1980a. User's Manual for the Plume Visibility Model (PLUVUE II). EPA 450/4-80-032. Environmental Research Laboratory, U.S. Environmental Protection Agency, Research Triangle Park, N.C.

EPA (Environmental Protection Agency). 1980b. Visibility Protection for Federal Class I Areas. 45 Fed. Reg. 80084-95 (Dec. 2, 1980).

EPA (Environmental Protection Agency). 1982. Analysis of New Source Review (NSR) Permitting Experience. EPA Contract No. 68-02-3174. Office of Planning and Resource Management, U.S. Environmental Protection Agency, Research Triangle Park, N.C.

EPA (U.S. Environmental Protection Agency). 1984. Approval and Promulgation of State Implementation Plans; Settlement of Litigation. 49 Fed. Reg. 20647- (May 16, 1984).

EPA (U.S. Environmental Protection Agency). 1985a. Developing Long-term Strategies for Regional Haze: Findings and Recommendations of the Visibility Task Force. U.S. Environmental Protection Agency, Washington, D.C. April.

EPA (U.S. Environmental Protection Agency). 1985b. Analysis of New Source Review (NSR) Permitting Experience--Part 2. EPA Contract No. 68-02-3515, 3816, and 6558. Office of Planning and

Program Evaluation, U.S. Environmental Protection Agency, Research Triangle Park, N.C.

EPA (U.S. Environmental Protection Agency). 1985c. Compilation of Air Pollutant Emission Factors, Vol. I: Stationary Point and Area Sources, 4th ed. EPA Publication AP-42. Office of Air Quality Planning and Standards, Research Triangle Park, N.C.

EPA (Environmental Protection Agency). 1986a. Analysis of New Source Review (NSR) Permitting Experience, Part 3. EPA Contract No. 68-02-3889. Office of Planning and Program Evaluation, Research Triangle Park, N.C.

EPA (Environmental Protection Agency). 1986b. Technical Support Document for Residential Wood Combustion. EPA-600/9-85-012. Office of Air Quality Planning and Standards, U.S. Environmental Protection Agency, Research Triangle Park, N.C. February.

EPA (Environmental Protection Agency). 1987a. Air Programs; Review of the National Secondary Ambient Air Quality Standards for Particulate Matter; Advance Notice of Proposed Rulemaking. 52 Fed. Reg. 24670 (July 1, 1987).

EPA (Environmental Protection Agency) 1987b. Approval and Promulgation of Implementation Plans; Vermont; Visibility in Federal Class I Areas; Lye Brook Wilderness. 52 Fed. Reg. 26,973 (July 17, 1987).

EPA (Environmental Protection Agency). 1987c. PM_{10} SIP (State Implementation Plan) Development Guideline. EPA 450/2-86-001. Office of Air Quality Planning and Standards, U.S. Environmental Protection Agency, Research Triangle Park, N.C. 163 pp.

EPA (Environmental Protection Agency). 1988a. Workbook for Plume Visual Impact Screening and Analysis. EPA 450/4-88-015. Office of Air Quality Planning and Standards, U.S. Environmental Protection Agency, Research Triangle Park, N.C. 202 pp.

EPA (Environmental Protection Agency). 1988b. In-Situ Emission Factors for Residential Wood Combustion. EPA-450/3-88-013. Office of Air Quality Planning and Standards, U.S. Environmental Protection Agency, Research Triangle Park, N.C. 41 pp.

EPA (Environmental Protection Agency). 1988c. Control of Open Fugitive Dust Sources. EPA 450/3-88-008. Office of Air Quality Planning and Standards, U.S. Environmental Protection Agency, Research Triangle Park, N.C. 349 pp.

EPA (Environmental Protection Agency). 1989a. Prevention of Significant Deterioration for Particulate Matter. 54 Fed. Reg. 41218 (Oct. 5, 1989).

EPA (Environmental Protection Agency). 1989b. Guidance Document for Residential Wood Combustion Emission Control Measures. EPA-450/2-89-015. Office of Air Quality Planning and Standards, U.S. Environmental Protection Agency, Research Triangle Park, N.C. 233 pp.

EPRI (Electric Power Research Institute). 1989. TAG^{TM} Technical Assessment Guide. Vol. I: Electricity Supply--1989 (Revision 6). EPRI Special Report P-6587-L. Electric Power Research Institute, Palo Alto, Calif. November.

EPRI (Electric Power Research Institute). 1990. Efficient Energy Use: Estimates of Maximum Energy Savings. Report CU-6746. Electric Power Research Institute, Palo Alto, Calif. March.

EPRI (Electric Power Research Institute). 1991a. Utility Coal Markets Under Acid Rain Legislation. Report IE-7110. Prepared by Energy Ventures Analysis, Inc. Electric Power Research Institute, Palo Alto, Calif. June.

EPRI (Electric Power Research Institute). 1991b. Economic Evaluation of Flue Gas Desulfurization Systems, Vol. 1. EPRI Report GS-7193. Prepared by United Engineers and Constructors, Inc. Electric Power Research Institute, Palo Alto, Calif. February.

Fehsenfeld, F.C., J.W. Drummond, U.K. Roychowdhury, P.J. Galvin, E.J. Williams, M.P. Buhr, D.D. Parrish. G. Hübler, A.O. Langford, J.G. Calvert, B.A. Ridley, F. Grahek, B.G. Heikes, G.L. Kok, J.D. Shetter, J.G. Walega, C.M. Elsworth, R.B. Norton, D.W. Fahey, P.C. Murphy, C. Hovermale, V.A. Mohnen, K.L. Demerjian, G.I. Mackay, and H.I. Schiff. 1990. Intercomparison of NO_2 measurement techniques. J. Geophys. Res. 95:3579–3597.

Ferm, M. 1979. Method for the determination of atmospheric ammonia. Atmos. Environ. 13:1385–1393.

Ferm, M., and A. Sjodin. 1985. A sodium carbonate coated denuder for determination of nitrous acid in the atmosphere. Atmos. Environ. 19:979–984.

Ferman, M.A., G.T. Wolff, and N.A. Kelly. 1981. The nature and sources of haze in the Shenandoah-Blue Ridge Mountains area. J. Air Pollut. Control Assoc. 31:1074–1082.

Fissan, H.J., C. Helsper, and H.J. Thielen. 1983. Determination of particle size distributions by means of an electrostatic classifier. J. Aerosol Sci. 14:354–357.

Fontijn, A., A.J. Sabadell, and R.T. Ronco. 1970. Homogeneous chemiluminescent measurement of nitric oxide with ozone. Anal. Chem. 42:575–579.

Foot, J.S., and C.G. Kilsby. 1989. Absorption of light by aerosol particles: An intercomparison of techniques and spectral observations. Atmos. Environ. 23:489–496.

Forrest, J., D.J. Spandau, R.L. Tanner, and L. Newman. 1982. Determination of atmospheric nitrate and nitric acid employing a diffusion denuder with a filter pack. Atmos. Environ. 16:1473–1485.

Freeburn, S.A. 1986. Benefit/Cost Analysis of Impact Reduction Alternatives for Prescribed Burning in Western Oregon. State of Oregon Department of Environmental Quality, Air Quality Division, Portland, Oreg. April.

Friedlander, S.K. 1973. Chemical element balances and identification of air pollution sources. Environ. Sci. Technol. 7:235–240.

Friedlander, S.K. 1977. Smoke, Dust and Haze: Fundamentals of Aerosol Behavior. New York: John Wiley and Sons.

Fuller, W.A. 1987. Measurement Error Models. New York: John Wiley and Sons.

Galasyn, J.F., J.F. Hornig, and R.H. Soderberg. 1984. The loss of PAH from quartz fiber high volume filters. J. Air Pollut. Control Assoc. 34:57–59.

Galloway, J.N., D.M. Whelpdale, and G.T. Wolff. 1984. The flux of sulfur and nitrogen eastward from North America. Atmos. Environ. 18:2595–2607.

Galvin, P., P.J. Samson, P.E. Coffey, and D. Romano. 1978. Transport of sulfate to New York state. Environ. Sci. Technol. 12:580–584.

GAO (General Accounting Office). 1990. Air Pollution: Protecting Parks and Wilderness From Nearby Pollution Sources. Report to the Chairman, Environment, Energy, and Natural Resources Subcommittee, Committee on Government Operations, House of Representatives. GAO/RCED-90-10. U.S. General Accounting Office, Washington, D.C. February.

Garber, R.W., P.H. Daum, R.F. Doering, T. O'Ottravio, and R.L.

Tanner. 1983. Determination of ambient aerosol and gaseous sulfur using a continuous flame photometric detector (FPD). 3. Design and characterization of a monitor for airborne applications. Atmos. Environ. 17:1381–1385.

Gebhart, K.A., and Malm, W.C. 1990. Source apportionment of particulate sulfate concentrations at three national parks in the eastern United States. Pp. 898–913 in Visibility and Fine Particles, C.V. Mathai, ed. A&WMA Publication No. TR-17. Air and Waste Management Association, Pittsburgh, Penn.

Gelbard, F. 1984. Computer-program and algorithm review, MAEROS. Aero. Sci. Technol. 3:117–118.

Gelbard, F., Y. Tambour, and J.H. Seinfeld. 1980. Sectional representations for simulating aerosol dynamics. J. Colloid Interface Sci. 76:541–556.

Georgi, B., H. Horvath, C. Norek, and I. Kreiner. 1987. Use of rare earth tracers for the study of diesel emissions in the atmosphere. Atmos. Environ. 21:21–28.

Gillette, D.A., and P.C. Sinclair. 1990. Estimation of suspension of alkaline material by dust devils in the United States. Atmos. Environ. 24A:1135–1142.

Gilpin, A. 1978. Air Pollution. St. Lucia, Queensland, Australia: University of Queensland Press.

Gins, J.D., D.H. Nochumson, and J.C. Trijonis. 1981. Statistical relationship between median visibility and conditions of worst-case impact on visibility. Atmos. Environ. 15:2451–2462.

Gleason, J.F., A. Sinha, and C.J. Howard. 1987. Kinetics of the gas-phase reaction $HOSO_2 + O_2 \rightarrow HO_2 + SO_3$. J. Phys. Chem. 91:719–724.

Goldberger, A.S. 1964. Econometric Theory. New York: John Wiley and Sons.

Gordon, G.E. 1980. Receptor models. Environ. Sci. Technol. 14:792–800.

Gordon, G.E. 1988. Critical review: Receptor models. Environ. Sci. Technol. 22:1132–1142.

Gray, H. 1986. Control of Atmospheric Fine Primary Carbon Particle Concentrations. EQL Report No. 23. California Institute of Technology, Environmental Quality Laboratory, Pasadena, Calif.

Gray, H.A., G.R. Cass, J.J. Huntzicker, E.K. Heyerdahl, and J.A.

Rau. 1986. Characteristics of atmospheric organic and elemental carbon particle concentrations in Los Angeles. Environ. Sci. Technol. 20:580-589.

Greenberg, J.P., and P.R. Zimmerman. 1984. Nonmethane hydrocarbons in remote in tropical, continental, and marine atmospheres. J. Geophys. Res. 89:4764-4778.

Groblicki, P.J., G.T. Wolff, and R.J. Countess. 1981. Visibility-reducing species in the Denver 'brown cloud' --I. Relationships between extinction and chemical composition. Atmos. Environ. 15:2473-2484.

Groblicki, P.J., S.H. Cadle, C.C. Ang, and P.A. Mulawa. 1983. Interlaboratory Comparison of Methods for the Analysis of Organic and Elemental Carbon in Atmospheric Particulate Matter. General Motors Research Report 4054. General Motors Research Laboratories, Warren, Mich.

Grosjean, D., and J.H. Seinfeld. 1989. Parameterization of the formation potential of secondary organic aerosols. Atmos. Environ. 23:1733-1747.

Gschwandtner, G., K.C. Gschwandtner, and K. Eldridge, C. Mann, and D. Mobley. 1985. Historic Emissions of Sulfur and Nitrogen Oxides in the United States from 1900-1980. EPA 600/7-85-009a. U.S. Environmental Protection Agency, Research Triangle Park, N.C.

Haagenson, P.L., Y.-H. Kuo, M. Skumanich, and N.L. Seaman. 1987. Tracer verification of trajectory models. J. Clim. Appl. Meteor. 26:410-426.

Hammerle, R.H., and W.R. Pierson. 1975. Sources and elemental composition of aerosol in Pasadena, Calif., by energy-dispersive X-ray fluorescence. Environ. Sci. Technol. 9:1058-1068.

Hänel, G. 1976. The properties of atmospheric aerosol particles as functions of the relative humidity at thermodynamic equilibrium with the surrounding moist air. Adv. Geophys. 19:73-188.

Hänel, G. 1987. Radiation budget of the boundary layer. II. Simultaneous measurement of mean solar volume absorption and extinction coefficients of particles. Beitr. z. Phys. Atmos. 60:241-247.

Hanrahan, P.L. 1981. Improved particulate dispersion modeling results: A new approach using chemical mass balance. Paper No. 81-64.5. In Proceedings of the 74th Annual Meeting and Exhibition of the Air Pollution Control Association, Pittsburgh, Penn.

Hansen, A.D.A., H. Rosen, T. Novakov. 1984. The aethalometer--an instrument for the real-time measurement of optical absorption by aerosol particles. Sci. Total Environ. 36:191–196.

Harley, R.A., S.E. Hunts, and G.R. Cass. 1989. Strategies for control of particulate air quality: Least cost solutions based on receptor-oriented models. Environ. Sci. Technol. 23:1007–1014.

Harrison, L. 1985. The segregation of aerosols by cloud-nucleating activity. Part II. Observation of an urban aerosol. J. Clim. Appl. Meteor. 24:312–321.

Harwood, J.E., and A.L. Kuhn. 1970. A colorimetric method for ammonia in natural waters. Water Res. 4:805–811.

Hasan, H., and C.W. Lewis. 1983. Integrating nephelometer response corrections for bimodal size distributions. Aerosol Sci. Technol. 2:443–453.

Hasan, H., and T.G. Dzubay. 1983. Apportioning light extinction coefficients to chemical species in atmospheric aerosol. Atmos. Environ. 17:1573–1581.

Hatakeyama, S., K. Izumi, T. Fukuyama, and H. Akimoto. 1989. Reactions of ozone with α-pinene and β-pinene in air: Yields of gaseous and particulate products. J. Geophys. Res. 94:13013–13024.

Heffter, J.L. 1980. Air Resources Laboratories Atmospheric Transport and Dispersion Model (ARL-ATAD). NOAA Technical Memorandum ERL ARL-81. NOAA Air Resources Laboratories, National Oceanic and Atmospheric Administration, Department of Commerce, Silver Spring, Md.

Hemeon, W.C.L., G.F. Haines, Jr., and H.M. Ide. 1953. Determination of haze and smoke concentrations by filter paper samplers. J. Air Pollut. Control Assoc. 3:22–28.

Henry, R.C. 1982. Stability analysis of receptor models that use least squares fitting. Pp. 141–157 in Receptor Models Applied to Contemporary Air Pollution Problems, S.L. Dattner and P.K. Hopke, eds. Proceedings No. SP-48. Air Pollution Control Association, Pittsburgh, Penn.

Henry, R.C. 1987. Current factor analysis receptor models are ill-posed. Atmos. Environ. 21:1815–1820.

Henry, R.C., and L. Matamala. 1990. A visual colorimeter for atmospheric research (VICAR). Color Research and Application 15:74–79.

Henry, R.C., C.W. Lewis, P.K. Hopke, and H.J. Williamson. 1984. Review of receptor model fundamentals. Atmos. Environ. 18:1507-1515.

Hering, S.V., and S.K. Friedlander. 1982. Origins of aerosol sulfur size distributions in the Los Angeles basin. Atmos. Environ. 16: 2647-2656.

Hering, S.V., and V.A. Marple. 1986. Low-pressure and micro-orifice impactors. Pp. 103-127 in Cascade Impactor: Sampling and Data Analysis, J.P. Lodge, Jr. and T.L. Chan, eds., American Industrial Hygiene Association, Akron, Ohio.

Hering, S.V., and P.H. McMurry. 1991. Response of a PMS LAS-X laser optical counter to monodisperse atmospheric aerosols. Atmos. Environ. 25:465-468.

Hering, S.V., S.K. Friedlander, J.J. Collins, and L.W. Richards. 1979. Design and evaluation of a new low pressure impactor. 2. Environ. Sci. Technol. 13:184-188.

Hidy, G.M., and S.K. Friedlander. 1972. The Nature of the Los Angeles Aerosol. Second International Union of Air Pollution Prevention Associations (IUAPPA) Clean Air Congress. International Union of Air Pollution Prevention Associations, Brighton, E. Sussex, England.

Hill, A.C. 1990. Measuring how landscape color changes affect aesthetic value. Pp. 570-581 in Visibility and Fine Particles, C.V. Mathai, ed. A&WMA Publication No. TR-17. Air and Waste Management Association, Pittsburgh, Penn.

Ho, W., G.M. Hidy, and R.M. Govan. 1974. Microwave measurements of the liquid water content of atmospheric aerosols. J. Appl Meteor. 13:871-879.

Hoffmann, M.R., and D.J. Jacob. 1984. Kinetics and mechanisms of the catalytic oxidation of dissolved sulfur dioxide in aqueous solution: An application to nighttime fog water chemistry. Pp. 101-172 in SO_2, NO and NO_2 Oxidation Mechanisms: Atmospheric Considerations, J.G. Calvert, ed. Boston: Butterworth Publishers.

Holgate, M.W., M. Kassas, and G.F. White. 1982. The World Environment, 1972-1982. Dublin, Ireland: Tycooly International Publishing, Ltd.

Hopke, P.K. 1985. Receptor Modeling in Environmental Chemistry. New York: Wiley-Interscience.

Hopke, P.K., and S.L. Dattner, eds. 1982. Receptor Models Applied to Contemporary Pollution Problems. Proceedings No. SP-48. Air and Waste Management Association, Pittsburgh, Penn.

Hopke, P.K., R.E. Lamb, and D.F. S. Natusch. 1980. Multielemental characterization of urban roadway dust. Environ. Sci. Technol. 14:164–172.

Hoppel, W.A., J.W. Fitzgerald, G.M. Frick, R.E. Larson, and E.J. Mack. 1990. Aerosol size distributions and optical properties found in the marine boundary layer over the Atlantic Ocean. J. Geophys. Res. 95(D-4):3659–3686.

Horvath, H., and K.E. Noll. 1969. The relationship between atmospheric light scattering coefficient and visibility. Atmos. Environ. 3:543–550.

Hosker, R.P., and S.E. Lindberg. 1982. Review: Atmospheric deposition and plant assimilation of gases and particles. Atmos. Environ. 16:889–910.

Houck, J.E., J.C. Chow, J.G. Watson, C.A. Simon, L.C. Pritchett, J.M. Goulet, and C.A. Frazier. 1989. Determination of Particle Size Distribution and Chemical Composition of Particulate Matter from Selected Sources in California, Vol. 1. Prepared by OMNI Environmental, Inc. and Desert Research Institute for California Air Resources Board, Sacramento, Calif.

Hov, O., and I.S.A. Isaksen. 1981. Generation of secondary pollutants in a power plant plume: A model study. Atmos. Environ. 15:2367–2376.

Hov, O., J. Schjoldager, and B.M. Wathne. 1983. Measurement and modeling of the concentrations of terpenes in coniferous forest air. J. Geophys. Res. 88:10679–10688.

Hudischewskyj, A.B., and C. Seigneur. 1989. Mathematical modeling of the chemistry and physics of aerosols in plumes. Environ. Sci. Technol. 23:413–421.

Hudischewskyj, A.B., P. Saxena, and C. Seigneur. 1987. Mathematical modeling of light scattering by atmospheric aerosols. Pp. 564–575 in Visibility Protection: Research and Policy Aspects, P.S. Bhardwaja, ed. APCA Publication No. TR-10. Air Pollution Control Association, Pittsburgh, Penn.

Huebert, B.J., G. Lee, and W.L. Warren. 1990. Airborne aerosol inlet passing efficiency measurement. J. Geophys. Res. 95(D-10):16369–

16381.

Hunter, S.C., and N.L. Helgeson. 1976. Control of Oxides of Sulfur from Stationary Sources in the South Coast Air Basin of California. KVB Inc., Tustin, Calif. Prepared for the State of California Air Resources Board (CARB), Sacramento, Calif.

Huntzicker, J.J., S.K. Friedlander, and C.I. Davidson. 1975. Material balance for automobile-emitted lead in Los Angeles Basin. Environ. Sci. Technol. 9:448–457.

Husar, R.B. 1986. Emissions of sulfur dioxide and nitrogen oxides and trends for Eastern North America. Pp. 48–92 in Acid Deposition: Long-Term Trends. Washington, D.C.: National Academy Press.

Husar, R.B. 1988. Trends of Seasonal Haziness and Sulfur Emissions over the Eastern United States. EPA 600/S3-89/062. Atmospheric Research and Exposure Assessment Laboratory, U.S. Environmental Protection Agency, Research Triangle Park, N.C. September.

Husar, R.B. 1990. Eastern United States Haze Trends During the 1978-82 Mini-Recession. EPA CR810351. Atmospheric Sciences Research Laboratory, U.S. Environmental Protection Agency, Research Triangle Park, N.C.

Husar, R.B., D.E. Patterson, J.M. Holloway, W.E. Wilson, and T.G. Ellestad. 1979. Trends of Eastern United States Haziness Since 1948. Proceedings of the Fourth Symposium on Atmospheric Turbulence, Diffusion, and Air Pollution. American Meteorological Society, Boston, Mass.

Husar, R.B., J.M. Holloway, D.E. Patterson, and W.E. Wilson. 1981. Spatial and temporal pattern of eastern United States haziness: A summary. Atmos. Environ. 15:1919-1928.

Husar, R.B., and W.E. Wilson. 1993. Emission trends: Haze and sulfar in the eastern United States. Environ. Sci. Technol. 27:12–16.

Hwang, H., and P.K. Dasgupta. 1986. Fluorometric flow injection determination of aqueous peroxides at nonomolar level using membrane reactors. Anal. Chem. 58:1521-1524.

Hwang, C.-S., K.G. Severin, and P.K. Hopke. 1984. A comparison of R-modes and Q-modes in target transformation factor analysis for resolving environmental data. Atmos. Environ. 18:345-352. Discussed by N.Z. Heidam, and D.Kronborg in "A comparision of R-mode and Q-mode in target transformation factor-analysis for resolving environmental data." Atmos. Environ. 19:1549-1551.

ICF Incorporated. 1991. Regulatory Impact Analysis of the Proposed Acid Rain Implementation Regulations. EPA Contract 68-DO-011102. Prepared for the Office of Atmospheric and Indoor Air Programs, Acid Rain Division, U.S. Environmental Protection Agency, Research Triangle Park, N.C. September 16.

Iyer, H.K., W.C. Malm, and R.A. Ahlbrandt. 1987. A mass balance method for estimating the fractional contribution of pollutants from various sources to a receptor site. Pp. 861–871 in Visibility Protection: Research and Policy Aspects, P.S. Bhardwaja, ed. APCA Publication No. TR-10. Air Pollution Control Association, Pittsburgh, Penn.

Jaenicke, R. 1980. Natural aerosols. Ann. N.Y. Acad. Sci. 338:317–329.

Japar, S.M. 1990. Discussion: Absorption of light by aerosol particles: An intercomparison of techniques and spectral observations. Atmos. Environ. 24A:979–981.

Javitz, H.S., J.G. Watson, J.P. Guertin, and P.K. Mueller. 1988a. Results of a receptor modeling feasibility study. J. Air Pollut. Control Assoc. 38:661–667.

Javitz, H.S., J.G. Watson, and N. Robinson. 1988b. Performance of the chemical mass balance model with simulated local-scale aerosols. Atmos. Environ. 22:2309–2322.

John, W., S.M. Wall, and J. Ondo. 1988. A new method for nitric acid and nitrate aerosol measurement using the dichotomous sampler. Atmos. Environ. 22:1627–1636.

John, W., S.M. Wall, J.L. Ondo, and W. Winklmayr. 1990. Modes in the size distribution of atmospheric inorganic aerosol. Atmos. Environ. 24A:2349–2360.

Johnson, C.E., J.V. Molenar, J.R. Hein, W.C. Malm. 1984. The use of a scanning densitometer to measure visibility-related parameters from photographic slides. Paper 84-60P.8. Presented at the 77th Annual Meeting of the Air Pollution Control Association, San Francisco, Calif., June 24-29.

Joos, E., A. Mendonca, and C. Seigneur. 1987. Evaluation of a reactive plume model with power plant plume data: Application to the sensitivity analysis of sulfate and nitrate formation. Atmos. Environ. 21:1331–1344.

Kahl, J.D., and P.J. Samson. 1986. Uncertainty in trajectory calculations due to low resolution meteorological data. J. Clim. Appl. Meteor. 25:1816–1831.

Kahl, J.D., and P.J. Samson. 1988. Uncertainty in estimating boundary-layer transport during highly convective conditions. J. Appl. Meteor. 27:1024–1035.

Kamens, R.M., H.E. Jeffries, M.W. Gery, R.W. Wiener, K.G. Sexton, and G.B. Howe. 1981. The impact of α-pinene on urban smog formation: An outdoor smog chamber study. Atmos. Environ. 15:969–982.

Kamens, R.M., M.W. Gery, H.E. Jeffries, M. Jackson, and E.I. Cole. 1982. Ozone isoprene reactions: Product formation and aerosol potential. Int. J. Chem. Kinet. 14:955–975.

Katz, M., and C. Chan. 1980. Comparative distribution of eight polycyclic aromatic hydrocarbons in airborne particulates collected by conventional high-volume sampling and by size fractionation. Environ. Sci. Technol. 14:838–843.

Keeler, G.J., and P.J. Samson. 1989. Spatial representativeness of trace element ratios. Environ. Sci. Technol. 23:1358–1364.

Keuken, M.P., C.A.M. Schoonebeek, A. van Wensveen-Louter, and J. Slanina. 1988. Simultaneous sampling of NH_3, HNO_3, HCl, SO_2, and H_2O_2 in ambient air by a wet annular denuder system. Atmos. Environ. 22:2541–2548.

Khalil, M.A.K., S.A. Edgerton, and R.A. Rasmussen. 1983. A gaseous tracer model for air pollution from residential wood burning. Environ. Sci. Technol. 17:555–559.

Kleindienst, T.E., P.B. Shepson, D.N. Hodges, C.M. Nero, R.R. Arnts, P.K. Dasgupta, H. Hwang, G.L. Kok, J.A. Lind, A.L. Lazrus, G.I. Mackay, L.K. Mayne, and H.I. Schiff. 1988. Comparison of techniques for measurement of ambient levels of hydrogen peroxide. Environ. Sci. Technol. 22:53–61.

Kleinman, M.T., B.S. Pasternack, M. Eisenbud, and T.J. Kneip. 1980. Identifying and estimating the relative importance of sources of airborne particulates. Environ. Sci. Technol. 14:62–65.

Kley, D., and M. McFarland. 1980. Chemiluminescence detector for NO and NO_2. Atmos. Technol. 12:63–69.

Klockow, D., B. Jablonski, and R. Niessner. 1979. Possible artifacts in filter sampling of atmospheric sulphuric acid and acidic sulphates. Atmos. Environ. 13:1655–1676.

Knapp, K.T., J.L. Durham, and T.G. Ellestad. 1986. Pollutant sampler for measurements of atmospheric acidic dry deposition. Environ. Sci. Technol. 20:633–637.

Kneip, T.J., M.T. Kleinman, and M. Eisenbud. 1972. Relative Contribution of Emission Sources to the Total Airborne Particulates in New York City. Third International Union of Air Pollution Prevention Associations (IUAPPA) Clean Air Congress. International Union of Air Pollution Prevention Associations, Brighton, E. Sussex, England.

Knudson, D.A. 1985. Estimated Monthly Emissions of Sulfur Dioxide and Oxides of Nitrogen for the 48 Contiguous States, 1975-1984. ANL/EES-TM-318. Argonne National Laboratories, U.S. Department of Energy, Washington, D.C.

Knutson, E.O., and K.T. Whitby. 1975. Aerosol classification by electric mobility: Apparatus, theory and applications. J. Aerosol Sci. 6:443–451.

Kok, G.L., T.P. Holler, M.B. Lopez, H.A. Nachtrieb, and M. Yuan. 1978. Chemiluminiscent method for determination of hydrogen peroxide in the ambient atmosphere. Environ. Sci. Technol. 12:1072–1076.

Kokkinos, A., J.E. Cichanowicz, R.E. Hall, and C.B. Sedman. 1991. Stationary combustion NO_x control: A summary of the 1991 symposium. J. Air Waste Manage. Assoc. 41:1252-1259.

Koenig, J., W. Funcke, E. Balfanz, B. Grosch, and F. Pott. 1980. Testing a high volume air sampler for quantitative collection of polycyclic aromatic hydrocarbons. Atmos. Environ. 14:609-614.

Koschmieder, H. 1924. Theorie der horizontalen sichtweite. Beitr. Phys. freien Atmos. 12:33–.

Kowalczyk, J.F., and B.J. Tombleson. 1985. Oregon's woodstove certification program. J. Air Pollut. Control Assoc. 35:619-625.

Kowalczyk, G.S., C.E. Choquette, and G.E. Gordon. 1978. Chemical element balances: Identification of air pollution sources in Washington, D.C. Atmos. Environ. 12:1143-1154.

Kuo, Y.-H, M. Skumanich, P.L. Haagenson, and J.S. Chang. 1985. The accuracy of trajectory models as revealed by the observing system simulation experiments. Mon. Wea. Rev. 113:1852-1867.

Lamb, R.G. 1983. A Regional Scale (1000 km) Model of Photochemical Air Pollution. Part 1: Theoretical Formulation. EPA 600/3-83-035. Environmental Sciences Research Laboratories, U.S. Environmental Protection Agency, Research Triangle Park, N.C.

Langford, A.O., P.D. Goldan, and F.C. Fehsenfeld. 1989. A molybdenum oxide annular denuder system for gas phase ambient ammonia measurements. J. Atmos. Chem. 8:359–376.

Larson, S.M., G.R. Cass, K.J. Hussey, and F. Luce. 1988. Verification of image processing based visibility models. Environ. Sci. Technol. 22:629–637.

Larson, S.M., and G.R. Cass. 1989. Characteristics of summer midday low-visibility events in the Los Angeles area. Environ. Sci. Technol. 23:281–289.

Latimer, D.A., and G.S. Samuelsen. 1978. Visual impact of plumes from power plants: A theoretical model. Atmos. Environ. 12:1455–1466.

Latimer, D.A., H. Hogo, and T.C. Daniel. 1981. The effects of atmospheric optical conditions on perceived scenic beauty. Atmos. Environ. 15:1865–1874.

Latimer, D.A., H.K. Iyer, and W.C. Malm. 1990. Application of a differential mass balance model to attribute sulfate haze in the Southwest. Pp. 819–830 in Visibility and Fine Particles, C.V. Mathai, ed. A&WMA Publication No. TR-17. Air and Waste Management Association, Pittsburgh, Penn.

Latin, H. 1985. Ideal versus real regulatory efficiency: Implementation of uniform standards and fine tuning regulatory reform. Stan. L. Rev. 37:1267–.

Lawson, D.R. 1980. Impaction surface coatings intercomparison and measurements with cascade impactors. Atmos. Environ. 14:195–200.

Lazrus, A.L., G.L. Kok, S.N. Gitlin, J.A. Lind, and S.E. McLaren. 1985. Automatic fluorometric method for hydrogen peroxide in atmospheric precipitation. Anal. Chem. 57:917–922.

Lazrus, A.L., G.L. Kok, J.A. Lind, S.N. Gitlin, B.G. Heikes, R.E. Shetter. 1986. Automated fluorometric method for hydrogen peroxide in air. Anal. Chem. 58:594–597.

LeBel, P.J., J.M. Hoell, J.S. Levine, and S.A. Vay. 1985. Aircraft measurements of ammonia and nitric acid in the lower troposphere. Geophys. Res. Lett. 12:401–404.

Lewin, E.E., and D. Klockow. 1982. Application of the TCM denuder for SO_2 collection. Pp. 54–61 in Proceedings of the 2nd European Symposium on Physico-Chemical Behavior of Atmospheric Pollutants,

B. Versino and H. Ott, eds. Dordrecht, Holland: D. Reidel Publishing Company.

Lewis, C.W., and E.S. Macias. 1980. Composition of size-fractionated aerosol in Charleston, West Virginia. Atmos. Environ. 14:185–194.

Lewis, C.W., and R.K. Stevens. 1985. Hybrid receptor model for secondary sulfate from a sulfur dioxide point source. Atmos. Environ. 19:917–924.

Lewis, C.W., R.E. Baumgardner, R.K. Stevens, and G.M. Russwurm. 1986. Receptor modeling study of Denver winter haze. Environ. Sci. Technol. 20:1126–1136.

Lewis, C.W., R.E. Baumgardner, R.K. Stevens, L.D. Claxton, and J. Lewtas. 1988. Contribution of woodsmoke and motor vehicle emissions to ambient aerosol mutagenicity. Environ. Sci. Technol. 22: 968–971.

Lin, C.-I., M. Baker, and R.J. Charlson. 1973. Absorption coefficient of atmospheric aerosols: A method for measurement. Appl. Opt. 12:1356–1363.

Lindgren, P.F., and P.K. Dasgupta. 1989. Measurement of atmospheric sulfur dioxide by diffusion scrubber coupled ion chromatography. Anal. Chem. 61:19–24.

Liroff, R. 1986. Reforming Air Pollution Regulation: The Toil and Trouble of EPA's Bubble. Washington, D.C.: Conservation Foundation.

Liu, B.Y.H., and D.Y.H. Pui. 1975. On the performance of the electrical aerosol analyzer. J. Aerosol Sci. 6:249–264.

Liu, M.-K., R.E. Morris, and J.P. Killus. 1984. Development of a regional oxidant model and application to the northeastern United States. Atmos. Environ. 18:1145–1162.

Lowenthal, D.H., and K.A. Rahn. 1988. Tests of regional elemental tracers of pollution aerosols. 2. Sensitivity of signatures and apportionments to variations in operating parameters. Environ. Sci. Technol. 22:420–426.

MacCracken, M.C., D.J. Wuebbles, J.J. Walton, W.H. Duewer, and K.E. Grant. 1978. The Livermore regional air quality model: I. Concept and development. J. Appl. Meteor. 17:254–272.

Maffiolo, G., Y. Leriquier, and J. Dubois. 1982. Comparative laboratory evaluation of four atmospheric sulfur dioxide analyzers. Pollut. Atmos. 94:89–94.

Magnotta, F., and H.S. Johnston. 1980. Photodissociation quantum yields for the NO_3 free radical. Geophys. Res. Lett. 7:769-772.

Malm, W.C. 1979. Considerations in the measurement of visibility. J. Air Pollut. Control Assoc. 29:1042-1052.

Malm, W.C., and M. Pitchford. 1989. The use of an atmospheric quadratic detection model to assess change in aerosol concentrations to visibility. Paper 89-67.3. Proceedings of the 82nd Annual Meeting and Exhibition of the Air and Waste Management Association, Anaheim, Calif., June 25-30.

Malm, W.C., K.K. Leiker, and J.V. Molenar. 1980. Human perception of visual air quality. J. Air Pollut. Cont. Assoc. 30:122-131.

Malm, W. C., K.K. Kelley, J. Molenar, and T. Daniel. 1981. Human perception of visual air quality (uniform haze). Atmos. Environ. 15:1875-1890.

Malm, W., J. Molenar, and L.L. Chan. 1983. Photographic simulation techniques for visualizing the effect of uniform haze on a scenic resource. J. Air Pollut. Cont. Assoc. 33:126-129.

Malm, W.C., G. Persha, R. Tree, R. Stocker, I. Tombach, and H. Iyer. 1987. Comparison of atmospheric extinction measurements made by a transmissometer, integrating nephelometer, and teleradiometer with natural and artificial black target. Pp. 763-782 in Visibility Protection: Research and Policy Aspects, P.S. Bhardwaja, ed. APCA Publication No. TR-10. Air Pollution Control Association, Pittsburgh, Penn.

Malm, W.C., H.K. Iyer, and K. Gebhart. 1990. Application of tracer mass balance regression WHITEX data. Pp. 806-818 in Visibility and Fine Particles, C.V. Mathai, ed. A&WMA Publication No. TR-17. Air and Waste Management Association, Pittsburgh, Penn.

Marcus, A.A. 1981. Improving Forest Productivity: Prescribed Burning in the Light of Clean Air Act Visibility Standards. Battelle Human Affairs Research Center, Seattle, Wash. November.

Margitan, J.J. 1984. Mechanism of the atmospheric oxidation of sulfur dioxide. Catalysis by hydroxyl radicals. J. Phys. Chem. 88:3314-3318.

Maroulis, P.J., A.L. Torres, A.B. Goldberg, and A.R. Bandy. 1980. Atmospheric SO_2 measurements on project Gametag. J. Geophys. Res. 85(C-12):7345-7349.

Marple, V.A., K.L. Rubow, and S.M.Behm. 1991. A microorifice uniform deposit impactor (MOUDI): Description, calibration, and use. Aerosol. Sci. Technol. 14:434-446.

Martin, L.R. 1984. Kinetic studies of sulfite oxidation in aqueous solution. Pp. 63-100 in SO_2, NO and NO_2 Oxidation Mechanisms: Atmospheric Considerations, J.G. Calvert, ed. Boston: Butterworth Publishers.

Marty, J.C., M.J. Tissier, and A. Saliot. 1984. Gaseous and particulate polycyclic aromatic hydrocarbons from the marine atmosphere. Atmos. Environ. 18:2183-2190.

McDermott, D.L., K.D. Reiszner, and P.W. West. 1979. Development of long-term sulfur dioxide monitor using permeation sampling. Environ. Sci. Technol. 13:1087-1090.

McDonald, J.E. 1958. The physics of cloud modification. Adv. Geophys. 5:223-303.

McDow, S.R., and J.J. Huntzicker. 1990. Vapor adsorption artifact in the sampling of organic aerosol: Face velocity effects. Atmos. Environ. 24:2563-2572.

McFarland, A.R., C.A. Ortiz, and C.E. Rodes. 1979. Characteristics of aerosol samplers used in ambient air monitoring. Paper presented at the 86th National Meeting of the American Institute of Chemical Engineering, Houston, Texas. April 1-5.

McGaughey, J.F., and S.K. Gangwal. 1980. Comparison of three commercially available gas chromatographic-flame photometric detectors in the sulfur mode. Anal. Chem. 52:2079-2083.

McMahon, T.A., and P.J. Denison. 1979. Empirical atmospheric deposition parameters: A survey. Atmos. Environ. 13:571-586.

McMurry, P.H., and J.C. Wilson. 1983. Droplet phase (heterogeneous) and gas phase (homogeneous) contributions to secondary ambient aerosol formation as functions of relative humidity. J. Geophys. Res. 88:5101-5108.

McMurry, P.H., and M.R. Stolzenburg. 1989. On the sensitivity of particle size to relative humidity for Los Angeles aerosols. Atmos. Environ. 23:497-507.

McMurry, P.H., and X.Q. Zhang. 1989. Size distributions of ambient organic and elemental carbon. Aerosol Sci. Technol. 10:430-437.

McNider, R.T. 1981. Investigation of the Impact of Topographic Circulations in the Transport and Dispersion of Air Pollutants. Ph.D. Dissertation. Department of Environmental Sciences, University of

Virginia, Charlottesville, Va. 210 pp.

McRae, G.J., W.R. Goodin, and J.H. Seinfeld. 1982. Development of a second generation mathematical model for urban air pollution-I. Model formulation. Atmos. Environ. 16:679–696.

Middleton, W.E.K. 1952. Vision Through the Atmosphere. Toronto: University of Toronto Press.

Middleton, P.B., and J.R. Brock. 1977. Modeling the urban aerosol. J. Air Pollut. Cont. Assoc. 27:771–775.

Middleton, P., and J.S. Chang. 1990. Analysis of RADM gas concentration predictions using OSCAR and NEROS monitoring data. Atmos. Environ. 24:2113–2125.

Middleton, P., and S. Burns. 1991. Denver Air Quality Modeling Study. In Proceedings of the Annual Air and Waste Management Association International Meeting in Vancouver, B.C., Canada. June 17-21.

Middleton, P., T.R. Stewart, and R.L. Dennis. 1983a. Modeling human judgments of urban visual air quality. Atmos. Environ. 17: 1015–1022.

Middleton, P., T.R. Stewart, R.L. Dennis, and D. Ely. 1983b. Implications of NCAR's urban visual air quality assessment method for pristine areas. Pp. 51–63 in Managing Air Quality and Scenic Resources in National Parks and Wilderness Areas, R.D. Rowe and L.C. Chestnut, eds. Boulder, Colo.: Westview Press.

Middleton, P., T.R. Stewart, D. Ely, and C.W. Lewis. 1984. Physical and chemical indicators of urban visual air quality judgments. Atmos. Environ. 18:861–870.

Middleton, P., T.R. Stewart, and J. Leary. 1985. On the use of human judgment and physical/chemical measurements in visual air quality management. J. Air Pollut. Cont. Assoc. 35:11–18.

Middleton, P., J.S. Chang, J.C. del Corral, and H. Geiss. 1988. Comparison of RADM and OSCAR precipitation chemistry. Atmos. Environ. 12:1195–1208.

Mie, G. 1908. Beitrage zur Optik trüber Medien speziell kolloidaler Metallösungen. Ann. Physik 25:377–445.

Miller M.S., S.K. Friedlander, and G.M. Hidy. 1972. A chemical element balance for the Pasadena aerosol. J. Colloid Interface Sci. 39:165–176.

Miller, D.F., D.E. Schorran, T.E. Hoffer, D.P. Rogers, W.H. White, and E.S. Macias. 1990. An analysis of regional haze using tracers of opportunity. J. Air Waste Management Assoc. 40:757-761.

Mills, T.J., P.B. Shinkle, and G.L. Cox. 1985. Direct Costs of Sylvacultural Treatments on National Forests, 1975-1978. U.S. Department of Agriculture (USDA) Forest Service Research Paper WO-40.

Miner, J.R. 1984. Use of natural zeolites in the treatment of animal wastes. Pp. 257-262 in Zeo-agriculture: Use of Natural Zeolites in Agriculture and Aquaculture, W.G. Pond and F.A. Mumpton, eds. Boulder, Colo: Westview Press.

Morandi M.T., J.M. Daisey, and P.J. Lioy. 1987. Development of a modified factor analysis/multiple regression model to apportion suspended particulate matter in a complex urban airshed. Atmos. Environ. 21:1821-1832.

Morris, R.E., S.D. Reynolds, M.A. Yocke, and M.K. Liu. 1987. The Systems Application, Inc. regional transport model: Current status and future needs. Air and Pollution Control Association (APCA) Conference on the Scientific and Technical Issues Facing the Post-1987 Ozone Control Strategies, Hartford, Conn.

Mozurkewich, M., and J.G. Calvert. 1988. Reaction probability of N_2O_5 on aqueous aerosols. J. Geophys. Res. 93:15889-15896.

Mueller, P.K., and G.M. Hidy. 1983. The Sulfate Regional Experiment: Report of Findings. EPRI EA-1901. Electric Power Research Institute, Palo Alto, Calif.

Mueller, P.K., D.A. Hansen, and J.G. Watson, Jr. 1986. The Subregional Cooperative Electric Utility, Department of Defense, National Park Service, and EPA Study (SCENES) on Visibility: An Overview. EPRI EA-4664-SR. Electric Power Research Institute, Palo Alto, Calif.

Mulawa, P.A., and S.H. Cadle. 1985. A comparison of nitric-acid and particulate nitrate measurement by the penetration and denuder difference methods. Atmos. Environ. 19:1317-1324.

Mumpower, J., P. Middleton, R.L. Dennis, T.R. Stewart, and V. Veirs. 1981. Visual air quality assessment: Denver case study. Atmos. Environ. 15:2433-2441.

Munn, R.E. 1973. Secular increases in summer haziness in the Atlantic provinces. Atmos. Environ. 11:156-161.

Murray, L.C., R.J. Farber, M. Zeldin, and W.H. White. 1990. Using statistical analyses to evaluate modulation in SO_2 emissions. Pp.

923–934 in Visibility and Fine Particles, C.V. Mathai, ed. A&WMA Publication No. TR-17. Air And Waste Management Association, Pittsburgh, Penn.

NAPAP (National Acid Precipitation Assessment Program, Office of the Director). 1991a. Sensitivity to Change. 1990 Integrated Assessment Report. National Acid Precipitation Assessment Program (NAPAP), Washington, D.C. November.

NAPAP (National Acid Precipitation Assessment Program, Office of the Director). 1991b. Future Projections. 1990 Integrated Assessment Report. National Acid Precipitation Assessment Program (NAPAP), Washington, D.C. November.

NCAR (National Center for Atmospheric Research). 1983. Regional Acid Deposition: Models and Physical Processes. NCAR/TN-214+STR. National Center for Atmospheric Research, Boulder, Colo. 386 pp. August.

NCAR (National Center for Atmospheric Research). 1985. The NCAR Eulerian Regional Acid Deposition Model. NCAR/TN-256+STR. National Center for Atmospheric Research, Boulder, Colo. 178 pp. June.

NCAR (National Center for Atmospheric Research). 1986. Preliminary Evaluation Studies with the Regional Acid Deposition Model (RADM). NCAR/TN-265+STR. National Center for Atmospheric Research, Boulder, Colo. 202 pp.

Nederbragt, G.W., A. Van Der Horst, and J. Van Duijn. 1965. Rapid ozone determination near an accelerator. Nature 206:87.

NERC (North American Electric Reliability Council). 1991. Reliability Assessment. Chapter on Clean Air Regulations. North American Electric Reliability Council, Princeton, N.J.

Newman, L., and C.M. Benkovitz. 1986. Comments on acid deposition in the western United States. Science 233:11–12.

NPS (National Park Service). 1988. Air Quality in the National Parks: A Summary of Findings from the National Park Service Air Quality Research and Monitoring Program. Natural Resources Report 88-1. Prepared by Energy and Resource Consultants, Inc. for the U.S. Department of the Interior, National Park Service, Air Quality Division, Denver, Colo. July.

NPS (National Park Service). 1989. National Park Service Report on the Winter Haze Intensive Tracer Experiment (WHITEX). Final

Report. Air Quality Division, National Park Service, Ft. Collins, Colo. December 4.

NRC (National Research Council, Commission on Natural Resources). 1975. Air Quality and Stationary Source Emission Control. Washington, DC.: U.S. Government Printing Office. 909 pp.

NRC (National Research Council). 1981. On Prevention of Significant Deterioration of Air Quality. Washington, D.C.: National Academy Press. 141 pp.

NRC (National Research Council). 1986. Acid Deposition: Long-Term Trends. Washington, D.C.: National Academy Press. 506 pp.

NRC (National Research Council). 1990. Haze in the Grand Canyon: An Evaluation of the Winter Haze Intensive Tracer Experiment. Washington, D.C.: National Academy Press. 97 pp.

NRC (National Research Council). 1991a. Toward a New National Weather Service. Washington, D.C.: National Academy Press. 67 pp.

NRC (National Research Council). 1991b. Rethinking the Ozone Problem in Urban and Regional Air Pollution. Washington, D.C.: National Academy Press. 489 pp.

NWS (National Weather Service). 1991. A National Weather Service Proposal for the Climate Data Continuity Project. Test and Evaluation Division, National Weather Service, Test National Oceanic and Atmospheric Administration, Sterling, Va. February.

Oppenheimer, M., C.B. Epstein, and R.E. Yuhnke. 1985. Acid deposition, smelter emissions, and the linearity issue in the western United States. Science 229:859-862.

Oren, C.N. 1988. Prevention of Significant Deterioration: Control-Compelling Versus Site Shifting. Iowa Law Review 74. 114 pp.

Oren, C.N. 1989. The protection of parklands from air pollution: A look at current policy. Harvard Environ. Law Rev. 13(2):313-422.

Orgill, M.M., and G.A. Sehmel. 1976. Frequency and diurnal variation of dust storms in the contiguous United States. Atmos. Environ. 10:813-825.

Ott, W.R. 1990. A physical explanation of the lognormality of pollutant concentrations. J. Air Waste Mgmt. Assoc. 40:1378-1383.

Ouimette, J.R., and R.C. Flagan. 1982. The extinction coefficient of multicomponent aerosols. Atmos. Environ. 16:2405-2419.

Pace, T.G., ed. 1986. Receptor Methods for Source Apportionment: Real World Issues and Applications. APCA Publication No. TR-5. Air Pollution Control Association, Pittsburgh, Penn.

Pace, T.G. 1990. Prescribed burning: Public policy issues. Paper 90-172.7. Presented at the 83rd Annual Meeting and Exhibition of the Air and Waste Management Association, Pittsburgh, Penn., June 24-29.

Pace, T.G., and J.G. Watson. 1987. Protocol for Applying and Validating the CMB Model. EPA 450/4-87-010. U.S. Environmental Protection Agency, Research Triangle Park, N.C.

Pandis, S.N., S.E. Paulson, J.H. Seinfeld, and R.C. Flagan. 1991. Aerosol formation in the photooxidation of isoprene and beta-pinene. Atmos. Environ. 25(A):997–1008.

Patterson, D.E., R.B. Husar, W.E. Wilson, and L.F. Smith. 1981. Monte Carlo simulation of daily regional sulfur distributions: Comparison with SURE sulfate data and visual range observations during August 1977. J. Appl. Meteor. 20:404–420.

Penndorf, R. 1957. Tables of the refractive index for standard air and the Rayleigh scattering coefficient for the spectral region between 0.2 and 20 μm and their application to atmospheric optics. Optic. Soc. Am. J. 47:176–182.

Penkett, S.A. 1982. Non-methane organics in the remote troposphere. Pp.329–355 in Atmospheric Chemistry, E.D. Goldberg, ed. New York: Springer-Verlag.

Penkett, S.A., B.M.R. Jones, K.A. Brice, and A.E.J. Eggleton. 1979. The importance of atmospheric ozone and hydrogen peroxide in oxidizing sulphur dioxide in cloud and rainwater. Atmos. Environ. 13:123–137.

Penner, J.E., and P.S. Connell. 1987. Pollutant transport study: Bay Area to North Central Coast air basin. California Air Resources Board, Sacramento, Calif.

Peters, J., and B. Seifert. 1980. Losses of benzo(a)pyrene under the conditions of high-volume sampling. Atmos. Environ. 14:117–120.

Pierson, W.R., and W.W. Brachaczek. 1983. Particulate matter associated with vehicles on the road, II. Aerosol Sci. Technol. 2:1–40.

Pierson, W.R., and W.W. Brachaczek. 1988. Coarse and fine-particle atmospheric nitrate and nitric acid gas in Claremont, California during the 1985 nitrogen species methods comparison study. Atmos.

Environ. 22:1665-1668.

Pilinis, C., and J.H. Seinfeld. 1988. Development and evaluation of an eulerian photochemical gas-aerosol model. Atmos. Environ. 22: 1985-2002.

Pitchford, M., and D. Allison. 1984. Lake Tahoe visibility study. J. Air Pollut. Control Assoc. 31:213-221.

Pitchford, M., and D. Joseph. 1990. IMPROVE Progress Report. EPA 450/4-90-008. Office of Air Quality Planning and Standards, U.S. Environmental Protection Agency, Research Triangle Park, N.C. May.

Pitchford, M., B.V. Polkowsky, M. R. McGown, W. C. Malm, J.V. Molenar, and L. Mauch. 1990. Percent change in extinction coefficient: A proposed approach for federal visibility protection strategy. Pp. 37-49 in Visibility and Fine Particles, C.V. Mathai, ed. A&WMA Publication No. TR-17. Air and Waste Management Association, Pittsburgh, Penn.

Placet, M., and D.G. Streets. 1987. Emission of acidic deposition precursors. Pp. 1, 53-57 in Volume 2, NAPAP Interim Assessment: Emissions and Controls. National Acid Precipitation Assessment Program (NAPAP), Washington, D.C.

Placet, M., R.E. Battye, F.C. Fehsenfeld, and G.W. Bassett. 1990. Emissions Involved in Acidic Deposition Process. State-of-Science/Technology Report 1, Volume I of the National Acid Precipitation Assessment Program (NAPAP), Washington, D.C. December.

Poirot, R.L., and P.R. Wishinski. 1986. Regional apportionment of ambient sulfate contributions to a remote site in northern Vermont. Pp. 239-250 in Receptor Methods for Source Apportionment: Real World Issues and Applications, T.G. Pace, ed. APCA Publication No. TR-5. Air Pollution Control Association, Pittsburgh, Penn.

Poirot, R.L., R.G. Flocchini, and R.B. Husar. 1990. Winter fine particle composition in the Northeast: Preliminary results from the NESCAUM monitoring network. Paper 90-84.5. Presented at the 83rd Annual Meeting and Exhibition of the Air and Waste Management Association, Pittsburgh, Penn., June 24-29.

Possanzini, M., A. Febo, and A. Liberti. 1983. New design of a high-performance denuder for the sampling of atmospheric pollutants. Atmos. Environ. 17:2605-2610.

Post, J.E., and P.R. Buseck. 1984. Characterization of individual particles in the Phoenix urban aerosol using electron-beam instruments. Env. Sci. Technol. 18:35–42.

Pratsinis, S.E., M.D. Zeldin, and E.C. Ellis. 1988. Source resolution of the fine carbonaceous aerosol by principal component-stepwise regression analysis. Environ. Sci. Technol. 22:212–216.

Pruppacher, H.R., and J.D. Klett. 1978. Microphysics of Atmospheric Clouds and Precipitation. Dordrecht, Holland: D. Reidel Publishing Co.

Quinn, P.K., and T.S. Bates. 1989. Collection efficiencies of a tandem sampling system for atmospheric aerosol particles and gaseous ammonia and sulfur dioxide. Environ. Sci. Technol. 23:736–739.

Raabe, O.G., D.A. Braaten, R.L. Axelbaum, S.V. Teague, and T.A. Cahill. 1988. Calibration studies of the DRUM impactor. J. Aerosol. Sci. 19:183–196.

Rahn, K.A. 1981. The manganese to vanadium ratio as a tracer of large-scale sources of pollution aerosol for the Arctic. Atmos. Environ. 15:1457–1464.

Rahn, K.A., and D.H. Lowenthal. 1985. Pollution aerosol in the Northeast: Northeastern-midwestern contributions. Science 228:275–284.

Rasmussen, R.A., and M.A.K. Khalil. 1988. Isoprene over the Amazon Basin. J. Geophys. Res. 93:1417–1421.

Reible, D.D, J.R. Ouimette, and F.H. Shair. 1982. Atmospheric transport of visibility degrading pollutants into the California Mojave Desert. Atmos. Environ. 16:599–613.

Reid, J., B.K. Garside, J. Shewchun, M. El-Sherbiny, and E.A. Ballik. 1978a. High sensitivity point monitoring of atmospheric gases employing tunable diode lasers. Appl. Opt. 17:1806–1810.

Reid, J., J. Shewchun, B.K. Garside, and E.A. Ballik. 1978b. High sensitivity pollution detection employing tunable diode lasers. Appl. Opt. 17:300–307.

Reischl, G.P. 1991. Measurement of ambient aerosols by the differential mobility analyzer method: Concepts and realization criteria for the size range between 2 and 500 nm. Aerosol. Sci. Technol. 14:5–24.

Reischl, G.P., and W. John. 1978. The collection efficiency of impaction surfaces: A new impaction surface. Staub-Reinhalt. Luft 38:55–58.

Reynolds, S.D., P.M. Roth, and J.H. Seinfeld. 1973. Mathematical modeling of photochemical air pollution. I. Formulation of the model. Atmos. Environ. 7:1033–1061.

Reynolds, S.D., L. Reid, M. Hillyer, J.P. Killus, T.W. Tesche, R.I. Pollack, G.E. Anderson, and J. Ames. 1979. Photochemical Modeling of Transportation Control Strategies: Model Development, Performance Evaluation, and Strategy Assessment. Report EF79-37 prepared by Systems Applications, Inc. for the Federal Highway Administration, Office of Research, U.S. Department of Transportation, Washington, D.C.

Rheingrover, S.W., and G.E. Gordon. 1988. Wind-trajectory method for determining compositions of particles from major air pollution sources. Aero. Sci. Tech. 8:29–61.

Richards, L.W. 1988. Sight path measurements for visibility monitoring and research. J. Air Pollut. Control Assoc. 38:784–791.

Richards, L.W., J.A. Anderson, D.L. Blumenthal, A.A. Brandt, J.A. McDonald, N. Walters, E.S. Macias, and P.S. Bhardwaja. 1981. The chemistry, aerosol physics, and optical properties of a western coal-fired power plant plume. Atmos. Environ. 15:2111–2134.

Richards, L.W., C.L. Blanchard, and D.L. Blumenthal. 1991. Navajo Generating Station Visibility Study, Final Report. Sonoma Technology Inc., Santa Rosa, Calif. November 15. Submitted to Prem Bhardwaja, Salt River Project, Phoenix, Ariz.

Ridley, B.A., and L.C. Howlett. 1974. An instrument for nitric oxide measurements in the stratosphere. Rev. Sci. Instrum. 45:742–746.

Ridley, B.A., M.A. Carroll, G.L. Gregory, and G.W. Sachse. 1988. NO and NO_2 in the troposphere: Technique and measurements in regions of a folded tropopause. J. Geophys. Res. 93(D-12): 15813–15830.

Rosen, H., A.D.A. Hansen, R.L. Dod, L.A. Gundel, and T. Novakov. 1982. Graphitic carbon in the urban environments and the Arctic. Pp. 273–294 in Particulate Carbon: Atmospheric Life Cycle, G.T. Wolff and R.L. Klimisch, eds. New York: Plenum Press.

Rosenthal, R. 1987. Judgment Studies: Design, Analysis, and Meta-analysis. Cambridge: Cambridge University Press.

Roth, P.M., S.D. Reynolds, T.W. Tesche, P.D. Gutfreund, and C. Seigneur. 1983. An appraisal of emission control requirements in the California South Coast air basin. Environ. Int. 9:549–571.

Russell, A.G., K.F. McCue, and G.R. Cass. 1988. Mathematical modeling of the formation of nitrogen-containing air pollutants. 1. Evaluation of an Eulerian photochemical model. Environ. Sci. Technol. 22:263-271.

Russell, A.G., K.F. McCue, and G.R. Cass. 1988. Mathematical modeling of the formation of nitrogen-containing air pollutants. 2. Evaluation of the effect of emission controls. Environ. Sci. Technol. 22:1336-1347.

Samson, P.J. 1978. Ensemble trajectory analysis of summertime sulfate concentration in New York state. Atmos. Environ. 12: 1889-1894.

Samson, P.J. 1980. Trajectory analysis of summertime sulfate concentrations in the northeastern United States. J. Appl. Meteor. 19: 1382-1394.

Sandberg, D.V. 1983. Research Leads to Less Smoke from Prescribed Fires. U.S. Department of Agriculture (USDA) Forest Service Pacific Northwest Forest and Range Experiment Station, Portland, Oreg.

Savoie, D.L., and J.M. Prospero. 1982. Particle size distribution of nitrate and sulfate in marine atmosphere. Geophys. Res. Let. 9: 1207-.

Saxena, P., and C. Seigneur. 1987. On the oxidation of SO_2 to sulfate in atmospheric aerosols. Atmos. Environ. 21:807-812.

Saxena, P., A.B. Hudischewskyj, C. Seigneur, and J.H. Seinfeld. 1986. A comparative study of equilibrium approaches to the chemical characterization of secondary aerosols. Atmos. Environ. 86:1471-1484.

SCAQMD (South Coast Air Quality Management District). 1978. Sulfur Dioxide/Sulfate Control Study. South Coast Air Quality Management District, El Monte, Calif.

Schere, K.L. 1986. Evaluation O_3 predictions from a test application of the EPA regional oxidant model. Proceedings of the 5th Joint Conference on Applications of Air Pollution Meteorology, November 18-21. Chapel Hill, N.C.

Schiff, H.I., G.I. Mackay, C. Castledine, G.W. Harris, and Q. Tran. 1986. Atmospheric measurements of nitrogen dioxide with a sensitive luminol instrument. Water Air Soil Pollut. 30:105-114.

Schiff, H.I., G.W. Harris, and G.I. Mackay. 1987. Measurement of atmospheric gases by laser absorption spectrometry. Pp. 274-288 in

The Chemistry of Acid Rain-Sources and Atmospheric Processes, R.W. Johnson and G.E. Gordon, eds. American Chemical Society Symposium Series 349. American Chemical Society, Washington, D.C.

Schwarz, F.P., H. Okabe, and J.A. Whittaker. 1974. Fluorescence detection of sulfur dioxide in air at the parts per billion level. Anal. Chem. 46:1024–1028.

Schwartz, S.E. 1984. Gas-aqueous reactions of sulfur and nitrogen oxides in liquid-water clouds. Pp. 173–208 in SO_2, NO and NO_2 Oxidation Mechanisms: Atmospheric Considerations, J.G. Calvert, ed. Boston: Butterworth Publishers.

Schwoeble, A.J., A.M. Dalley, B.C. Henderson, G.S. Casuccio. 1988. Computer-controlled SEM and microimaging of fine particles. J. Meteor. 40:11–14.

Seaman, N.L., and R. Stauffer. 1989. Development of Four-dimensional Data Assimilation for Regional Dynamic Modeling Studies, Final Report. CR-814068-01-0. Atmospheric Research and Exposure Assessment Laboratory, U.S. Environmental Protection Agency, Research Triangle Park, N.C. 102 pp.

Seber, G.A.F. 1977. Linear Regression Analysis. New York: John Wiley and Sons.

Sehmel, G.A. 1980. Particle and gas dry deposition: A review. Atmos. Environ. 14:983–1011.

Seigneur, C. 1982. A model of sulfate aerosol dynamics in atmospheric plumes. Atmos. Environ. 16:2207–2228.

Seigneur, C., and P. Saxena. 1990. Status of Subregional and Mesoscale Models, Vol. 1: Air Quality Models. EN-6649, Vol. 1, Research Project 2434-6. Final report prepared by Bechtel Environmental Inc. for Pacific Gas and Electric Company, San Ramon, Calif. and Electric Power and Research Institute, Palo Alto, Calif. January.

Seigneur, C., P. Saxena, and A.B. Hudischewskyj. 1982. Formation and evolution of sulfate and nitrate aerosols in plumes. Sci. Total Environ. 23:283–292.

Seigneur, C., T.W. Tesche, P.M. Roth, and M.K. Liu. 1983. On the treatment of point source emissions in urban air quality modeling. Atmos. Environ. 17:1655–1676.

Seigneur, C., H. Hogo, and C.D. Johnson. 1984. Comparison of teleradiometric and sensitometric techniques for visibility measurements. Atmos. Environ. 18:227–233.

Seigneur, C., A.B. Hudischewskyj, J.H. Seinfeld, K.T. Whitby, E.R. Whitby, J.R. Brock, and H.M. Barnes. 1986. Simulation of aerosol dynamics: A comparative review of mathematical models. Aerosol. Sci. Technol. 5:205–222.

Seinfeld, J.H. 1986. Atmospheric Chemistry and Physics of Air Pollution. New York: Wiley Interscience.

Shah, J.J., T.J. Kneip, and J.M. Daisey. 1985. Source apportionment of carbonaceous aerosol in New York City by multiple linear regression. J. Air Pollut. Control Assoc. 35:541–544.

Shah, J.J., R.L. Johnson, E.K. Heyerdahl, and J.J. Huntzicker. 1986. Carbonaceous aerosol at urban and rural sites in the United States. J. Air Pollut. Control Assoc. 36:254–257.

Shannon, J.D. 1981. A model of regional long-term average sulfur atmospheric pollution, surface removal, and net horizontal flux. Atmos. Environ. 15:689–702.

Shareef, G.S., W.A. Butler, L.A. Bravo, and M.B. Stockton. 1988. Air Emissions Species Manual. Volume II: Particulate Matter Species Profiles. EPA 450/2-88-002. U.S. Environmental Protection Agency, Research Triangle Park, N.C.

Shaw, R.W., Jr., R.K. Stevens, and J. Bowermaster, J.W. Tesch, and E. Tew. 1982. Measurements of atmospheric nitrate and nitric acid: The denuder difference experiment. Atmos. Environ. 16:845–853.

Sheffield, A.E., and G.E. Gordon. 1986. Variability of composition from ubiquitous sources: Results from a new source-composition library. Pp. 9–22 in Receptor Methods for Source Apportionment: Real World Issues and Applications, T.G. Pace, ed. APCA Publication No. TR-5. Air Pollution Control Association, Pittsburgh, Penn.

Shi, B., J.D. Kahl, Z.D. Christidis, and P.J. Samson. 1990. Simulation of the three-dimensional distribution of tracer during the Cross-Appalachian Tracer Experiment. J. Geophys. Res. 95:3693–3703.

Shum, Y.S., W.D. Loveland, and E.W. Hewson. 1975. The use of artificial activable trace elements to monitor pollutant source strengths and dispersal patterns. J. Air Pollut. Control Assoc. 25:1123–1128.

Sisler, J.F., and W.C. Malm. 1990. Relationship of trends in regional sulfur dioxide emissions to particulate sulfate concentrations in the southwestern United States. Pp. 750–762 in Visibility and Fine Particles, C.V. Mathai, ed. A&WMA Publication No. TR-17. Air and

Waste Management Associton, Pittsburgh, Penn.

Slanina, J., M.P. Keuken, and C.A.M. Schoonebeek. 1987. Determination of sulfur dioxide in ambient air by a computer-controlled thermodenuder system. Anal. Chem. 59:2764-2766.

Slemr, F., G.W. Harris, D.R. Hastie, G.I. Mackay, and H.I. Schiff. 1986. Measurement of gas phase hydrogen peroxide in air by tunable diode laser absorption spectroscopy. J. Geophys. Res. 91: 5371-5378.

Sloane, C.S. 1982. Visibility trends. 2. Mideastern United States. Atmos. Environ. 16:2309-2322.

Sloane, C.S. 1983. Optical properties of aerosols: Comparisons of measurements with model calculations. Atmos. Environ. 17: 409-416.

Sloane, C.S. 1984. Optical properties of aerosols of mixed composition. Atmos. Environ. 18:871-878.

Sloane, C.S. 1986. Effect of composition on aerosol light scattering efficiencies. Atmos. Environ. 20:1025-1073.

Sloane, C.S. 1988. Forecasting visibility impairment: A test of regression estimates. Atmos. Environ. 22:2033-2046.

Sloane, C.S., and G.T. Wolff. 1985. Prediction of ambient light scattering using a physical model responsive to relative humidity: Validation with measurements from Detroit, Michigan. Atmos. Environ. 19:669-680.

Sloane, C.S., and W.H. White. 1986. Visibility: An evolving issue. Environ. Sci. Technol. 20:760-766.

Small, M., M.S. Germani, A.M. Small, W.H. Zoller, and J.L. Moyers. 1981. Airborne plume study of emissions from the processing of copper ores in southeastern Arizona. Environ. Sci. Technol. 15: 293-299.

Solorzano, L. 1969. Determination of ammonia in natural waters by the phenolhypochlorite method. Limnol. Oceanogr. 14:799-801.

Söderlund, R., and B.H. Svensson. 1976. The global nitrogen cycle. In Nitrogen and Sulfur-Global Cycles, R. Söderlund and B.H. Svensson, eds. SCOPE Report No. 7. Ecol. Bull. (Stockholm) 22:23-73.

South, D.W., T.E. Emmel, J.L. Gillette, C.L. Saricks, and J.T. Waddell. 1990. Technologies and Other Measures for Controlling Emissions: Performance, Costs, and Applicability. State-of- Science/ Technology Report 25, Volume IV. National Acid Precipitation

Assessment Program (NAPAP), Washington, D.C. January.

State of Vermont. 1986. Implementation Plan for the Protection of Visibility in the State of Vermont. Vermont Agency of Natural Resources, Division of Air Pollution Control, Montpelier, Vt. May.

State of Oregon, Department of Environmental Quality. 1986. Visibility Protection Program for Oregon's Class I Areas. OAR 340-20-052. Air Quality Division, Oregon Department of Environmental Quality, Portland, Oreg.

State of Oregon, Department of Environmental Quality. 1988. Field Burning Program Annual Report for 1988. Air Quality Division,, Oregon Department of Environmental Quality, Portland, Oreg. June.

State of Oregon, Department of Environmental Quality. 1990. Final Recommendations of the Oregon Visibility Advisory Committee. Air Quality Division, Oregon Department of Environmental Quality, Portland, Oreg. December 2.

State of Washington, Department of Ecology. 1983. Revisions to the Washington State Implementation Plan: Washington State's Visibility Protection Program. State of Washington Department of Ecology, Office of Hazardous Substances and Air Quality Control, Olympia, Wash. October.

State of Washington, Department of Natural Resources. 1989. Smoke Management Annual Report. State of Washington Department of Natural Resources, Olympia, Wash.

Stevens, R.K., and T.G. Pace. 1984. Review of the mathematical and empirical receptor models workshop (Quail Roost II). Atmos. Environ. 18:1499-1506.

Stevens, R.K., and C.W. Lewis. 1987. Hybrid receptor modeling. In Extended Abstracts for the Fifth Joint Conference on Applications of Air Pollution Meteorology with Air Pollution Control Association, November 18-21, Chapel Hill, N.C. Published by the American Meteorological Society, Boston, Mass.

Stevens, R.K., T.G. Dzubay, C.W. Lewis, and R.W. Shaw, Jr. 1984. Source apportionment methods applied to the determination of the origin of ambient aerosols that affect visibility in forested areas. Atmos. Environ. 18:261-272.

Stewart, D.A., and M.-K. Liu. 1981. Development and application of a reactive plume model. Atmos. Environ. 15:2377-2394.

Stewart, D.A., P. Middleton, and D. Ely. 1983. Urban visual air quality judgments: Reliability and validity. J. Environ. Psych. 3: 129-145.

Stewart, T.R., P. Middleton, M. Downton, and D. Ely. 1984. Judgments of photographs vs. field observations in studies of perception and judgment of the visual environment. J. Environ. Psych. 4:-283-302.

Stockwell, W.R., and J.G. Calvert. 1983. The mechanism of the hydroxyl-sulfur dioxide reaction. Atmos. Environ. 17:2231-2235.

Szymanski, W.W., and B.Y.H. Liu. 1986. On the sizing accuracy of laser optical particle counters. Part. Character. 3:1-7.

Tang, I.N. 1980. Deliquescence properties and particle size change of hygroscopic aerosols. Pp. 153-167 in Generation of Aerosols and Facilities for Exposure Experiments, K. Willeke, ed. Ann Arbor, Mich.: Ann Arbor Science Publishers, Inc.

ten Brink, H.M., S.E. Schwartz, and P.H. Daum. 1987. Efficient scavenging of aerosol sulfate by liquid-water clouds. Atmos. Environ. 21:2035-2052.

Thornton, D.C., A.R. Driedger, III, and A.R. Bandy. 1986. Determination of part-per-trillion levels of sulfur dioxide in humid air. Anal. Chem. 58:2688-2691.

Thurston, G.D., and J.D. Spengler. 1985. A quantitative assessment of source contributions to inhalable particulate matter pollution in metropolitan Boston, Massachusetts. Atmos. Environ. 19:9-26.

Torrens, I.M. 1990. Developing clean coal technologies. Environment 32:10-15; 28-33.

Torrens, I.M., and P.T. Radcliffe. 1990. SO_2 control in the 90s: An EPRI perspective. {p. 5-23 in Proceedings of the 1990 SO_2 Control Symposium, Vol. 1. EPRI Report GS-6963. Electric Power Research Institute, Palo Alto, Calif. September.

Trexler, E. 1990. Evaluation of Costs to Control SO_2 Emissions from Coal-Fired Power Plants Within 200 Miles of Class I PSD Areas. Draft report prepared by PEI Associates for the U.S. Department of Energy, Office of Fossil Energy, Washington, D.C. June.

Trijonis, J.C. 1972. Ph.D. Dissertation. An Economic Air Pollution Control Model: Application to Photochemical Smog in Los Angeles County in 1975. Division of Engineering and Applied Science, California Institute of Technology, Pasadena, Calif.

Trijonis, J.C. 1974. Economic air pollution control model for Los

Angeles County in 1975. Environ. Sci. 8:811-826.

Trijonis, J. 1979. Visibility in the Southwest: An exploration of the historical data base. Atmos. Environ. 13:833-843.

Trijonis, J. 1982a. Visibility in California. J. Air Pollut. Control Assoc. 32:165-169.

Trijonis, J. 1982b. Existing and natural background levels of visibility and fine particles in the rural East. Atmos. Environ. 16:2431-2446.

Trijonis, J., G. Richard, K. Crawford, R. Tan, and R. Wada. 1975. An Implementation Plan for Suspended Particulate Matter in the Los Angeles Region. TRW, Inc. report under contract number EPA/68-02-1384 by TRW, Inc., submitted to the Environmental Protection Agency, Region IX, San Francisco, Calif.

Trijonis, J.C., K. Yuan, and R.B. Husar. 1978. Visibility in the Northeast: Long-term Visibility Trends and Visibility/Pollutant Relationships. EPA 600/3-78-075. U.S. Environmental Protection Agency, Research Triangle Park, N.C.

Trijonis, J.C., M. Pitchford, M. McGown and others. 1987. Preliminary extinction budget results from the RESOLVE program. Pp. 872-883 in Visibility Protection: Research and Policy Aspects, P.J. Bhardwaja, ed. APCA Publication No. TR-10. Air Pollution Control Association, Pittsburgh, Penn.

Trijonis, J., M. McGown, M. Pitchford, D. Blumenthal, P. Roberts, W. White, E. Macias, R. Weiss, A. Waggoner, J. Watson, J. Chow, and R. Flocchini. 1988. Final Report for the RESOLVE Project: Visibility Conditions and Causes of Visibility Degradation in the Mojave Desert of California. NWC TP 6869. Naval Weapons Center, China Lake, Bloomington, Calif.

Trijonis, J., W. Malm, M. Pitchford, and W.H White. 1990. Visibility: Existing and Historical Conditions--Causes and Effects. State-of-Science/Technology Report 24, Volume III of the National Acid Precipitation Assessment Program (NAPAP), Washington, D.C. October.

Tuncel, S.G., I. Olmez, J.R. Parrington, G.E. Gordon, and R.K. Stevens. 1985. Composition of fine particle regional sulfate component in Shenandoah Valley. Environ. Sci. Technol. 19:529-537.

Turner, J.R., and S.V. Hering. 1987. Greased and oiled substrates as bounce-free impaction surfaces. J. Aerosol Sci. 18:215-224.

Turpin, B.J., and J.J. Huntzicker. 1991. Secondary formation of organic aerosol in the Los Angeles basin: A descriptive analysis of organic and elemental carbon concentrations. Atmos. Environ. 25(A):207–215.

U.S. Congress, House of Representatives. 1977. Clean Air Act Amendments of 1977--Report by the Committee on Interstate and Foreign Commerce. Report No. 95-294. Washington, D.C.: U.S. Government Printing Office.

U.S. Congress, House of Representatives. 1981. Health and the Environment Miscellaneous, Part 5. Hearings before the Subcommittee on Health and the Environment of the House Committee on Energy and Commerce. Washington, D.C.: U.S. Government Printing Office.

U.S. Congress, Senate. 1977. Clean Air Act Amendments of 1977-- Report of the Committee on Environment and Public Works. Report No. 95-127. Washington, D.C.: U. S. Government Printing Office.

USDA (U.S. Department of Agriculture, Forest Service). 1988. Managing Competing and Unwanted Vegetation: Final Environmental Impact Statement Characterizing and Management of Risk. Pacific Northwest Region, Portland, Oreg. November.

USDA (U.S. Department of Agriculture, Forest Service). 1989. The Effects of Forest Fire Smoke on Firefighters. Proposal by USDA Forest Service and Johns Hopkins University. U.S. Forest Service Pacific Northwest Region, Portland, Oreg.

Valaoras, G., J.J. Huntzicker, and W.H. White. 1988. On the contribution of motor vehicles to the Athenian nephos: An application of factor signatures. Atmos. Environ. 22:965–972.

van Dijk, C.A., S.T. Sandholm, D.D. Davis, and J.D. Bradshaw. 1989. NH ($b^1\Sigma^+$.) deactivation/reaction rate constants for the collisional gases H_2, CH_4, C_2H_6, Ar, N_2, O_2, H_2O, and CO_2. J. Phys. Chem. 93:6363–6367.

Venkatram, A., P.K. Karamchandani, and P.K. Misra. 1988. Testing a comprehensive acid deposition model. Atmos. Environ. 22:737–747.

Vincent, J.H. 1989. Aerosol Sampling: Science and Practice. Chichester, N.Y.: Wiley.

Vossler, T.L., R.K. Stevens, R.J. Paur, R.E. Baumgardner, and J.P. Bell 1988. Evaluation of improved inlets and annular denuder systems to measure inorganic air pollutants. Atmos. Environ. 22:1729–1736.

Wall, S.M., W. John, and J.L. Ondo. 1988. Measurement of aerosol size distributions for nitrate and major ionic species. Atmos. Environ. 22:1649-1656.

Walstad, J.D., S.R. Radosevich, and D.V. Sandberg, eds. 1990. Natural and Prescribed Fire in Pacific Northwest Forests. Corvallis, Oreg.: Oregon State University Press.

Wang, H.-C., and W. John. 1987. Comparative bounce properties of particle materials. Aeros. Sci. Technol. 7:285-299.

Wang, H.-C., and W. John. 1988. Characteristics of the Berner impactor for sampling inorganic ions. Aeros. Sci. Technol. 8:157-172.

Warren, G.J., and G. Babcock. 1970. Portable ethylene chemiluminescence ozone monitor. Rev. Sci. Instrum. 41:280-282.

Watson, J.G. 1979. Chemical Element Balance Receptor Model Methodology for Assessing the Sources of Fine Total Suspended Particulate Matter in Portland, Oregon. Ph.D. Dissertation, Oregon Graduate Center, Beaverton, Oreg.

Watson, J.G., J.A. Cooper, and J.J. Huntzicker. 1984. The effective variance weighting for least squares calculations applied to the mass balance receptor model. Atmos. Environ. 18:1347-1356.

Watson, J.G., J.C. Chow, and C.V. Mathai. 1989. Receptor models in air resources management: A summary of the A&WMA International Specialty Conference. J. Air Pollut. Control Assoc. 39:419-426.

Watson, J.G., N.F. Robinson, J.C. Chow, R.C. Henry, B. Kim, T. Nguyen, E.L. Meyer, and T.G. Pace. 1990a. Receptor Model Technical Series. Volume 3. CMB User's Manual, Version 7.0. EPA 450/4-90-004. Office of Air Quality Planning and Standards, U.S. Environmental Protection Agency, Research Triangle Park, N.C. 127 pp.

Watson, J.G., N.F. Robinson, J.C. Chow, R.C. Henry, B.M. Kim, T.G. Pace, Q. Nguyen, and E.L. Meyer. 1990b. EPA/DRI (Desert Research Institute) 7.0 Chemical Mass Balance Modeling Software. Environ. Software 5:38-49.

Watson, J.G., J.C. Chow, and T.G. Pace. 1991. Chemical mass balance. Pp. 83-118 in Receptor Modeling for Air Quality Management, P.K. Hopke, ed. New York: Elsevier Scientific Publishing Company.

Went, F.W. 1960. Blue hazes in the atmosphere. Nature 187:-

641–643.

Wesolowski, J.J., W. John, W. Devor, T.A. Cahill, P.H. Feeney, G. Wolfe, and R. Flocchini. 1977. Collection surfaces of cascade impactors. Pp. 121–131 in X-Ray Fluorescence Analysis of Environmental Samples, T. Dzubay, ed. Ann Arbor, Mich.: Ann Arbor Science Publishers, Inc.

Westberg, H., and B. Lamb. 1985. Ozone Production and Transport in the Atlanta, Georgia Region. EPA 600/3-85/013. U.S. Environmental Protection Agency, Research Triangle Park, N.C. 244 pp.

Whitby, K.T., and W.E. Clark. 1966. Electro-aerosol particle counting and size distribution measuring system for the 0.015 to 1.0 μm size range. Tellus 13:573–586.

White, W.H. 1977. NO_x-O_3 photochemistry in power plant plumes: Comparison of theory with observation. Environ. Sci. Technol. 11:995–1000.

White, W.H. 1986. On the theoretical and empirical basis for apportioning extinction by aerosols: A critical review. Atmos. Environ. 20:1659–1672.

White, W.H. 1989a. Heteroscedasticity and the standard errors of regression estimates. Pp. 226–241 in Receptor Models in Air Resources Management, J.G. Watson, ed. A&WMA Publications No. TR-14. Air and Waste Management Association, Pittsburgh, Penn.

White, W.H. 1989b. MLRCLS: A BASIC routine for multiple regression by corrected least squares. Pp. 269–275 in Receptor Models in Air Resources Management, J.G. Watson, ed. A&WMA Publication No. TR-14. Air and Waste Management Association, Pittsburgh, Penn.

White, W.H. 1990. The components of atmospheric light extinction: A survey of ground-level budgets. Atmos. Environ. 24:2673–2680.

White W.H., and P.T. Roberts. 1977. On the nature and origins of visibility-reducing aerosols in the Los Angeles air basin. Atmos. Environ. 11:803–812.

White, W.H., and D.E. Patterson. 1981. On the relative contributions of NO_2 and particles to the color of smoke plumes. Atmos. Environ. 15:2097–2104.

White, W.H., and E.S. Macias. 1987a. On measurement error and the empirical relationship of atmospheric extinction to aerosol composition in the non-urban West. Pp. 783–794 in Visibility Protection:

Research and Policy Aspects, P.S. Bhardwaja, ed. APCA Publication No. TR-10. Air Pollution Control Association, Pittsburgh, Penn.

White, W.H., and E.S. Macias. 1987b. Uncertainties in the atmospheric transparency inferred from teleradiometry of natural targets. Pp. 499–509 in Visibility Protection: Research and Policy Aspects, P.S. Bhardwaja, ed. APCA Publication No. TR-10. Air Pollution Control Association, Pittsburgh, Penn.

White, W.H., and E.S. Macias. 1989. Carbonaceous particles and regional haze in the western United States. Aerosol Sci. Technol. 10:111–117.

White, W.H., and E.S. Macias. 1990. Light scattering by haze and dust at Spirit Mountain, Nevada. Pp. 914–922 in Visibility and Fine Particles, C.V. Mathai, ed. A&WMA Publication No. TR-17. Air and Waste Management Association, Pittsburgh, Penn.

White, W.H., and E.S. Macias. 1991. Chemical-mass balancing with ill-defined sources: Regional apportionment in the California desert. Atmos. Environ. 25:1547–1557.

White, W.H., S.L. Heisler, R.C. Henry, G.M. Hidy, and I. Straughan. 1978. The same-day impact of power plant emissions on sulfate levels in the Los Angeles air basin. Atmos. Environ. 12:779–784.

White, W.H., D.E. Patterson, and W.E. Wilson, Jr. 1983. Urban exports to the nonurban troposphere. Results from Project MISTT. J. Geophys. Res. 88:10745–10752.

White, W.H., C. Seigneur, D.W. Heinold, M.W. Eltgroth, L.W. Richards, P.T. Roberts, P.S. Bhardwaja, W.D. Conner, and W.E. Wilson, Jr. 1985. Predicting the visibility of chimney plumes: An intercomparison of four models with observations at a well-controlled power plant. Atmos. Environ. 19:515–528.

White, W.H., C. Seigneur, D.W. Heinold, L.W. Richards, W.E. Wilson, Jr., and P.T. Roberts. 1986. Radiative transfer budgets for scattering and absorbing plumes: Measurements and model predictions. Atmos. Environ. 20:2243–2257.

White, W.H., E.S. Macias, D.F. Miller, D.E. Schorran, T.E. Hoffer, and D.P. Rogers. 1990. Regional transport of the urban workweek: Methylchloroform cycles in the Nevada-Arizona desert. Geophys. Res. Let. 17:1081–1084.

Whitten, G.Z., H. Hogo, and J.P. Killus. 1980. The carbon-bond mechanism: A condensed kinetic mechanism for photochemical smog. Environ. Sci. Technol. 14:690–700.

Wiebe, H.A., K.G. Anlauf, E.C. Tuazon, A.M. Winer, H.W. Biermann, B.R. Appel, P.A. Solomon, G.R. Cass, T.G. Ellestad, K.T. Knapp, E. Peake, C.W. Spicer, and D.R. Lawson. 1990. A comparison of measurements of atmospheric ammonia by filter packs, transition-flow reactors, simple and annular denuders and Fourier transform infrared spectroscopy. Atmos. Environ. 24A:1019–1028.

Williams, M.D., E. Treiman, and M. Wecksung. 1980. Plume blight visibility modeling with a simulated photograph technique. J. Air Pollut. Cont. Assoc. 30:131–134.

Wilson, W.E., Jr. 1981. Sulfate formation in point source plumes: A review of recent field studies. Atmos. Environ. 15:2573–2581.

Wilson, J.C., and P.H. McMurry. 1981. Secondary aerosol formation in the Navajo power plant plume. Atmos. Environ. 15:2329–2340.

Winchester, J.W., and G.D. Nifong. 1971. Water pollution in Lake Michigan by trace elements from pollution aerosol fallout. Water, Air Soil Poll. 1:50–64.

Winklmayr, W. 1987. Ph.D. Dissertation. University of Vienna, Vienna, Austria.

Wolff, G.T. 1984. On the nature of nitrate in coarse continental aerosols. Atmos. Environ. 18: 977–982.

Wolff, G.T., R.J. Countess, P.J. Groblicki, M.A. Ferman, S.H. Cadle, and J.M. Muhlbaier. 1981. Visibility-reducing species in the Denver brown cloud--II. Sources and temporal patterns. Atmos. Environ. 15:2485–2502.

Wolff, G.T., N.A. Kelly, and M.A. Ferman. 1982. Source regions of summertime ozone and haze episodes in the eastern United States. Water, Air, Soil Pollut. 18:65–81.

Wolff, G.T., C.M. Stroup, and D.P. Stroup. 1983. The coefficient of haze as a measure of particulate elemental carbon. J. Air Pollut. Control Assoc. 33:746–750.

Young, J.R., C. Ellis, and G.M. Hidy. 1988. Deposition of air-borne acidifiers in the western environment. J. Environ. Qual. 17:1–26.

Zhang, X.Q. 1990. Measurements of size resolved atmospheric aerosol chemical composition with impactors: Data integrity and applications to visibility. Ph.D. Dissertation, Department of Mechanical Engineering, University of Minnesota, Minneapolis, Minn.

Zhang, X.Q., and P.H. McMurry. 1987. Theoretical analysis of evaporative losses from impactor and filter deposits. Atmos. Environ. 21:1779–1790.

Zhang, D., Y. Maeda, and M. Munemori. 1985. Chemiluminescence method for direct determination of sulfur dioxide in ambient air. Anal. Chem. 57:2552–2555.

Zimmerman, D., W. Tax, M. Smith, J. Demmy, and R. Battye. 1988a. Anthropogenic Emissions Data for the 1985 NAPAP Inventory. EPA 600/7-88-002. Environmental Protection Agency, Washington, D.C. 295 pp.

Zimmerman, P.J., J.P. Greenberg, and C.E. Westberg. 1988b. Measurements of atmospheric hydrocarbons and biogenic emission fluxes in the Amazon boundary layer. J. Geophys. Res. 93:1407–1416.

Appendix A

Scientific Background Information

ATMOSPHERIC CHEMISTRY AND SECONDARY PARTICLE PRODUCTION

Sulfate Aerosol Chemistry

Sulfate (SO_4^{2-}) airborne particles are predominantly composed of sulfuric acid (H_2SO_4) and its salts, NH_4HSO_4, $(NH_4)_2SO_4$, etc. Only a small fraction of the particles are emitted directly as particulate matter from the sources; most are produced in the atmosphere by the oxidation of sulfur-containing gases emitted into the air by both natural and anthropogenic sources. Sulfur dioxide (SO_2) is by far the most important gas precursor to SO_4^{2-} in the atmosphere over North America. The compounds dimethyl sulfide (CH_3SCH_3), hydrogen sulfide (H_2S), dimethyl disulfide (CH_3SSCH_3), and related species are derived from natural sources and undergo rapid oxidation in the troposphere to provide a small background source of SO_2, which amounts to about 6% of the SO_2 found over North America (Placet and Streets, 1987). However, the major sources of SO_2 are anthropogenic. Most important is the combustion of sulfur-containing fuels (primarily coal) by electric utility boilers, which account for almost three-fourths of SO_2 emissions. Emissions from smelters and other combustion sources account for most of the remainder.

Gas-phase and liquid-phase oxidation processes are important in sulfate production. The major gas-phase oxidation pathway involves oxidation by the hydroxyl radical (HO), although other transient species can

315

also contribute (Calvert and Stockwell, 1984). Ozone (O_3) is a major source of the HO radical through its photodecomposition to the highly reactive, excited oxygen atom, $O(^1D)$, which reacts with water vapor:

$$O_3 + h\upsilon(\lambda < 305nm) \rightarrow O(^1D) + O_2, \qquad (A\text{-}1)$$

$$O(^1D) + H_2O \rightarrow 2HO. \qquad (A\text{-}2)$$

The major elementary reactions that occur in the homogeneous atmospheric oxidation of SO_2 are well established (Stockwell and Calvert, 1983; Calvert and Stockwell, 1984; Margitan, 1984; Calvert et al., 1985; Gleason et al., 1987). The first of those, Reaction A-3, involves the HO radical formed in Reaction A-2:

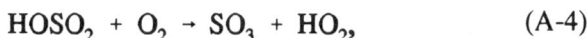

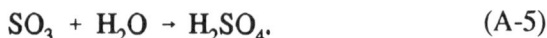

$$HO + SO_2 \ (+M) \rightarrow HOSO_2 \ (+M), \qquad (A\text{-}3)$$

$$HOSO_2 + O_2 \rightarrow SO_3 + HO_2, \qquad (A\text{-}4)$$

$$SO_3 + H_2O \rightarrow H_2SO_4. \qquad (A\text{-}5)$$

Reactions A-4 and A-5 occur rapidly following the rate-determining step (A-3). When coupled with Reaction A-6, Reactions A-3 and A-4 constitute the elements of a $HO\text{-}HO_2$ radical chain propagation sequence; that is, although an HO radical is removed in Reaction A-3, another is regenerated in Reaction A-6:

$$HO_2 + NO \rightarrow HO + NO_2. \qquad (A\text{-}6)$$

H_2SO_4 formed in Reaction A-5 condenses on airborne particles because of its low vapor pressure. Field data and theoretical analyses show that secondary SO_4^{2-} produced by gas-phase reactions tend to accumulate on particles smaller than about 0.3 μm in diameter (Wilson and McMurry, 1981; Hering and Friedlander, 1982; McMurry and Wilson, 1983; John et al., 1990).

SO_4^{2-} particles also can be produced by the oxidation of dissolved SO_2 (largely present as bisulfite ion, HSO_3^{-1}) within cloud or fog droplets (Penkett et al., 1979; Martin, 1984; Schwartz, 1984; Hoffmann and Jacob, 1984; Calvert et al., 1985). The major oxidants, H_2O_2 and O_3, are the secondary products formed in gas-phase reactions initiated by the irradiation of air masses containing NO, NO_2, and hydrocarbons. Those compounds, as well as molecular oxygen in the presence of certain transition metal catalysts, can oxidize bisulfite to sulfuric acid and ultimately to the various partially neutralized species (principally ammonium bisulfate) that are the common aerosol SO_4^{2-} (Hoffmann and Jacob, 1984).

The formation of sulfates by liquid-phase reactions occurs in several steps. First, airborne particles (cloud condensation nuclei) are activated to produce much larger (1-50 μm) fog or cloud droplets (Pruppacher and Klett, 1978). The SO_2 and oxidant then dissolve in the droplet and react to produce SO_4^{2-}. When the droplets evaporate, the resulting airborne particle is larger than the original particle due to the addition of the new SO_4^{2-}.

Recent work has shown that two kinds of SO_4^{2-} particle size distributions are found in the accumulation mode (Hering and Friedlander, 1982; McMurry and Wilson, 1983; John et al, 1990). Under dry conditions, when gas-phase reactions are likely to dominate, secondary SO_4^{2-} tend to accumulate in particles smaller than about 0.3 μm. In contrast, under humid conditions, when aerosol processing by clouds or fog is likely, SO_4^{2-} particles tend to range in size from 0.5 to 1.0 μm. The investigators also have shown theoretically that SO_4^{2-} produced by gas-phase reactions accumulate in particles smaller than those produced by liquid-phase reactions. Thus, both theory and atmospheric observations show a close coupling between SO_4^{2-} particle size distributions and the chemical mechanism of sulfur oxidation.

Nitrate Aerosol Chemistry

Nitrate airborne particles also can be produced by several mechanisms. One major mechanism of nitrate formation involves the gas-phase reaction of NO_2 with HO to produce nitric acid (Calvert et al., 1985):

$$HO + NO_2\ (+M) \rightarrow HNO_3\ (+M). \qquad (A\text{-}7)$$

Nitric acid also can be formed in heterogeneous chemistry, which occurs largely at night. That involves the reaction of gaseous dinitrogen pentoxide (N_2O_5) with aqueous aerosols:

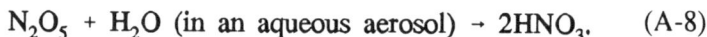

$$N_2O_5 + H_2O\ \text{(in an aqueous aerosol)} \rightarrow 2HNO_3. \qquad (A\text{-}8)$$

About one in every 10-20 collisions between an aqueous aerosol and the N_2O_5 gas-phase molecule results in nitric acid generation (Mozurkewich and Calvert, 1988). N_2O_5 is formed through the reaction involving NO_2 and O_3:

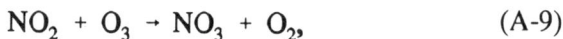

$$NO_2 + O_3 \rightarrow NO_3 + O_2, \qquad (A\text{-}9)$$

$$NO_3 + NO_2\ (+M) \rightarrow N_2O_5\ (+M). \qquad (A\text{-}10)$$

During daylight hours the sequence of Reactions A-9, A-10, and A-8 is relatively unimportant since the transient NO_3 radical is kept at very low concentrations through its decomposition by sunlight (Magnotta and Johnston, 1980).

The gaseous nitric acid can react at basic airborne particle surfaces to form nitrate salts (Seinfeld, 1986). For example, a particle containing calcium carbonate, can neutralize the nitric acid to produce calcium nitrate. It follows that nitrate size distributions depend, in part, on the size distributions of the particles on which they react. Reactions with sea salt or soil dust material tend to produce coarse particle nitrates (particle diameter > 2 μm).

Ammonium Aerosol Chemistry

Gaseous ammonia (NH_3), is the principal gas-phase neutralizing species in the atmosphere, and, as such, its distribution in the particle phase is closely linked to that of the predominant acid aerosol component, SO_4^{2-}. Consequently, most of the particulate NH_4^+ is found in the submicron-particle-size fraction. Ammonium or acid SO_4^{2-} are believed to be the most abundant inorganic components of cloud condensation nuclei. Nonetheless, the acidic aerosol components are seldom fully neutralized by NH_4^+, even in relatively remote areas; consequently, stoichiometric ratios that approximate NH_4HSO_4 are common.

When the levels of NH_3 and HNO_3 are sufficiently high (e.g., each above about 1 ppbv at 298 K), ammonium nitrate (NH_4NO_3) can be formed. Ammonium nitrate is often found in submicron particles in locations such as Denver or Los Angeles (Wall et al., 1988; Pierson and Brachaczek, 1988). For example, John et al. (1990) have shown that particulate nitrate measured in Los Angeles accumulates in three distinct modes with mass mean aerodynamic diameters of 0.2 $\pm$ 0.1 μm, 0.7 $\pm$ 0.2 μm, and 5 $\pm$ 1 μm. However, little submicron ammonium nitrate is typically found in continental air, where ammonia concentrations are low and acid sulfate concentrations are high. When ammonium nitrate reacts with the acidic particles, such as sulfuric acid, the ammonium is retained and the nitric acid is released. The nitric acid tends to react with and be retained by coarse particle cations, including calcium and sodium (Wolff, 1984).

Organic Aerosol Chemistry

Secondary organic particle formation is poorly understood in comparison to the formation of SO_4^{2-} and nitrates. There are several reasons for that. First, the problem is far more complex, involving many organic gas-phase particle precursors, each with a variety of possible particle formation pathways and products. Little is known about the contributions of primary and secondary organics and of anthropogenic and natural sources to ambient loadings, especially for remote areas.

Second, methods for sampling and analysis of organic gases and particulate matter are less well developed than those for nitrates and SO_4^{2-}. The sampling of organic materials can be very difficult; organic particles that are collected on filters can lose volatile components during the collection period, and reactive particulate organics can undergo chemical changes after collection on filters, in flasks, or on adsorption media. Problems are especially severe in pristine areas where low concentrations lead to large relative errors because of sampling artifacts.

The suggested pathways for secondary organic particle formation primarily involve gas-phase reactions with ozone, the hydroxyl radical (OH), and the nitrate radical (NO_3). Grosjean and Seinfeld (1989) have summarized what is known about the potential for airborne particle formation of organic gases. They show that alkenes and aromatic hydrocarbons tend to be effective particle producers in photochemical systems, but alkanes and carbonyl compounds are relatively ineffective. Within any given chemical class, the particle-formation potential tends to increase with increasing molecular weight and increasing number of polar groups (such as nitrate, hydroxyl, etc.), because those products are less volatile.

The contributions of terpenes and isoprene emitted by plants to secondary organic particles are uncertain. However, it is important to attempt to judge the extent of that contribution, as it may contribute to the background particle loadings and to light scattering in national parks and wilderness areas. The particle-formation potential of the monoterpenes was recognized as early as 1960 (Went, 1960). More recent studies by Kamens et al. (1981, 1982), Hatakeyama et al. (1989), and Pandis et al. (1991) show that a significant fraction of the α- and ß-pinenes, which react with O_3 or HO radicals in the atmosphere, can lead to organic particle formation, but isoprene forms very little particulate matter under similar conditions (Pandis et al., 1991).

Based on the study of the time dependence of the number and size distribution of organic airborne particles formed in isoprene and ß-pinene photooxidation at low, ambient concentrations, estimates were made by Pandis et al. (1991) to show that the potential organic particulate formed from the monoterpenes could be significant in three types of environments:

• In poorly ventilated urban areas such as Los Angeles, which has extensive urban landscaping and brush-covered hills, natural hydrocarbons might be responsible for up to 50% of secondary organic particles.

• In urban areas such as Atlanta, which has extensive wooded areas, natural sources could produce approximately 27 Mg of organic airborne particles per day, probably an order of magnitude greater than the anthropogenic secondary organic particles formed.

• In highly wooded areas, typical of many national parks and wilderness areas, 2,000 $\mu g/m^2$-hr of reactive organic hydrocarbons are emitted throughout the day (Hov et al., 1983). Pandis et al. (1991) have estimated that stagnant conditions in those areas could lead to organic particle concentrations of 30-39 $\mu gC/m^3$ on the second or third night of stagnant conditions.

Although such estimates suggest that secondary organic compounds can contribute significantly to visibility impairment, there is little experimental evidence that permits quantitative estimates of the relative contributions of primary and secondary compounds to organic particle formation on a national scale. Gray et al. (1986) found that, on a long-term average, organic carbon particles in Los Angeles originate largely from primary emissions, even during the summer when there is extensive secondary particle formation by photochemical reactions. However, during smog episodes, secondary organic particles can be the largest contributor to secondary organic levels. Turpin and Huntzicker (1991) have reported that secondary particles constitute about 70% of the Los Angeles organic particle fraction during the afternoon on smoggy summer days. Similar studies have not been done in Class I areas.

Because organic carbon often contributes significantly to optical extinction in such areas (e.g., Ouimette and Flagan, 1982), present understanding of its primary and secondary origins should be improved. That will require improvements in the sampling and analysis of carbon-containing particulate constituents (see Appendix B) as well as a more satisfactory understanding of atmospheric processes.

METEOROLOGY

In some areas, visibility degradation often is associated with the transport of plumes, both individually and collectively as a polluted air mass, from specific source regions. The magnitude of the visibility degradation is controlled in part by the degree of dispersion of plumes. To assess the sources of pollutants that affect a region and to predict the degree of visibility degradation that might be expected, accurate measurements of a number of atmospheric characteristics are needed.

Atmospheric transport can be estimated by (1) interpolating wind data obtained from meteorological stations; (2) extrapolating the measured wind field by means of a system of hydrodynamic equations, incorporating the appropriate initial and boundary conditions; or (3) using physical or chemical tracers characteristic of a specific source (or types of sources). The transport of visibility-reducing plumes often occurs within a layer located a few hundred meters or more above the surface of the earth. Consequently, wind data are required not only at the surface but also throughout the transport layer. The following sections outline the measurements needed to estimate plume transport through the interpolation of observations.

Surface-Wind Measurements

Surface-wind measurements are the easiest to obtain and can be useful in determining plume transport. Surface winds are most likely to be representative of the flow of air aloft in the afternoon when thermally driven convection couples the surface wind to the winds aloft. Normally, in pollution dispersion studies, unmanned meteorological stations are used to measure surface winds. The stations are set up in a relatively dense network. Wind data from the network are integrated with those obtained from existing weather service stations (such as those commonly located at airports), which provide information on a larger scale. Remote systems, such as the NCAR PAM stations, automatically collect information on wind speed and direction, humidity, and solar radiation. The data are periodically transmitted to satellites for relay to a primary data collection site. The stations can be powered by solar cells for use at remote locations.

Estimates of atmospheric transport based solely on surface winds are subject to interpolation errors in the horizontal direction and to extrapolation errors in the vertical direction. Horizontal interpolation errors are particularly sensitive to topographic features; such errors are a major problem in visibility studies performed in national parks and wilderness areas, because those areas are often located in mountainous regions. Consequently, surface-wind networks should be regarded as a necessary, but not a sufficient, component in studies that evaluate pollutant transport.

Non-Surface-Wind Measurements

Balloon Studies

Balloon studies are commonly used to measure winds above the surface. Measurements are made with tethered balloons, free-rise balloons, and constant altitude balloons. Tethered balloons, with attached tether-sondes, can be used to measure winds, temperature, and humidity up to altitudes of about 2 km during light wind conditions; however, these balloons become unstable at wind speeds above roughly 8 m/sec. Free-rise balloons, including instrumented units (e.g., rawinsondes) and those that are tracked optically (pibals), can be flown in all weather conditions.

Constant altitude balloons (tetroons) ride at a fixed altitude and are used to simulate the trajectory of an air mass. Tetroons can deviate from their preassigned altitude, particularly during convection, thereby reducing their being representative of atmospheric transport. The accuracy of tetroons is also difficult to verify, as there is no independent measure of transport.

Although balloons are generally less expensive than other techniques, they are more labor-intensive. Another shortcoming of balloons is that they provide little information about the magnitude of atmospheric turbulence, an important factor in pollution dispersion.

Tower Studies

Instrumented towers are used to obtain information about atmospheric transport and turbulence. The measurements are usually limited to heights of a few tens of meters; therefore, tower winds generally are not representative of the winds throughout the plume layer. However, towers can furnish data on the variation in wind direction and speed with height; those data provide a measure of the dispersing capacity of the atmosphere.

Remote Sensing

Radar and lidar techniques can be used to provide data for wind measurements at rates that are orders-of-magnitude smaller than those available from operational measurement systems. Given the need for wind fields with high temporal resolution, particularly in areas of complex terrain, this technology is promising. However, it can only measure winds at one location, so spatial interpolation is still necessary. Also, the technique is expensive and requires a technically well-trained staff. With the continuing advances in computers and hardware, however, the cost of remote sensing systems will inevitably decrease.

Ideally, a meteorological measurement program would include the use of remote sensing to measure temporal variations of winds above the surface. Remote sensing would be complimented by balloons, either tethered or free-rise, depending on the problem being studied.

Moisture Measurements

The extinction of light by hygroscopic airborne particles generally increases with increasing relative humidity because of the increase in particle size. Consequently, detailed information about humidity must be obtained during field programs. The relative humidity (or the dew point temperature) should be measured at all surface meteorological stations and in vertical profiles from balloons as well. Such measurements are usually inexpensive and can provide information for testing the ability of meteorological models to reproduce the vertical and spatial extent of water vapor transport; that information also is needed for modeling the wet removal of pollutants.

Appendix B

Measurement Methods

GAS-PHASE CHEMICAL MEASUREMENTS

Certain trace gases are important in airborne particle formation, light scattering, and light absorption in the troposphere. The key gases are sulfur dioxide (SO_2), nitric oxide (NO), nitrogen dioxide (NO_2), ozone (O_3), ammonia (NH_3), hydrogen peroxide (H_2O_2), and the non-methane hydrocarbons (see Appendix A). A knowledge of their concentration distributions is essential for applying modeling methods and is necessary for developing mechanistic models. There generally are accepted methods for measuring those gases in the atmosphere. The methods are considered briefly in this appendix.

Gaseous Sulfur Dioxide

Sulfur dioxide is the precursor to sulfuric acid (H_2SO_4), ammonium bisulfate (NH_4HSO_4), and ammonium sulfate ($(NH_4)_2SO_4$). In most regions, (SO_4^{2-}) are the single most important class of particles responsible for visibility degradation. Methods for measuring SO_2 that have proven successful in the field are flame photometric instruments (e.g., Maroulis et al., 1980; McGaughey and Gangwal, 1980; Garber et al., 1983; Thornton et al., 1986); fluorescence analyzers (Schwarz et al., 1974); pulsed fluorescence analyzers (Maffiolo et al., 1982); single photon laser-induced fluorescence (Bradshaw et al., 1982); mass spectrometric analysis (Driedger et al., 1987); filter pack systems (e.g., Quinn and Bates, 1989) coupled with colorimetric methods or ion chromatography; permeation sampling (McDermott et al., 1979); diffusion scrubber (Lindgren and Dasgupta, 1989); diffusion denuder (Possanzini et al., 1983; Keuken et al., 1988; Vossler et al., 1988); thermo-denuder systems (e.g., Slanina et al., 1987); and chemiluminescence (e.g., Zhang et al., 1985).

When SO_2 is present at very low concentrations (< 0.1 ppb), a sample can be accumulated by cryo-trapping an airstream over a period of several minutes; the trapped SO_2 in the sample is then separated by gas chromatography and detected by one of the methods named above. Commercial instrumentation is available, but the instruments often must be modified to improve their sensitivity in relatively clean air.

Oxides of Nitrogen

The oxides NO and NO_2, collectively referred to as NO_x, are important in the generation of nitric acid (HNO_3) in the troposphere. HNO_3 can react with various bases, such as NH_3, to form light-scattering airborne particles, such as NH_4NO_3. NO is commonly analyzed through the chemiluminescence of excited NO_2 molecules formed in the NO-O_3 reaction (Ridley and Howlett, 1974; Ridley et al., 1988). Commercial instruments are available for NO, but they must be modified to measure NO at low levels encountered in relatively clean air (less than 0.1 ppb) (Dickerson et al., 1984).

The other common oxide of nitrogen, NO_2, is also important in airborne particle formation. NO_2 is a strong absorber of visible and ultraviolet light and can thereby contribute to haze. However, because of its high reactivity and relatively short lifetime, NO_2 does not normally contribute significantly to haze in remote areas; it is a problem only in areas close to sources.

NO_2 can be analyzed in air by several instrumental methods: photolysis and chemiluminescence (Kley and McFarland, 1980); tunable diode laser spectrometers (Reid et al., 1978a,b; Schiff et al., 1987); and chemiluminescence from the NO_2 luminol reaction (Schiff et al., 1986, 1987). A field intercomparison of the three methods has shown their utility and their limitations (Fehsenfeld et al., 1990). In some commercial instruments, NO_2 is estimated by first using NO-O_3 chemiluminescence to measure NO in an ambient air sample; a sample of ambient air is then subjected to a reduction cycle (with heated molybdenum oxide catalyst) that reduces NO_2 to NO. The concentration of NO_2 in the air is estimated by taking the difference between the two measurements. However, that procedure is subject to errors because the reduction cycle will generate NO from nitrogen compounds other than NO_2 (e.g., peroxyacetyl nitrate, PAN). The NO_2 luminol method is subject to small interferences from PAN and O_3.

Ozone

Ozone is a critical gas because it is responsible for initiating much of the chemistry that leads to the generation of secondary airborne particles. O_3 can be measured easily by several available techniques, including chemiluminescence from the reaction with either ethylene (Nederbragt et al., 1965; Warren and Babcock, 1970) or nitric oxide (Fontijn et al., 1970) and absorption spectroscopy in the 253.7 nm region (Bowman and Horak, 1972; de Vera et al., 1974). Commercial instruments for each of those methods are available; the sensitivity of some of the instruments is sufficient for use in remote "background" areas.

Ammonia

Ammonia gas is the only major basic gas-phase substance in the troposphere; as such it is important in the generation of sulfate and nitrate particles through the neutralization of sulfuric and nitric acid. NH_3 can be measured by a variety of techniques: photo-fragmentation and laser-induced fluorescence (van Dijk et al., 1989); tungstic oxide (WO_3) denuder sampling followed by chemiluminescence detection (Braman et al., 1982, 1986; LeBel et al., 1985); molybdenum oxide annular denuder sampling followed by chemiluminescence detection (Langford et al., 1989); citric acid coated denuder sampling followed by liquid extraction and ion chromatographic analysis (R.B. Norton, pers. comm., Aeronomy Lab, NOAA, Boulder, Colo., 1991, modified from Ferm, 1979); and filter pack sampling followed by liquid extraction and colorimetric (Solorzano, 1969; Harwood and Kuhn, 1970) or ion chromatographic analysis.

Hydrogen Peroxide

Hydrogen peroxide gas is important in the generation of sulfuric acid through its oxidation of $SO_2(HSO_3^-)$ in fog and cloud water. The subsequent evaporation of fog or cloud droplets results in the formation of particles of sulfuric acid or its salts. H_2O_2 can be measured by several techniques: scrubbing coupled with fluorometric detection (Lazrus et al., 1985, 1986); diffusion scrubbing with fluorescence detection (Dasgupta et al., 1986; Hwang and Dasgupta, 1986); tunable diode laser absorption spectrometry (Slemr et al., 1986); and glass impinger collection followed by luminol detection (Kok et al., 1978). Those techniques have been recently compared (Kleindienst et al., 1988); the results provide guidance about the suitability of each for specific environmental conditions.

Gas-Phase Organic Carbon Compounds

Volatile organic compounds (VOCs) are present in great variety in the atmosphere. Both natural and anthropogenic emissions are important sources of VOCs. Many of them play a key role in atmospheric chemistry because their oxidation products react with the oxides of nitrogen to generate ozone; indeed, in many polluted regions, high concentrations of ozone are attributable largely to those reactions (NRC, 1991b). The hydrocarbons usually are analyzed with gas chromatographic separation and flame ionization detection of the individual components (Greenberg and Zimmerman, 1984; Westberg and Lamb, 1985; Christian and Reilly, 1986; Zimmerman et al., 1988b; Rasmussen and Khalil, 1988). Mass spectrometers to identify components following chromatography have proved to be useful (Penkett, 1982).

Conclusions

It is often difficult to use conventional techniques to analyze air samples for gaseous substances in remote areas such as national parks and wilderness areas because of the complexity of the equipment required and the lack of electrical power. Portable battery-operated systems have been designed to detect NO_2, NO, O_3, and SO_2. Alternatively, certain simple filter systems can be used readily in remote areas, or the air can be collected in stainless steel, electropolished canisters and returned to the laboratory for analysis where more complex but more accurate methods can be used. In the latter case, care must be taken to avoid loss of trace gases during storage in the canisters (Greenberg and Zimmermann, 1984).

Many of the commercial instruments designed to monitor impurities at elevated concentrations in stack gases are not sensitive enough to monitor the same impurities at the low concentrations usually encountered in the atmosphere. The analysis of atmospheric trace gases requires highly specialized equipment which satisfies rather specific and demanding requirements:

• The instrument should minimize interferences from other components of the atmosphere.
• The sensitivity of the instrument should be sufficient to track a given trace gas at the very low levels commonly present in the atmosphere.
• The stability of the instrument should be good enough to allow many hours of unattended operation.

In the review of the alternative instrumentation for the analysis of particle precursor gases given above, the types of equipment available today which satisfy these stringent requirements have been discussed very briefly. Further guidance on the choice of instrumentation to satisfy particular needs can be gained from the references cited in this section.

MEASUREMENTS OF
ATMOSPHERIC PARTICLES

Airborne particle measurements are critical in any visibility program for several reasons:

• Fine particles (i.e., smaller than 2.5 μm in diameter) are responsible for most visibility impairment. The measurement of fine-particle mass concentrations can provide a first-order means of tracking trends in visibility impairment and therefore can be used as a basis for regulatory action.

• The chemical composition of the aerosol yields information about the amount of extinction due to each pollutant (e.g., sulfates, organic carbon, and soil dust).

• When used with receptor models, the chemical composition of the aerosol can provide clues about its origin. Apportioning extinction to various groups of sources is an essential step in targeting emission controls.

For those reasons, airborne particle measurements are a major component of any field study of visibility impairment.

Particle Size Distribution

Single-particle electro-optical counters and electrical mobility analyzers are the most commonly used instruments for measuring size distributions of atmospheric particles. An extensive review of instrumentation is given in ACGIH (1989).

Single-Particle Optical Counters

Single-particle optical counters (OPCs) measure the amount of light scattered by individual particles as they flow through a volume illuminated by a laser or white light source. The minimal detection limit for OPCs varies with instrument design but typically is in the range of 0.1-0.5 μm. The response of a given OPC to a given particle depends on the characteristics of the light source (wavelength distribution and polarization) and on the geometry of optics used to collect the scattered light. For homogeneous spherical particles of known refractive index, typically there is good agreement between measured and predicted OPC response and particle size.

There are several major difficulties in using OPCs to measure size distributions of airborne particles. If a large fraction of the measured particles are nonspherical, generally it is not possible to measure size distribution accurately. Also, the refractive index of sampled atmospheric particles generally is unknown. To relate measurements to size distributions, a refractive index must be assumed. Indeed, measurements have shown that atmospheric particles of a given size can contain an external mixture of particles having two or more refractive indexes (Hering and McMurry, 1991; Covert et al., 1990). Even when particles are chemically alike, particles of different sizes can produce an identical OPC response (e.g., Szymanski and Liu, 1986). That phenomenon is especially likely for particles in the optically sensitive 0.5-1.0 μm diameter range. Finally, heating within the instrument can drive off volatile components, especially water; that will lead to a measured size distribution that is smaller than that in the ambient air and, consequently, to an erroneous estimate of the effect of the particles on visibility (Biswas et al., 1987).

These difficulties limit the use of OPCs for measuring airborne particle size distributions. Nevertheless, because the OPC provides real-time information and can operate in the optically sensitive 0.1-1.0 μm size range, the OPC is one of the most convenient instruments available for measuring of physical size distributions.

Electrical Mobility Analyzers

The electrical mobility of a particle in an electric field depends on the particle charge, size, and shape. The size and electrical mobility of spherical particles with a known charge are uniquely related. Therefore, electrical mobility measurements can be used to determine the diameters of such particles.

There are two types of electrical mobility analyzers. The electrical aerosol analyzer (EAA) (Whitby and Clark, 1966; Liu and Pui, 1975) measures the electrical mobility distribution of airborne particles that have been charged by exposure to a unipolar cloud of small ions. The unipolar charger is used to put the maximal possible charge on the particles. One limitation of the EAA is that particles of a given size emerging from the charger contain a distribution of charges rather than a unique charge. Therefore, there is no unique relation between particle size and electrical mobility. For particles smaller than about 0.1-0.3

μm, various charges do not present a major problem because the range of charges on particles of that size is narrow. However, for larger particles charge distribution broadens and the size sensitivity of electrical mobility decreases. It is difficult, therefore, to obtain accurate information on the size distribution of particles larger than about 0.3 μm. Because OPCs perform best for particles larger than 0.3 μm and EAAs perform best for smaller particles, both instruments are often used to measure overall particle size distributions.

The differential mobility analyzer (DMA), also referred to as the electrostatic classifier, can be used to determine atmospheric particle size distributions from measurements of electrical mobility distributions (Liu and Pui, 1975; Knutson and Whitby, 1975; Fissan et al., 1983; Hoppel et al., 1990). Two major differences between the DMA and the EAA are that (1) particles entering the DMA are typically charged with a bipolar ion charger, in which particles are exposed to an ion cloud containing both positive and negative ions; and (2) the DMA detects particles in a narrow mobility window rather than particles with mobilities below some minimal value, as is done with the EAA. The net effect of the differences is that size distributions of particles smaller than about 0.5 μm can be measured far more precisely with the DMA than the EAA. Recent improvements in the design of such instruments permit accurate size distribution measurements of particles as small as 0.002 μm in diameter (Winklmayr, 1987; Reischl, 1991). Although the DMA has received little application in measuring atmospheric particles, it undoubtedly will be used increasingly.

Particle Mass and
Chemical Size Distributions

Filters are used often for obtaining bulk aerosol samples. A wide variety of filter sampling materials is available (e.g., Teflon, nylon, polypropylene, polystyrene, cellulose fiber, glass fiber, and polycarbonate). Filter samples are useful for obtaining a measure of gross particle concentrations and composition. Such measurements are important for flux data, but they do not provide information about chemical speciation as a function of size that is required for many applications. Also, filter samples are likely to have many artifacts that are generated by reactions of the sampled substances with the sample matrix or that are among the atmospheric constituents themselves. That subject is discussed more fully later in this appendix.

Cascade impactors are the most commonly used instruments for measuring the size-resolved chemical composition of airborne particles. A cascade impactor consists of a series of impactor stages. The top stage removes the largest particles, and subsequent stages remove successively smaller particles. Impactors classify particles according to their aerodynamic diameter, which depends on particle shape and density. Although most impactors have a lower size cut (i.e., minimum particle size that can be collected by inertial impaction) of about 0.3-0.5 μm in aerodynamic diameter, impactors with size cuts as small as 0.05 μm have been developed (Hering et al., 1979; Berner et al., 1979; Hering and Marple, 1986; Cahill et al., 1987). Because most atmospheric particle mass consists of particles larger than 0.05 μm, such impactors can collect virtually all the particulate mass.

Impactors are highly versatile devices and cover a wide range of particle sizes and sampling rates. With appropriate substrate materials, impactor deposits can be analyzed for a large variety of aerosol components. Analytical techniques that have been applied to impactor deposits include ion chromatography, flash volatilization, elemental and organic carbon analysis, x-ray fluorescence, and proton-induced x-ray emission.

Although impactors and filter systems are used widely in visibility studies, problems with their performance must be thoroughly understood, to avoid compromising the acquired data.

Particle Bounce

There are several common problems with using of cascade impactors for atmospheric particles. Particles bouncing off the collection stage and collecting downstream is the most troubling problem (Dzubay et al., 1976; Wesolowski et al., 1977; Reischl and John, 1978; Lawson, 1980; Wang and John, 1987). Particle bounce leads to overestimates of concentrations in the small-size ranges. Bounce can be reduced or eliminated by coating the impaction substrates with a sticky substance (Turner and Hering, 1987; Wang and John, 1988), although greases can interfere with the subsequent chemical analysis.

Particle Shape and Density

Particle density must be known to compute the actual diameter from the aerodynamic diameter. However, atmospheric ammonium nitrate particles are not always spherical densities are not always known, and particles of a given size might have more than one density. Thus, computed diameters are often based on assumptions that can lead to considerable error.

Chemical Changes in the Sample

Chemical changes can occur while sampling with filters or cascade impactors and can lead to measurement errors. The artifacts include evaporation of semivolatile components during sampling, adsorption or absorption of gases onto sampling substrates, and changes in particle composition during sampling due to reactions among particles or with the gas stream.

Evaporative losses occur if the equilibrium partial pressure at the surface of the particulate deposit exceeds the gas-phase partial pressure above the deposit. Evaporation can result from the pressure drop through the sampler and from changes in temperature or aerosol composition during sampling. The theory is that evaporative loss is likely to be high when the ratio of the gas-phase concentration to the particulate concentration at the sampler inlet is large and that the extent of evaporation depends on whether the semivolatile compound is adsorbed, absorbed, or condensed on the particles (Zhang, 1990). Substances that can volatilize during sampling include HNO_3 from acidified NO_3^- particles, NH_3 from basic NH_4^+ particles, and certain organics.

Evaporative losses of particulate nitrates have been investigated in both laboratory and field experiments with impactors and filters. The laboratory studies (Wang and John, 1988) involved parallel sampling of ammonium nitrate particles with a Berner impactor (size cuts ranging from 0.08 to 16 μm) and a Teflon filter. Both samplers were followed by nylon filters to collect evaporated nitric acid. Losses from the impactor were 3-7% at 35°C and 18% relative humidity, and losses from the filter were 81-95% under the same conditions. The result that evaporative losses from the filter exceeded those from the impactor is consistent with theoretical predictions (Zhang and McMurry, 1987). The atmospheric measurements (Wall et al., 1988) involved sampling Los Angeles airborne particles with a Berner impactor located downstream of a nitric acid denuder parallel with a dichotomous sampler with nylon filters. The dichotomous sampler is known to sample particulate nitrate quantitatively (John et al., 1988). It was found that losses from the impactor were less than 10%. There have been no similar experiments in pristine regions where low particulate nitrate loadings might lead to relatively large evaporative-loss errors in impactors.

Because evaporative losses of nitrates from filters tend to be large, measurement methods for particulate nitrate must include collection and measurement of nitric acid that evaporates from particles during sampling. One common sampling method involves using a nitric acid denuder to remove gas-phase nitric acid upstream of the particulate filter. The particulate filter is then followed by an adsorber to collect nitric acid that evaporates from the particles during sampling. Discussions of sampling methods for nitrates and other inorganics are given by Lewin and Klockow (1982), Possanzini et al. (1983), Eatough et al. (1985), Ferm and Sjodin (1985), Mulawa and Cadle (1985), Knapp et al. (1986), Keuken et al. (1988), and Vossler et al. (1988).

The importance of evaporative sampling losses is not as well understood for organics as for nitrates. There is evidence that trace organic compounds, such as polycyclic aromatic hydrocarbons (PAH) and polychlorinated biphenyls (PCB), can evaporate from aerosol deposits during sampling (Commins and Lawther, 1957; De Wiest and Rondia, 1976; Katz and Chan, 1980; Koenig et al., 1980; Peters and Seifert, 1980; Galasyn et al., 1984; Marty et al., 1984; Coutant et al., 1988). Eatough et al. (1990) used a sampling system that included a denuder, filter, and sorbent bed for measuring organic particles. Their sampling scheme permitted determination of positive sampling artifacts (associated with vapor adsorption on filter media) and negative artifacts (associated with evaporative losses of vapors). They concluded that both positive and negative artifacts occurred but that negative artifacts tend to dominate. Negative artifacts as large as 2.6 $\mu gC/m^3$ were collected at the Grand Canyon. Their results require confirmation by other measurement methods; however, they suggest that current methods for measuring atmospheric organic carbon may be seriously flawed.

Many investigators have reported that the adsorption of organic vapors on quartz filters can lead to a substantial positive sampling of arti-facts for particulate organic carbon. For example, Cadle et al. (1983) found that when two quartz fiber filters were used in a series, the amount of carbon collected on the second filter was at least 15% of the amount collected on the first filter. Because particles could not penetrate through the first filter, the carbon on the backup filter was attributed to vapor adsorption. McDow and Huntzicker (1990) found that quartz backup filters collected more organic carbon when they followed Teflon prefilters than when they followed quartz prefilters, presumably because quartz prefilters are more effective than Teflon prefilters at removing adsorbing vapors, thereby reducing the vapor exposure of the afterfilter.

McMurry and Zhang (1989) used a cascade impactor with a cut point of about 0.1 μm to collect size-resolved atmospheric particle samples for chemical analysis; particles smaller than 0.1 μm were collected on a quartz afterfilter. Samples were analyzed for organic and elemental carbon, for various elements (by x-ray fluorescence), and for nitrate and sulfate ions. Measurements were made in Los Angeles and at pristine sites such as the Grand Canyon and Glen Canyon. Except for organic carbon, only a small fraction of the particle species were found on the afterfilter. However, a large portion (40-70%) of the collected organic carbon was found on the quartz afterfilter; the amount was higher at remote sites where particulate organic carbon concentrations were lower. McMurry and Zhang argued that much of the carbon on the afterfilter was due to the adsorption of carbon-containing gases. That suggests that gas adsorption on quartz filters can lead to major uncertainty in deter-mining the particulate organic carbon concentrations in samples collected in visibility monitoring networks, where concentrations often are low but can be a large fraction of the fine particle mass.

Chemical reactions that involve the deposited particles can also lead to sampling artifacts. For example, Klockow et al. (1979) found that sulfuric acid particles could not be extracted accurately from certain filter materials because of chemical reactions of the acid with the filter substrate. Brorström et al. (1983) found that particle-associated PAH was degraded when air containing NO_2 and O_3 was passed through the filter; the rate of PAH degradation increased with increasing acidity of the aerosol deposits.

Sampling Inlets

An overview of aerosol sampling and sampling inlets has been given by Vincent (1989). Because of the strong size dependence of radiation-particle interactions, it is important to use samplers with well-characterized sampling inlets for visibility research and monitoring. In recent years, considerable effort has been made in developing inlets with a well-defined sampling efficiency for coarse particles. There have been several reasons for the effort. First, the size-dependent sampling efficiency of the hi-vol sampler, the standard reference method for measuring ambient total suspended particulate concentrations, was found to depend on wind speed and direction (McFarland et al., 1979). Therefore, the characteristics of the sampled particles were not well defined. There also has been an interest in developing inlets that collect only particles of the size range that can be inhaled by humans and, hence, might harm human health.

Recent work has shown that particle deposition in aircraft sampling inlets may be a much more severe problem than was previously believed. Huebert et al. (1990) found that 50-90% of sampled atmospheric particles deposited within aircraft inlets of various designs. Deposition losses occurred both for coarse particles and for fine-particle non-sea-salt sulfates. The poor efficiency of the sampling is not understood; theories predict high efficiency, especially for fine particles. Based on those observations, the conclusion of Huebert and co-workers was that much of the aircraft data for atmospheric particle concentrations probably underestimate the actual ambient concentrations by factors of 2-10. However, other intercomparisons (e.g., Daum et al., 1987; Baumgardner et al., 1992) have shown relatively good agreement between aircraft and ground measurements. A workshop was recently convened to investigate sampling characteristics for airborne particles (Baumgardner et al., 1992). It was concluded that "at present, there is insufficient airborne experimental data to validate any existing inlet models." Because of the importance of airborne sampling in visibility research, the committee supports the recommendation from this workshop that more research on inlet sampling efficiencies be conducted, especially for submicron particle sampling with aircraft.

Electron Microscopic Analyses
Of Single Particles

Filter and impactor samples can be analyzed to determine the average composition of all collected particles. Such analyses, however, provide no information about particle-to-particle variations in composition. Recent work has shown that atmospheric particles are externally mixed to some extent; that is, particles in a given size range comprise of distinct types of materials (Covert and Heintzenberg, 1984; Harrison, 1985; McMurry and Stolzenburg, 1989; Covert et al., 1990; Hering and McMurry, 1991). To understand optical properties of the atmosphere containing those particles, detailed knowledge is needed of the composition of the individual particles and the relative concentration of each in the ensemble (White, 1986). Recent advances in electron beam technology make single-particle imaging and analysis a useful technique for obtaining such information.

Aden and Buseck (1983) reported on data-analysis procedures that provide quantitative ($\pm 10\%$) elemental analyses by energy dispersive spectrometry for particles as small as 0.1 μm. With light-element detectors, elements having molecular weights as light as carbon can be analyzed. Post and Buseck (1984) used these methods to analyze 8,000 particles from the urban Phoenix atmosphere and found that most fell into the categories of minerals, sulfur (presumably sulfate), and lead from automobiles. Their microscope was not equipped with a light-element detector, so carbon-containing particles were not identified. Schwoeble et al. (1988) reported on a computer-controlled scanning electron microscope (SEM) that automatically locates particles on a substrate and digitally records the particle image and elemental composition. Automated microscopy facilitates the analysis of a statistically significant number of particles at a modest cost.

There are some problems with electron microscopy. Volatile substances such as water, nitrates, and some organics might evaporate when irradiated by the electron beam, and all important elements cannot be analyzed. The environmental SEM (Danilatos and Postle, 1982; Danilatos, 1988) eliminates some of those problems. The electron gun and optics operate under high vacuum, but the samples are exposed to a positive gas pressure (greater than 5 torr). The gas phase above the sample can include water vapor, so water is not necessarily lost during pump down, as occurs in most other electron microscopes. The presence of the gas also eliminates the need to coat the sample with a conducting film. Recent developments suggest that electron microscopy shows promise for providing information on particle properties that is not provided by most common analysis methods.

Organic and Elemental Carbon Analysis

The total carbon content of filter or impactor samples can be measured accurately with a variety of techniques. For visibility measurements, however, it is important to distinguish between organic and "elemental" or "black" carbon. The different chemical forms tend to originate from different sources and have different optical effects.

A variety of analytical schemes have been developed to distinguish between organic and elemental carbon. In all the methods, the samples are exposed to a carrier gas (typically oxygen, air, or helium) in an oven where the particulate carbon is converted to a gas, which is measured by an appropriate detector. Particulate organic carbon tends to be released at lower oven temperatures than does elemental carbon, and that temperature dependence is used to distinguish between them. With these techniques, the pyrolysis of organic carbon within the furnace can lead to overestimates of the ratio of elemental to organic carbon. Some techniques use optical absorption to correct for pyrolysis.

Interlaboratory comparisons (Groblicki et al., 1983; Countess, 1990) showed a wide variation in ratios of organic to elemental carbon for identical samples. Samples included atmospheric aerosols, diesel and unleaded gasoline exhaust, soot, and organic particles collected from a smog chamber. Laboratories participating in those studies included most of those involved in the analysis of samples from visibility monitoring networks. The inconsistency of results among laboratories suggests that organic and elemental carbon analysis is a major source of uncertainty in aerosol composition for samples collected in visibility monitoring networks.

Water Content of Airborne Particles

Water often constitutes a substantial portion of atmospheric particles. For example, water accounts for 63% of the mass of ammonium sulfate droplets at 85% relative humidity and 20°C. Because water causes substantial changes in the size of hygroscopic particles, it plays a central role in determining the visibility effect of airborne particles. However, because water is so volatile, the water content of airborne particles is difficult to measure.

Several methods have been used to measure aerosol water content. Hänel (1976) and co-workers used impactors to collect particles in one or two size cuts with geometric diameters larger than 0.3-0.5 μm. An electronic microbalance was then used to measure the mass for relative humidities ranging from 0 to 95%; densities were determined with a specially designed gas pycnometer. The mass and density measurements were used to determine the average dependence of size on relative humidity. Measurements of airborne particles collected in urban, desert, and mountainous regions showed that water uptake by these diverse particles was highly variable.

Ho et al. (1974) used microwave resonance to measure the water content of samples collected on glass fiber filters. Measurements were made in Southern California during Fall 1972. The particulate water content was found to increase from 10% of the airborne particle mass at 50% relative humidity to 40% at 70% relative humidity.

McMurry and Stolzenburg (1989) used a tandem differential mobility analyzer (TDMA) to measure the sensitivity of particle size to relative humidity for particles in Los Angeles. With this technique, atmospheric particles of a known size are segregated using a differential mobility analyzer (DMA). The final size after humidification or dehumidification is determined to within 1-2% using a second DMA. McMurry and Stolzenberg found that particles in the 0.05-0.5 μm range often contained both hygroscopic and nonhygroscopic fractions. The hygroscopics to nonhygroscopic ratio varied substantially from day to day. For particles in the hygroscopic fraction, particle diameters increased by factors ranging from 1.12 $\pm$ 0.05 at 0.05 μm, to 1.46 $\pm$ 0.02 at 0.5 μm as relative humidities were increased from 50% to 90%.

Continuous Measurements

The approaches described above for measuring the chemical composition of airborne particles all require the collection of samples over extended sampling periods. Useful information often can be obtained with more refined temporal resolution. For example, measurements that can be taken continuously (e.g., light extinction or scattering and concentrations of various gases, including SO_2) often show large variations during the sampling periods needed for collecting particulate samples. Concentrations of particles almost certainly undergo similar large variations. Continuous measurements of particulate chemical concentrations could provide important information about atmospheric transport and transformations and about the effects of specific point sources on visibility at a receptor site.

Continuous measurements of particulate composition would have provided useful information in the studies reported by White et al. (1990) and Miller et al. (1990). They used hourly information on the concentrations of methylchloroform and perchloroethylene to document the effect of urban plumes on receptor sites in the southwestern United States. They found that concentrations of those gases at desert receptor sites followed a diurnal pattern that mimicked industrial emission patterns in distant urban areas. By coupling those data with wind trajectory analyses, it was possible to identify likely source regions. If continuous real-time particulate data had also been available, it would have been possible to obtain valuable information about long-range transport of particulate species and about the effects of such species on visibility impairment at the remote receptor sites.

Some work has been done to develop continuous or nearly continuous detectors for some particulate substances, such as sulfur (Garber et al., 1983), or for elements analyzed by techniques such as PIXE (proton-induced x-ray emission) (e.g., Cahill et al., 1987). However, such instruments typically are not used for routine monitoring because of cost, complexity, limitations in sensitivity, and the need for specialized analysis facilities. More attention should be given to developing techniques for continuous measurements of particulate composition.

OPTICAL MEASUREMENTS

The appearance of a distant object viewed through the atmosphere is affected by several factors: the amount and color of light emitted by the object (initial radiance); the transmittance of that light from the object to the observer; and the scattering of ambient light into the sight path by the atmosphere (path radiance) (Duntley et al., 1957; Malm, 1979; Richards, 1988). Because the initial radiance, transmittance, and path radiance are sensitive to the wavelength of the light, one must know how those factors depend on wavelength before visibility can be characterized fully. Middleton (1952) provides a comprehensive discussion of atmospheric visibility and its measurement.

Transmittance is determined by the average extinction coefficient between the observer and the object. The extinction coefficient is the sum of the scattering and absorption coefficients for both gases and particles. Although transmittance only partly characterizes the visibility along a particular sight path, it is an inherent atmospheric aerosol air-quality factor and is unaffected by time of day, viewing angle, or characteristics of the natural illumination.

Path radiance is affected by variables that are not specifically related to air quality. For example, under identical air-quality conditions, the path radiance along a given sight path will be greater if viewed toward the sun than if viewed away from the sun, because particles tend to scatter light more strongly in the forward direction. Ground reflectance and the extent of cloud cover also affect path radiance. For example, the path radiance decreases in the presence of clouds because less skylight is available for scattering into the sight path.

Methods for measuring atmospheric optical properties can be classified as either point or sight-path techniques. Point measurement techniques determine the contributions of gases and particles to scattering and absorption at the measurement location. That information can be used to determine the local extinction coefficient (and therefore the local atmospheric transmittance). Because the composition of the atmosphere may vary between an observer and a distant object, the local transmittance may not characterize accurately the average transmittance along a given sight path. Sight-path methods include direct measurements of transmittance over a long path, measurements of radiance along a given sight path, and photography. Photographic and radiance measurements are influenced by factors such as skylight and ground reflectance and, therefore, usually do not provide information on atmospheric optical properties. In the remainder of this section, the most commonly used point and sight-path methods are briefly reviewed and discussed.

Point-Measurement Methods

Both gases and particles contribute to optical extinction. Evaluating the contributions of gases to light scattering and absorption is relatively straightforward. All gases scatter light. The scattering of light by atmospheric gases is dominated by the most abundant kinds (i.e., N_2, O_2, and CO_2), and, thus, light scattering by gases can be determined largely by the total air pressure. Because the relationship between air scattering and total air pressure is well known (Penndorf, 1957), the contribution of light scattering by air to b_{ext} can be determined easily.

NO_2 is the only pollutant gas in the atmosphere that absorbs visible light and is present in sufficient concentrations to affect visibility. The specific absorption of NO_2 is well established, so extinction can be determined from concentration measurements.

The remainder of this section focuses on light scattering and absorption by particles. These processes are much more complex for particles than for gases.

Scattering-Coefficient Measurements: Integrating Nephelometry

The integrating nephelometer measures the total amount of light scattered by an aerosol sample. The most commonly used instrument of this type (see Charlson et al., 1974) draws a sample into an enclosed, dark chamber where it is illuminated. A detector measures light scattered at angles ranging from near forward to near backward. To determine the contribution of gases and electronic noise to the scattering signal, the instrument's light-scattering response to filtered air is measured periodically. The contribution of particles to scattering is then determined by taking the difference. When equipped with a photon-counting detector, the integrating nephelometer can measure particle light-scattering coefficients of less than 0.1 Mm^{-1}, a value equal to about 1% of the light-scattering coefficient of particle-free air at normal atmospheric pressure.

Because of its potential for high accuracy, portability, and moderate cost, the nephelometer has been used widely for measurements of light-scattering coefficients. There are, however, several sources of measurement error with this instrument. First, the contribution of coarse particles (particle diameter greater than about 5 μm) to scattering is underestimated because they tend to deposit at the inlet. Second, the optics do not permit measurement of light scattered between 0 and 7 degrees; however, coarse particles scatter strongly in the forward direction (Ensor and Waggoner, 1970; Hasan and Lewis, 1983). Another limitation of the nephelometer is that heating within the instrument can reduce the aerosol relative humidity during measurement. The light-scattering coefficient of atmospheric particles is sensitive to relative humidity, especially for values exceeding 60-70%, because the amount of water absorbed by particles (and therefore particle size) depends on humidity (Charlson et al., 1978).

Some of those problems can be dealt with relatively easily with available instruments. For example, the humidity's effects can be minimized by heating the inlet air so that the relative humidity is reduced to a low or constant value. Improved designs also will minimize or eliminate many of the problems. A nephelometer is being developed that uses solar power to operate at remote locations (J. Persha, pers. comm., Optec, Inc., 1990). Because of their many advantages, it is believed that nephelometers will play a major role in visibility programs.

Absorption Coefficient Measurements

Elemental carbon is apparently the dominant light absorbing particle component (e.g., Rosen et al., 1982; Adams et al., 1990). Particle absorption coefficients can be determined either by using in situ techniques, such as photoacoustic spectroscopy (Adams, 1988), or by measuring the extent to which light is absorbed by particles that have been collected on a filter (e.g., Lin et al., 1973; Clarke, 1982; Hänel, 1987). Salient features of these techniques are outlined below.

Photoacoustic Spectroscopy

Photoacoustic spectroscopy measures the absorption coefficients of suspended particles in real time (Adams, 1988). The air stream, from which NO_2 has been removed, is drawn into an acoustic cell where it is illuminated by light that is modulated at the resonant frequency of the cell. Light energy absorbed by the particles heats the carrier gas, which expands and then contracts according to the modulation frequency of the light. The associated pressure variation is a sound wave whose intensity can be measured with a microphone. Photoacoustic spectroscopy appears to be the best available technique for measuring particle absorption coefficients, but it requires skilled personnel and complex equipment and, consequently, is not suitable for routine regulatory monitoring.

Filter Techniques

Filter techniques are the most common methods used for measuring particle absorption coefficients. Because light transmittance through filters is affected by scattering and absorption, the effects of scattering, including multiple scattering, must be accounted for. If light interacts with more than one particle as it passes through the filter, the apparent absorption coefficient will exceed the correct value. Also, filter techniques are problematic because the optical properties of deposited particles may be different from those of airborne particles, especially if the particles undergo chemical reactions on the filter, either through exposure to gases passing through the filter or through contact with other particles on the surface of the filter.

Lin et al. (1973) developed the integrating plate technique for measuring absorption coefficients of particle deposits on filters. With this method, an opal glass plate is located between the filter and the optical detector. Because the opal glass is a diffuse reflector, light scattered by particles in the forward direction is detected with the same efficiency as light that enters the glass directly. If backward scattering is small in comparison to absorption, changes in filter transmittance before and after particle collection can be attributed to particle absorption. Lin et al. concluded that both backward scattering and multiple scattering did not contribute significantly to errors in their measurements. Clarke (1982) modified that technique to reduce some of the errors.

Hänel (1987) argued that multiple scattering and backward scattering led to significant measurement errors of absorption coefficients by previous investigators using filter techniques. He developed an approach for measuring absorption coefficients that permitted accounting for forward, backward, and multiple scattering from collected particles. Reported values for absorption coefficients with his approach are somewhat smaller than values determined with other techniques.

Foot and Kilsby (1989) compared particle absorption coefficients measured with a filter technique and with a photoacoustic technique. They used laboratory particles with known properties and found that agreement between the two methods was $\pm 15\%$. However, as was pointed out by Japar (1990), the uncertainties are likely to be greater in filter measurements with atmospheric particles that scatter strongly.

The earliest and simplest method of measuring light absorption by particles on filters is the coefficient of haze (COH) technique (Hemeon et al., 1953). The aethelometer described by Hansen et al. (1984) is an updated and more sensitive absorption measurement that operates on a similar principle. These techniques measure the light attenuation caused by an aerosol sample on a filter; no integrating plate is used to correct for light scattering. Wolff et al. (1983) found a good correlation between COH and concentrations of elemental carbon, and Campbell et al. (1989) found good correlations between COH and absorption measurements with an integrating plate and integrating sphere. Because these instruments can provide continuous, near real-time data, they are used often for monitoring. Quantification is achieved by calibrating against a more accurate absorption measurement standard.

Sight-Path Techniques

Instruments that are used for sight-path measurements include teleradiometers (telephotometers), scanning densitometers, and transmissometers. Although their strengths and limitations have been known for many years, only during the past 10-15 years have they come into widespread use for visibility research. Teleradiometers measure the radiance (light intensity per solid angle) along a given sight path. The most common approach to teleradiometric measurements involves measuring the radiances from a dark target and from the background sky adjacent to the target. The apparent contrast between the target and sky is obtained from the radiances and can be used to determine the visual range under ideal measurement conditions.

Atmospheric conditions that must be met to determine visual range from contrast measurements are well known (see Allard and Tombach,

1981; Richards, 1988). In particular, the initial contrast of the target against the background sky must be known, and the sky radiance at the target and at the measurement point must be equal. Factors that lead to spatial variations in sky radiance include nonuniform lighting conditions associated with clouds or variations in ground reflectance and variations in extinction along the sight path. Allard and Tombach (1981) have described "nonstandard" viewing conditions that should be recorded when telephotometric contrast measurements are used to infer visual range. During routine contrast measurements made with telephotometers in the Western Regional Air Quality Study, standard observing conditions were uncommon enough to raise concerns for the representativeness of data (White and Macias, 1987b). Therefore, although telephotometers can be used to obtain accurate measurements of sky-target contrast, the measurement conditions necessary for use of contrast data are seldom met.

There are measurement schemes for which teleradiometers can be used to measure atmospheric optical properties unambiguously, even under nonstandard viewing conditions (Richards, 1988). Typically, more than one teleradiometer is needed, and absolute calibrations or artificial targets might be required. Such measurement methods have been used in field research but have not been implemented in routine monitoring networks.

Photographic slides commonly are used to monitor visibility. A scanning densitometer is used to measure the contrast between an image of a dark target and background skylight. (Johnson et al., 1984). The technique involves measuring light transmission through the slide by focusing on a small area (typically 25-100 μm^2) of the target and of skylight and using those data in conjunction with known properties of the film to determine the sky-target contrast. A considerable library of slides has been acquired by the visibility monitoring networks of the National Park Service and the U.S. Forest Service. This technique provides a low-cost approach for routine visibility monitoring.

The use of slides for visibility measurements has limitations similar to those encountered with teleradiometers. Even though the apparent target-sky contrast can be measured, the information can be used only to infer atmospheric optical properties under ideal viewing conditions. Because ideal conditions are seldom encountered, slide data have limited utility for visibility monitoring. Nonetheless, slides do provide a useful record of visibility; studies have quantified the relationship between slide-based assessments of visibility and other types of measurements (see further discussion in this appendix).

Transmissometers measure the transmittance of light from an artificial light source over a measured path to a detector. Path lengths of about 15 km typically are required to obtain accurate data in pristine air. The average extinction coefficient along the path can be determined directly from the transmittance measurements. Because the measured light propagates through the open atmosphere, transmissometers measure the optical effects of the unperturbed airborne particle. That ensures that accurate measurements can be made at high relative humidities when the particle water content is large and that measurements include the effects of both fine and coarse particles. Because transmittance is a property of the aerosol along the sight path, transmissometric data are not affected by path radiance.

An essential difficulty of any long-path in situ method is the absence of any definitive field calibration. Because the contents of the sample volume cannot be controlled, airborne particles of known characteristics cannot be administered (e.g., ambient Rayleigh scattering), as is routinely done with the nephelometer and other enclosed instruments. The optical elements can be tested at close range, but that introduces potentially large geometric and alignment errors and in any case produces a calibration point at zero extinction, well below the minimal (Rayleigh) value encountered in the atmosphere. In particular, the potentially substantial effects of turbulence on the beam ("blooming") are invisible to such tests, although various schemes for minimizing such errors have been proposed (e.g., Malm et al., 1987; Richards, 1988). Problems such as environmental degradation of optical surfaces, drifting source intensity, and misalignment can introduce significant hidden scaling errors in the field, even though a unit tests out before installation. As transmissometers are only now passing from selective research to widespread operational use, few data are available on their overall accuracy and reliability.

Several other problems are encountered with transmissometers. Care must be taken to filter out data that are acquired when clouds, rain, and fog obscure the sight path. In practice, that is often difficult to do. Transmissometers are also much more expensive than other monitoring instruments. Despite the limitations, solar-powered transmissometers are routinely used in remote locations, and they have been the instrument of choice in the NPS/IMPROVE monitoring network.

Remote Sensing Techniques

Satellites can be used to estimate the total column loading of particles in the atmosphere on the basis of the upwelling irradiances from the Earth's surface. These techniques suffer from a number of problems. First of all, algorithms must be used to convert the irradiances to equivalent aerosol concentrations. Because of the complex nature of the interaction of radiation with particles, algorithms are in an early stage of development; consequently, measurements are subject to considerable error. Another operational problem is that it is difficult to distinguish between clear (cloud-free) regions and regions where there is broken (or sub-visible) cloud; contamination with even a small amount of cloud can lead to highly erroneous data. Also, in order to make accurate estimates of particle irradiances, it is necessary to accurately measure the albedo of the underlying surface; however, the surface albedo can vary greatly with time and location. Finally, satellites can only provide an estimate of the total column loading of particles; for the purpose of visibility studies, one needs to know the concentration in the atmospheric boundary layer close to the surface. Thus, while satellites can be useful for characterizing the large scale distribution of haze events, at this time they can not be used in quantitative visibility studies.

Summary

Point measurements and sight-path measurements are two distinct alternatives for visibility measurement. Point measurements can be used to determine the extinction coefficient, an inherent air-quality factor, and they can be used in conjunction with particulate and gaseous substance measurements to determine the contributions of different substances to extinction. The average atmospheric extinction coefficient over a given sight path can be determined with a transmissometer. However, extinction does not completely characterize visibility. For example, from extinction data alone, one cannot estimate the clarity with which detailed features can be resolved in a particular landscape scene or correct for changes in the perceived color of the scene.

Radiances measured directly with teleradiometers or indirectly with scanning densitometers on photographic slides can be used to provide more direct information on what is seen by an observer. Information of that kind, however, provides information on atmospheric optical properties only under restricted measurement conditions that are seldom encountered in practice. Data from these instruments can provide valuable insights into air-quality levels necessary to achieve a desired visual quality for a particular view, but they are of limited use for routine compliance monitoring, because measurements are strongly affected by factors unrelated to air quality.

On balance, point-measurement techniques appear to be preferable to sight-path methods for routine monitoring of atmospheric visibility. Point measurements permit measurement of both scattering and absorption coefficients, thereby providing important information on the contributions of each to extinction. Comparisons between optical properties and chemical composition measured at the same point can be used to infer the contributions of various substances to extinction, and point measurements are less prone to the complications of clouds, fog, or precipitation that can occur along sight paths.

The most important point-measurement instrument is the integrating nephelometer, which measures the light-scattering coefficient. Instruments of that kind can be designed to be sensitive to the range of scattering coefficients found in both pristine and polluted air, are easy to install and calibrate, and can be reasonably priced. Furthermore, there are absolute and unambiguous calibration methods for integrating nephelometers, which is not the case for long-path methods. Major drawbacks of integrating nephelometers include the unavailability of a commercial, electronically up-to-date instrument, the tendency of the instrument to heat sampled airborne particles, thereby underestimating scattering coefficients on high-humidity days, and inaccurate measurements of coarse-particle scattering due to particle losses at the instrument inlet.

We recommend that high-sensitivity integrating nephelometry be used for routine haze monitoring. To do that a modern, commercial integrating nephelometer must be developed. The instrument should be sensitive to the range of scattering coefficients encountered in the atmosphere, should be self-powered, and should be designed to minimize errors associated with sample heating and coarse-particle sampling losses.

TRACERS IN THE ENVIRONMENT

A "tracer" element or compound is a substance with unique characteristics, allowing its positive identification at very low concentrations. Tracers can be added in small amounts to a known substance of larger amounts whose subsequent physicochemical behavior needs to be understood. The goal is to add a detectable marker whose presence is so insignificant that the basic behavior of the larger system is unaffected even when the marker mimics the reaction pathways of the known species. In everyday life, the laundry mark on a shirt serves later to identify the origin of the shirt without significantly changing its appearance; the brand on a cow serves a similar purpose. In scientific application, the ideal tracer is one that has the same chemical form and properties as the material to be traced and that is distributed physically in the same manner.

In many laboratory situations, the tracer of choice is often a radioactive isotope of the same element whose stable isotopes are of interest—e.g., $^{14}CO_2$ with radioactive ^{14}C as a tracer for ordinary CO_2, which consists chiefly of $^{12}CO_2$ (99%) plus a small percentage of $^{13}CO_2$. If CO_2 might be coming simultaneously from two power stations, the addition of $^{14}CO_2$ to one of them allows evaluation of the contributions of the two power stations to the CO_2 observed at a downwind site through measurement of its ^{14}C content. Once the three isotopic versions of CO_2 are thoroughly mixed, all three behave similarly (although not in precisely the same way because of the slight differences in mass of the carbon atom), and measurement of the ^{14}C content of a sample provides information about the other two forms of CO_2. In some situations, the substance of concern carries its tracer with it as a natural component, as with atmospheric CO_2, which contains a small fraction (about 1 part in 10^{12}) of $^{14}CO_2$ formed by the natural process of cosmic ray bombardment of the atmosphere. When atmospheric CO_2 is incorporated into a growing plant by photosynthesis, the presence of the trace of $^{14}CO_2$ can be quantitatively demonstrated through the radioactivity of the ^{14}C in the plant.

The ideal conditions for using a tracer material may not be attainable (e.g., tracers for sulfur within coal), and near-substitutes must be found. For example, artificial radioisotopes are frequently the tracer of choice in controlled laboratory situations, but their release into the open environment usually is usually not acceptable because of the lack of control over the subsequent pathways of the radioactivity. During coal combustion, the sulfur atoms in coal are converted into SO_2, which is later changed chemically into sulfuric acid (H_2SO_4). Frequently, when H_2SO_4 is found in the atmosphere (e.g., as a component in acid rain), questions arise as to the source of the sulfur, because H_2SO_4 from many sources becomes essentially indistinguishable once mixed. Under those conditions, a tracer, which has accompanied the sulfur from a particular source, can be essential in determining whether the H_2SO_4 in question has come from that source. The available isotopic tracers for sulfur are the radioactive isotopes (especially ^{35}S with a half-life of 90 days) and sulfur isotopic mixtures, which have been enriched in stable ^{36}S. (The most abundant stable isotope is ^{32}S.) Neither of those is acceptable for tracing the path of fossil-fuel sulfur in the atmosphere; the amounts of ^{35}S needed are large enough that release of radioactivity in that quantity into the environment would not be approved, and the sensitivity of detection for enriched ^{36}S is sufficiently low to make such experiments impractical. Even if those factors were acceptable, sulfur atoms within coal exist in a chemical environment very hard to simulate, so that one could not be assured that the conversion to SO_2 would proceed at the same rate for the coal-bound sulfur and its isotopic tracer.

When no acceptable isotopic tracers are available, other similar tracers can be identified. The element selenium is just below sulfur in the periodic table of the elements, which suggests that the chemical behavior of selenium would be similar to that of sulfur. However, selenium and sulfur differ significantly in their chemical behavior under many atmospheric conditions, so their subsequent pathways might be different, thereby undermining the purpose of the tracer.

Another tracer procedure for following the SO_2 formed during coal combustion is the addition of another gaseous component intended to mix with SO_2 and to follow its physical path, even if it is not capable of undergoing a chemical change analogous to that of SO_2 changing to H_2SO_4. Two such gaseous tracer species are an isotopically labeled form of methane, fully deuterated CD_4, which can be readily separated from the naturally occurring CH_4 by mass spectrometry, and several perfluorocarbons, compounds not found naturally. The compound CD_4 is extremely rare in nature, where the ratio of D to H ratio is about 1:6,500, making the ratio of CD_4 to CH_4 about $1:10^{15}$. When such a gaseous tracer is introduced into the stack of a power plant and mixed with the SO_2 from the burned coal, one can anticipate that the gaseous tracer and SO_2 will be carried by winds to the same locations. If the tracer has arrived at a particular location, then it is likely that the SO_2 arrived at the same time—unless it was previously converted to H_2SO_4. In that case, the H_2SO_4 is likely to arrive coincidentally with the tracer, unless the H_2SO_4 has interacted further and has been chemically or physically removed from the atmosphere. Thus, when CD_4 is added to a known power-plant stack and some of it is measured downwind, the inference is strong that the SO_2 emitted from the same stack on the same day also arrived at the site as either SO_2 or H_2SO_4.

PERCEPTION MEASUREMENT

The effect of haze on visibility is an important issue in local and regional air-quality management. In the past, visibility was primarily the concern of aviation and military operations where the most important aspect of visibility was visual range—that is, the greatest distance at which an object could be discerned against the background sky. In that context, visual range has been quantified and related to light extinction, and it is used extensively to define visibility at airports.

In contrast, much of the present concern about visibility is related to the aesthetic damage from air pollution—that is, the impact on the perceived form, texture, and color of scenic features (Trijonis et al., 1990). In that regard, visibility degradation is distinct from nonvisual effects of air quality, such as health effects or economic effects. A judgment of visibility is an aesthetic judgment; depending on the judgment of an observer, atmospheric conditions can degrade the aesthetic quality of a scene (for example, when plumes or urban haze obscure a mountain backdrop) or enhance it (as when haze adds interest to a landscape) (Stewart et al., 1983).

The broader conception of visibility serves as the basis for regulation of stationary sources in the Clean Air Act. However, the definition of visibility impairment in the Clean Air Act is vague and qualitative at best. It is difficult to obtain quantitative cause-and-effect relationships when the effects are not clearly defined.

The ability to make quantitative connections between optical properties of the atmosphere and human judgments of visibility is still in the developmental stage because of the complexity of the physical and psychological phenomena. To quantify visibility impairment, an index must be developed that can incorporate the complexity of those phenomena; the index also must be understandable and useful to the general public and policy makers as well as to scientific researchers. Because impairment is based largely on human judgments of the visual environment, the human element must be incorporated in the development of such an index. In addition, the index must be based on properties of the physical environment that can be readily measured and monitored to enable enforcement of air-quality standards.

Human Response to
Visual Air Quality

The human response to visual air quality (VAQ) can be measured by a variety of techniques: (1) judgments of VAQ made in the field by experienced observers; (2) judgments of VAQ made from photographs by experienced observers; (3) judgments of VAQ made in the field by random passersby; (4) judgments of visual range made at airports by trained observers; and (5) judgments of selected perceptual cues that are components of the overall judgment of VAQ.

Field Judgments by
Experienced Observers

Field judgments of overall VAQ by experienced observers provide the most direct measurement of VAQ. The relationship between experienced judgments and those obtained by other techniques can be used to assess the suitability of the latter as indicators of VAQ (Middleton et al., 1985).

The use of experienced observers to rate VAQ has been tested under a wide variety of field conditions (Mumpower et al., 1981; Middleton et al., 1983b; Stewart et al., 1983; Middleton et al., 1984). Stewart et al. (1983) summarized those studies; they concluded that field observations made by experienced observers can be used as a basis for studying VAQ and for monitoring trends of VAQ in an urban area. Similar studies have been conducted for pristine areas (Malm et al., 1980, 1981; Latimer et al., 1981).

Judgments of Photographs

Judgments of VAQ from photographs were found to be highly correlated with judgments made in the field when the photographs were taken (Stewart et al., 1984). Although VAQ tends to be judged slightly worse in photographs than in the field, the relative differences in VAQ estimates for different scenes are about the same whether the estimates are based on photographs or made in the field.

As with field observations, the judgments of photographs of the same scenes by many observers can be averaged to decrease the variance in the responses. Tests show that 50-100 photographs can be judged in one sitting. Because photographic assessments of VAQ can readily accommodate a large number of observers, VAQ estimated from photographs can be more reliable than that provided by field observations, where the number of observers must be restricted because of time limitations and the expense of traveling from site to site. Photographic judgments also can be used as an alternative to field judgments for assessing the relationship of other measurement techniques to VAQ. One example is the recent visibility standards development process for Colorado, which used judgments of photographs as the technique for determining acceptable levels of VAQ (Ely et al., 1991).

Field Judgments by Passersby

Another alternative to field judgments by experienced observers is to conduct on-site interviews of passersby to obtain their judgment of the VAQ at the time and location of the interview. Stewart et al. (1983) found that, when the judgments of a number of passersby were averaged, the resulting average was highly correlated with judgments of trained observers. However, the judgments of passersby were more variable, so that 17 passersby would have to be interviewed to equal the statistical reliability of results obtained from three trained observers. Given limitations of reliability and cost and the restrictions on usable sites, on-site interviews do not appear to be a viable alternative to using experienced field observers or slide-based judgments.

Airport Visibility

Airport visual range is routinely estimated by trained meteorological observers. (Visual range is defined as the maximal distance at which a large black object can be perceived on the horizon.) Even though viewing conditions were well defined and the viewers well-trained, airport visual range was only weakly correlated to judgments of overall VAQ made by experienced observers in a study by Middleton et al. (1984). That can be explained, in part, by the fact that visual range is only one of the many factors that constitute VAQ. Indeed, airport observers are specifically trained to focus solely on visual range and to carefully exclude other visibility-related factors that might affect their judgment of visual range. Another deficiency of airport visual range data is that airports usually are located on the outskirts of cities; the public's perception of VAQ is based on observations during everyday life, which is largely spent in the cities themselves. In spite of the weak relationship between airport visibility and field judgments of VAQ, airport data have one positive feature that all other VAQ techniques lack—a long historical record. Airports have been recording visibility for decades. Such records are valuable for assessing long-term visibility trends and have been used extensively for that purpose (Trijonis et al., 1990).

Perceptual Cues

Human judgments of visual air quality are a composite of judgments of perceptual cues associated with scene characteristics, such as the color of the air, the clarity of objects at a distance, and the existence of borders between clear and discolored air (Stewart et al., 1983). Several visual indexes that are related to individual cues have been proposed as surrogates for the overall judgment of VAQ. Those indexes, which have been summarized in Trijonis et al. (1990), include prevailing visibility, apparent contrast, equivalent contrast, average landscape contrast, modulation depth, blue-red ratio, color difference delta E, and "just noticeable difference" (JND) (Malm et al., 1980, 1981; Latimer et al., 1981; Malm and Pitchford, 1989; Hill, 1990). Prevailing visibility and apparent contrast are based on observations of a specific target or element in a scene. Apparent contrast is computed from measurements of target and background radiances. Equivalent contrast, average landscape contrast, and modulation depth account for the spatial structure throughout a scene and require the measurement of radiance at many positions. Measurements of color differences involve adaptations of colorimeters optimized for color matching in the natural environment (Henry and Matamala, 1990). JND is defined as the change in the input stimulus (e.g., contrast) required for an observer to perceive that change 70% of the time.

The usefulness of the different indexes depends on how closely the index capture the overall judgment of VAQ in a particular setting and a particular time. In some cases, the index might be a major component of the overall judgment, and, in other cases, a small component. Additional field testing of the indexes is needed to determine their usefulness.

Summary

Human judgments of VAQ and perceptual cues are being used to help establish standards (Ely et al., 1991) and to document public assessment of VAQ in conjunction with measurements of physical, chemical, and optical properties (e.g., Malm et al., 1981; Middleton et al., 1984). More current measurements of light scattering and extinction, for example, are used to monitor changes in visibility. Periodic assessment of the public response to VAQ changes is determined by special programs. Measurements of human response are used to determine the selected thresholds for acceptable visibility.

Appendix C

Source Identification and Apportionment Models

SPECIATED ROLLBACK MODELS

The speciated rollback model is a simple, spatially averaged mathematical model that disaggregates the major airborne particle components into chemically distinct groups that are contributed by different types of sources (Trijonis et al., 1975, 1988).

A linear rollback model is based on the assumption that ambient concentrations C above background are directly proportional to total emissions E in the region of interest. Stated as a formula, $C - C_b = kE$, where C_b is the background concentration due to emissions other than E (i.e., to emissions outside the region of interest; natural sources, even inside the region, are usually included in this background term). The constant of proportionality, k, is determined over a historical time period when both concentrations, C and C_b, and regional emissions, E, are known. With that information, new concentration estimates can be derived for proposed changes in emission levels. For an inert pollutant, the only assumption required for the model to be exactly correct at all points in the region is that the relative spatial distribution of emissions remains fixed despite the changes in emissions.[1] That assumption be-

---

[1]As with all models, there is an assumption that the concentrations apply to some given meteorology and given averaging time. In a temporal sense, the rollback model has the requirement that the temporal (e.g., diurnal) pattern of emissions remains fixed.

comes less restrictive when the model is applied to larger, three-dimensional, and spatially averaged concentrations rather than to concentrations at individual points in a geographic region. Thus, the model is especially useful for spatially averaged problems, such as regional haze.

A speciated rollback model for airborne particles is an aggregation of several separate rollback models for each individual chemical component of the atmospheric particle complex. In almost all cases, the anthropogenic materials in the dry particle mass almost entirely consist of five components: sulfates, organics, elemental carbon, nitrates, and crustal material (e.g., soil dust and road dust). Organics can be further subdivided into primary organic and secondary organic particles. In the simplest case, it is assumed that linear rollback models can relate each primary particle component (elemental carbon, crustal material, and primary organics) to its regionwide emission level and each secondary aerosol component (sulfates, nitrates, and secondary organics) to the emission level of its controlling gas phase precursor (e.g., SO_2, NO_x, NH_3, and VOC).

In considering the *ambient* nature of airborne particles (not only the measured dry fine-particle mass), particle-bound water is an additional important component. Certain chemical constituents of anthropogenic particles—such as sulfates, nitrates, and some organics—have an affinity for water. The constituents acquire water vapor from the atmosphere and form a liquid phase at relative humidities well below the 100% level normally associated with condensation. If the concentration of hygroscopic particles (i.e., those that retain water) is reduced, there is a corresponding reduction in the particle-bound water. Accordingly, water retention is usually incorporated into the rollback models for the hygroscopic airborne particles (i.e., sulfate-bound water is assumed to change in proportion to sulfate concentrations at a particular relative humidity, nitrate-bound water in proportion to nitrate concentrations, etc.). For example, if one is considering a rollback model for visibility effects, the total light-extinction contribution from nonbackground sulfate particles plus sulfate-associated water is assumed to change in proportion to SO_x emissions.

The speciated rollback model incorporates a very restrictive assumption in addition to the assumption about the spatial homogeneity of emission changes. The restrictive assumption is that there is only one controlling precursor for each secondary airborne particle component and that transformation and deposition processes are completely linear with respect to the precursor (i.e., that transformation rates and deposition velocities are independent of pollutant concentrations).

Various complexities can be added to the speciated rollback model. First, rollback models can be disaggregated by particle-size fraction (e.g., coarse versus fine particles) as well as by chemical composition. Second, additional distinctions can be made between primary and secondary particles (e.g., separate rollback models can be formulated for primary versus secondary sulfate particles). Third, rather than using proportional relationships, nonlinearities in transformation processes can be approximately accounted for by assuming nonlinear functional relationships between emitted precursors and their atmospheric reaction products. Finally, the model can be disaggregated spatially by including separate transfer coefficients for different source areas or stack heights. The latter two modifications are ways of relaxing some of the restrictive assumptions of the rollback technique.

Four types of information are needed to implement a speciated rollback model:

• Data on airborne particle concentrations disaggregated by components of the particles;
• Knowledge or assumptions regarding the controlling precursor for each secondary airborne particle component;
• Emission inventories for the important source categories of each airborne particle component and each gaseous precursor substance;
• Knowledge or assumptions regarding background concentrations (due to sources other than those that are in the inventory) for each component of the airborne particles and each gaseous precursor substance.

One of the advantages of the rollback model is that this type of information is often obtained as a first step in any reasonable and practical control plan; therefore, rollback models often can be formulated and applied at an early stage in the source attribution process.

RECEPTOR-ORIENTED MODELS BASED ON CHEMICAL SIGNATURES

Receptor modeling is an active and developing field of research that has given rise to many different approaches and techniques. The following discussion provides a taxonomical overview, identifying some recurrent themes and attempting to clarify the relationships among various models. We then focus on two models (i.e., chemical mass balance and regression analysis) that are used so often that a more detailed discussion of their formulation is warranted.

A variety of techniques extract information on the types of sources contributing to a given airborne particle sample on the basis of the particles' chemical composition. All the techniques are conceptually based on the same underlying model (Friedlander, 1977; Henry et al., 1984; Hopke, 1985; Gordon, 1988):

$$c_{it} = \sum_j f_{ijt} S_{jt}. \qquad \text{(C-1)}$$

In Equation C-1, the subscripts i, j, and t index ambient aerosol characteristics, emissions sources, and sampling intervals, respectively. The terms c_i, S_j, and f_{ij} are defined as follows:

c_i is the ith characteristic of the airborne particles at the receptor site. That characteristic is typically the mass concentration (particle mass per unit air volume) of the particles or particle component. The characteristic also can be an air pollutant effect, such as the light-extinction coefficient (extinction cross-section per unit air volume) (Pitchford and Allison, 1984) or mutagenicity (revertants per unit air volume) (Lewis et al., 1988). For simplicity, our discussion will take the c_i to be the mass concentration of the ith particle component.

S_j is the ambient mass concentration (effluent mass per unit air volume) of the total effluent contributed by the jth emissions source at the receptor site. This contribution often is referred to as the source strength. A source can be defined as a specific industrial facility, such as the Navajo Generating Station (NGS) at Page, Arizona (NPS, 1989; NRC, 1990), a generic category, such as soil dust (Friedlander, 1973), or a geographic area, such as the midwestern United States (Rahn and Lowenthal, 1985).

f_{ij} is the mass fraction (particle mass per effluent mass) as measured at the receptor site, of particle component i in the effluent of the jth source. The sequence $f_{1j}, f_{2j}, \ldots, f_{nj}$ is referred to as the jth source's profile, or its chemical signature or fingerprint. For conserved chemicals, it may be possible to measure f_{ij} at the source (Core, 1989a). For substances produced or destroyed in the atmosphere, measurements at the source may have limited value (NRC, 1990).

All the techniques have as their objective the estimation of the source contributions, $f_{ij} \, S_{ij}$.

A class of methods referred to as chemical mass balances (CMB) can be applied to the solution of Equation C-1 when the source characteristics f_{ij} are known. The simplest case involves a conserved substance i, which is emitted by a unique source j. Such a tracer can be endemic, such as lead in Los Angeles automobile exhaust during the early 1970s (Miller et al., 1972), or inoculated, such as deuterated methane (CD_4) that was injected into the effluent of the Navajo Generating Station during the Winter Haze Intensive Tracer Experiment (WHITEX) (NPS, 1989; NRC, 1990). For the unique tracer, Equation C-1 simplifies to $c_{it} = f_{ijt}S_{jt}$, which can be solved directly for the source strength in terms of the measured ambient concentration of the tracer and the mass fraction of the tracer in the source's effluent: $S_{jt} = c_{it}/f_{ijt}$. The jth source's contribution to another conserved substance, one that may be emitted by multiple sources, then can be calculated from the mass ratio measured at the jth source:

$$c_{i't} = f_{i'jt}S_{jt} = c_{it}(f_{i'jt}/f_{ijt}). \qquad (C-2)$$

In most cases sources are distinguished by overall chemical profiles rather than by unique individual substances. Such situations are typically modeled in terms of n conserved substances that are wholly accounted for by the emissions of $m \leq n$ sources. If the chemical profiles are linearly independent, then the system given by Equation C-1 ($i = 1, \ldots, n$) can be solved for source strengths S_j ($j = 1, \ldots, m$) in terms of the measured ambient concentrations c_i and the source characteristics f_{ij}. To minimize the effects of measurement error, the number of substances is usually taken to exceed the number of sources ($n > m$), in which case an overdetermined solution is estimated by weighted least-squares fitting procedures (Watson et al., 1984). Useful information sometimes can be obtained even when there are more sources than substances (White and Macias, 1991). Given an estimate of source strengths, source contributions can be derived for any conserved substance, whether it is one of the n markers used in the solution or one with additional sources. If there are many more measured chemical substances than sources, then the comparison of modeled concentrations with observed ambient concentrations of all chemical substances can provide a valuable internal check on model consistency (Friedlander, 1973; Kowalczyk et al., 1978).

The CMB model possesses some attractive properties as a tool for apportioning conserved characteristics of the ambient airborne particles. Unlike the statistical approaches discussed below (e.g., factor analysis), the CMB model can be applied to individual ambient samples. More critically, it is an easily understood and easily scrutinized model that is straightforwardly derived from physical principles, and it contains no unmeasured quantities. The CMB's deterministic character carries a cost, however; it requires comprehensive prior information on the identities and chemical characteristics of all important sources that contribute to the ambient aerosol.

When multiple ambient samples are available, a class of methods referred to as factor analysis offers empirical insights into the identities and characteristics of major sources. The basic idea behind factor analysis is that the ambient concentrations of various conserved chemical substances should correlate with each other if they have a common source. That idea can be seen in the simplest case, where j is the only source of substances i and i', and the source characteristics $f_{ijt} = f_{ij}$ and $f_{i'jt} = f_{i'j}$ are stable from one sample to the next. According to Equation C-1, the only source of variability in the ambient concentrations $c_{it} = f_{ij}S_{jt}$ and $c_{i'} = f_{i'j}S_{jt}$ is then the common source strength S_{jt}. The two concentrations should therefore correlate, both being high when source j is present and both being low when source j is absent; moreover, their standard deviations should be proportional to the substances' abundance at the source. Inverting that logic, one can hypothesize that substances that are highly correlated in ambient air have a common source, and one can infer the chemical signature of the source from ambient measurements alone.

Factor analysis provides a framework for partially extending the simple reasoning outlined above to situations with multiple sources. A sequence of p ambient measurements, each characterizing n substances, can be represented as a cloud of p points in n-dimensional space ("Q-mode analysis") or n points in p-dimensional space ("R-mode analysis") (Hwang et al., 1984). In either representation all points should, according to Equation C-1, lie within model and measurement error of the m-

dimensional hyperplane determined by the chemical profiles of the m distinct emissions sources. The algebra of factor analysis allows the dimensionality and orientation of this hyperplane to be estimated from the data. The source profiles themselves can be recovered in the special case where each substance has a unique source (via "VARIMAX rotation") or when the profiles are approximately known already (via "target transformation") (Hopke, 1985). Factor analysis thus serves to validate and refine the source information used in the CMB model. The set of source profiles cannot be recovered uniquely without some such prior knowledge, because the set constitutes only one of an infinite number of possible coordinate systems (Henry, 1987).

In the context of visibility studies, the models of CMB and factor analysis are critically limited by their restriction to airborne particle characteristics that are conserved during transport from source to receptor. As discussed in Chapter 4 of this report, the extinction cross-section of the ambient aerosol is contributed largely by secondary particulate matter, which is not directly emitted by any source, and is inflated by liquid water whose abundance is determined by ambient relative humidity conditions. The optical characteristics of source emissions are thus a function of atmospheric transport and transformation, and are highly variable relative to the tracer substances used in CMB and factor analysis. The optically relevant portion of a source's profile at the receptor site consequently cannot be determined by direct measurements of its emissions but must be estimated by source-oriented modeling or by regression analysis of ambient data.

Linear regression analysis is a well-established (Seber, 1977; Draper and Smith, 1981) and well-studied (Belsley et al., 1980; Fuller, 1987) class of procedures for estimating unknown coefficients in linear relationships from multiple observations of the dependent and independent variables. Equation C-1, adapted to account for the sulfate concentration, adapted to, for example, is a linear relationship in which the source-specific ratios of sulfate to effluent are unknown parameters. Given measured ambient sulfate concentrations c_{it}, and ambient effluent concentrations S_{jt} derived from CMB analyses of conserved substances, regression analysis can generate estimates of the average sulfate-to-effluent ratios f_{ij} at the receptor. The regression estimates are determined by optimizing the agreement between measured and modeled sulfate values.

All the foregoing analyses require the existence of chemical signatures for at least some of the sources of interest in a given application. To be broadly useful, such signatures must be distinctive, stable, and measurable. Because they minimize collinearity problems in the solution of the system described by Equation C-1, the most helpful signatures involve substances predominantly attributable to a single major source or source category. Table 5-2 lists examples of substances that have been used as endemic markers and the sources to which they are usually attributed. Endemic tags also have been identified for some airsheds that are rich in a distinctive source type (Rahn, 1981; Miller et al., 1990). Unique signatures can be created by inoculating targeted sources with substances that are otherwise scarce in the atmosphere (e.g., unusual perfluorocarbons, deuterated methane, or sulfur hexafluoride). Such artificial tags have been applied to specific sources (Shum et al., 1975; Georgi et al., 1987; NPS, 1989) and to airsheds (Reible et al., 1982; Haagenson et al., 1987). Artificial tracers often are used to elucidate airflow patterns. In such studies, the tracer is typically released in discrete puffs. In contrast, tracers used to support receptor modeling should be released over a sustained period to avoid ambient samples that contain an unknown proportion of tagged and untagged effluent. To study particle fluxes, it is necessary that (1) the ratio of tracer injected into the stack to stack gas-particle loading is constant, and (2) the dispersive and depositional characteristics of the tracer and primary particles emitted from the stack are similar.

The distinctiveness of an endemic source signature depends in general on its context. Several of the tracer-substances' source attributions in Table 5-2, for example, must be considered unreliable in the southwestern United States, where copper smelters are important sources of vanadium, arsenic, sellenium, and lead (Small et al., 1981). In actual applications, it can be difficult to verify the attribution of a signature to a specific source. At the large distances over which sources can contribute to regional haze, there may be many sources for any endemic tracer. A useful multi-substance signature based on characteristic substance ratios rather than characteristic substances per se, must preserve its distinctiveness over all combinations of all potential sources of any of the signature's constituents.

Fluctuations in source signatures can produce significant uncertainty in source apportionment. There is some variability, often undocumented, in the composition of emissions from any individual source. The SO_2-to-NO_x ratio in the Navajo Generating Station's emissions varies by about 20% (Richards et al., 1981), for example, and the sulfur-to-selenium ratios in two samples taken during WHITEX differed by a similar amount (NPS, 1989). Additional variability is introduced by the diversity in the chemical composition of emissions from the individual sources that make up a given source category, especially at the large distances relevant to regional haze. The selenium-to-aluminum ratios in fine-particle emissions from coal-fired power plants can vary by 70% (Sheffield and Gordon, 1986), for example, even for facilities located in the same geographic region and using similar particle control technology. Figure 5-1 shows copper smelters within the same geographic region to have widely varied chemical signatures. Even suspended soil dust varies significantly in composition from site to site (Cahill et al., 1981).

No chemical signature is of value unless it can be identified at ambient concentrations. In many national parks and wilderness areas, this requirement places heavy demands on measurement technology. For example, Table 5-1 shows that most of the tracers identified in Table 5-2, including vanadium, manganese, nickel, arsenic, selenium, bromine, and lead, were not quantified routinely by the trace-element monitoring network operated by the National Park Service between 1979 and 1986. The main problem is that a source's impact on visibility through the atmospheric formation of secondary particle components does not lessen with distance in proportion to the dilution of its primary emissions. Moreover, the detectability of chemical signatures does not necessarily improve with the progress of technology, because analytical advances complete with improved emissions controls. As one example, the decrease in automotive lead emissions since the mid-1970s has clearly outpaced increases in analytical sensitivity, making it difficult to use lead as a tracer for automotive aerosol emissions.

Recent use of CMB calculations and regression analysis as part of the visibility impairment study contained in the National Park Service's WHITEX report (NPS, 1989) has focused particular attention on those two source apportionment methods. For this reason, an extended discussion of both models follows.

CMB Models

The CMB model was first proposed by Winchester and Nifong

(1971), Hidy and Friedlander (1972), Kneip et al. (1972), and Friedlander (1973). It has been applied widely to apportionment of sources of primary particulate emissions on local and regional scales, to groundwater problems, and to apportionment of sources of VOCs and air toxics and of sources contributing to light extinction (Cooper and Watson, 1980; Hopke and Dattner, 1982; Hopke, 1985; Pace, 1986; Gordon, 1980, 1988; Watson et al., 1989). The CMB model has been used widely in the regulatory community (EPA, 1987c), and many validation studies have been completed with the model (Stevens and Pace, 1984).

The current state of the art limits the model's regulatory application to particulate matter that is directly emitted to the atmosphere. The ability of the CMB model to apportion airborne particle concentration or light extinction to sources is limited to categories of sources with dissimilar source profiles, because of the assumptions inherent in the model and because of its inability to resolve sources of secondary particles.

The first-order principles of the CMB model have been described (Watson et al., 1991), and assumptions implicit in its application have been documented in the literature (Watson et al., 1991). The sensitivity of the model to deviations from modeling assumptions has been examined in two studies, both of which were designed to determine if the CMB model could be used in regulatory settings (Stevens and Pace, 1984; Javitz et al., 1988a,b).

CMB source apportionment was first used as a basis for regulatory action by the state of Oregon, when in 1977, it sponsored the Portland Aerosol Characterization Study (PACS). PACS was the first large-scale, successful receptor modeling study specifically designed to support State Implementation Plan revisions to attain EPA's Total Suspended Particulate NAAQS (National Ambient Air Quality Standard). The study spawned much of the receptor modeling technology that is in use today (Watson, 1979). The source apportionment results developed during PACS were applied by the staff of the Oregon Department of Environmental Quality in the first joint applications of receptor and dispersion modeling (Hanrahan, 1981; Core et al., 1982).

Concurrent with the revision of the NAAQS for particulate matter, EPA released several guidance documents to state regulatory agencies that supported the use of the CMB model as a technical basis for PM_{10} control strategies (Pace and Watson, 1987; EPA, 1987c). (PM_{10} refers to particles less than 10 μm in diameter.) EPA has continued to support state air regulatory agencies' application of the CMB model by continued development of software (Watson et al., 1991). Source profile information also is being gathered (Core et al., 1984; Shareef et al., 1988; Core, 1989a; Houck et al., 1989).

Theory of the CMB Model

Watson et al. (1990a,b) have described the theoretical basis of the CMB model in several publications.

The CMB model consists of a least-squares estimate of the solution to a set of linear equations that expresses each concentration of a chemical species at a receptor air-monitoring station as a linear sum of the products of source-profile species at the receptor site multiplied by source contributions. The source profile (i.e., the fractional amount of each chemical species in the emissions from each source type) and the ambient concentrations of each species measured at the receptor site with appropriate uncertainty estimates serve as input data to the model. The output consists of the ambient airborne particle mass increment and the amount of each chemical substance contributed by each source type. The model calculates values for the contributions from each source and the uncertainties associated with those source contributions. Input data on uncertainties are used both to weight the importance of input data on chemical species concentrations when computing the solution and to calculate the uncertainties associated with the source contributions.

Derivation and Solutions

The concentration of a conserved pollutant measured at a receptor air-monitoring site during a sampling period of length T due to a source j with a constant emission rate E_j is

$$S_j = D_j \times E_j, \qquad (C-3)$$

where D_j is a dispersion factor depending on wind velocity u, atmospheric stability, and location of source j with respect to a receptor (x).

All these factors vary over time, so the dispersion factor D must be an integral over a specified time period. Various analytical expressions for D have been proposed based on solutions to equations that describe atmospheric transport, but none have completely captured the complex and turbulent nature of atmospheric dispersion. A major advantage of the CMB model is that an exact knowledge of D is not required. Instead the CMB model replaces Equation C-3 above with an equation of the form of Equation C-1, which was described earlier.

If the number of source types that contribute to the airborne particle mass is less than or equal to the number of aerosol chemical features measured, then Equation C-1 can be solved for the unknown source contributions, the S_j's by a variety of methods. These include tracer, linear programming, ordinary least-squares solutions, ridge regression, weighted least-squares solutions, and effective variance least-squares solutions (Britt and Luecke, 1973; Henry, 1982). An estimate of the uncertainty associated with the source contributions is an integral part of several of those solution methods.

The CMB software in current use by EPA applies the effective variance solution because it makes use of all available chemical measurements, it estimates the uncertainties of the source contribution estimates, and it yields the most reasonable solutions because it preferentially weights those chemical species with the higher precision in both source and receptor measurement. The effective variance solution is derived by minimizing the weighted sums of the squares of the differences between the measured and calculated values of c_i and f_{ij}.

CMB Model Assumptions

Assumptions implicit in the version of the CMB model recommended by EPA include

(1)   The composition of the source emissions is not changed by transformation or deposition as the plume is dispersed downwind to the receptor, and the composition is constant over the time period of ambient and source sampling.

(2)   Chemical species do not react with each other—i.e., conversion of gases to particles and reactions between particles do not occur for the

species used for fitting a solution to the CMB equations. The chemical species are assumed to be linearly additive.

(3)   All sources with a potential for large contributions to pollutant concentrations at the receptor site have been identified, and their emissions have been characterized.

(4)   The relative chemical composition profiles that describe the emissions sources are linearly independent of each other.

(5)   The number of sources or source groups is less than or equal to the number of chemical species measured.

(6)   Measurement errors are not seriously correlated from one constituent to another and are not seriously biased (the calculation proceeds as though all correlations and biases vanish).

Assumptions 1 through 6 are fairly restrictive and will probably never be fulfilled totally. Fortunately, model validation studies using synthetic data sets have shown that deviations from the assumptions often can be tolerated by the model (within practical applications), although as the deviations increase, the uncertainties in the source contribution estimates also increase.

CMB Model Validation Studies

Validation studies (Stevens and Pace, 1984) have shown that the CMB model typically can resolve the separate contributions of five or six major emission sources to the ambient primary airborne particle mass. In simulations of a local airshed containing seven major sources—airborne soil dust, a coal-fired power plant, sea salt, a steel mill, a lead smelter, a municipal incinerator, and background aerosol—the CMB model was able to allocate the contribution of the coal-fired power plant to within an average relative uncertainty of $\pm 50\%$ and contributions of the oil-fired power plant to within an average of $\pm 20\%$, even though the standard error of each source profile's daily fluctuation was 25% and the standard error of the airborne particle measurements was 10% (Javitz et al., 1988a). The studies focused on urban settings and excluded secondary sulfate contributions from the sources.

The studies indicate that, as the composition of the source emissions varies (Assumption 1), errors in the estimated source contributions also vary in direct proportion to the magnitude of the bias. If the errors are random, the magnitude of the estimated error of the source contributions decreases as the difference between the number of species and sources increases. For that reason, use of the maximal possible number of fitting species in the model is encouraged.

Few studies have been performed on the basis of Assumption 2 (linear summation of species), but because errors introduced by the conversion of gases to particles and the reactions among particles are not necessarily linear, the model's ability to apportion secondary particles (taken as the quantity of ammonium, nitrate, sulfate, or organics that remains unexplained following apportionment of primary emissions) correctly among primary sources might be suspect. As chemical-reaction mechanisms and, in particular, the distribution of organic-reaction products become better understood, it might be possible to produce "fractionated" source profiles that can be used to apportion reactive species approximately among sources.

Regarding Assumption 3 (inclusion of correct profiles for all sources), model sensitivity studies have shown that (1) underestimation of the number of sources will have little effect on the calculated source contributions if prominent species contributed by unidentified sources are excluded from the calculation procedure; (2) if the number of sources is overestimated the contributions of those that are included will be underestimated because of the common properties of the emissions contributed by both included and excluded sources; (3) if major source types present in the airshed are excluded from the analysis, the calculated-to-measured ratio of the fitting species and other fitting criteria will prove unsatisfactory; and (4) if the number of sources is overestimated, the standard errors of the source contributions increase and the sources that are not present in the airshed are estimated to have smaller contributions than the standard errors of those estimates.

The linear independence requirement (Assumption 4) has been directly addressed in EPA's CMB software with inclusion of Henry's singular value decomposition analysis (Henry, 1982). When a model solution is reached that consolidates similarity clusters (groups of sources that cannot be resolved by the model), then the likelihood of significant errors due to collinearity is greatly reduced.

With regard to Assumption 5, the true number of individual sources contributing to receptor substances is usually much larger than the number of chemical species that can be measured. It is therefore common practice to group sources that have similar chemical composition into composite source types. For example, wind-entrained soil dust, emissions from a rock crusher, paved-road dust, and agricultural-tilling dust often are grouped together into a "geologic" source type represented by a source profile that has the chemical composition of soil dust.

There are no results from validation studies currently available to judge the effect of deviations from Assumption 6 (randomness, normality, and uncorrelated nature of measurement uncertainties), but simulations (sensitivity studies) as well as theoretical development, support the focuses and relative emphases implied by the wording of that assumption.

Apportionment of Light Extinction

Since the CMB model only apportions airborne particle mass concentrations among source types, further calculations must be completed to use those source contribution estimates to apportion contributions to light extinction among source types. If the particle mass contribution at a receptor site is known from CMB calculations and the chemical composition of the particles is known, then the concentration of each chemical substances (typically primary SO_4^{2-} and NO_3^-, organics, light-absorbing carbon, and fine-particle soils) contributed by the source can be calculated. Because the physical and chemical characteristics of the substance vary as a function of the environmental conditions (especially humidity), particle size, and particle shape, additional knowledge of the light-extinction efficiency (extinction-to-mass ratio coefficient in units of meters squared per gram) of each particle type (under assumed particle and atmospheric conditions) must be attained to calculate the contribution to the extinction coefficient associated with primary emissions from each source. Those extinction efficiency values are typically derived from the literature. Care must be taken to ensure that the extinction efficiency values represent the actual particle size, morphology, and humidity conditions that exist in the atmosphere under study.

Regression Analysis

Regression analysis generates empirical relationships of the form

$$c_{it} = \sum_j f_{ij} S_{jt} \qquad\qquad (C\text{-}4)$$

between ambient concentration c and source strengths S_j, the terms being as defined following Equation C-1. Equation C-4 has no t subscript on the factors f; conventional regression analysis addresses only the average relationship of c_i to the S_j, not the variations in this relationship from observation to observation. The subscript i on the mass or effect concentration c_i usually can be dropped with no loss of clarity. In the terminology of regression analysis, c is the response, or dependent, variable; the S_j are the regressor, or independent, variables; and the f_j are the regression coefficients.

In practice, the regressors are often taken to be variables that are proportional to source strengths, rather than the source strengths S_j themselves. The corresponding proportionality constants are then incorporated into the empirical regression coefficients. For example, tracer substances attributed to unique source types are commonly used directly as regressors, in which case the reciprocal of the tracer concentration in the emissions becomes an implicit factor of the regression coefficient. Apportionments based on such regressions do not depend on that possibly unknown factor, because it cancels out in the calculation of absolute contributions when the coefficient is multiplied by tracer concentration instead of source strength.

Background

Multiple regression has been widely used to apportion total particle mass among different types of emission sources. The most common approach has been to use tracer concentrations directly as regressor variables (Kleinman et al., 1980; Cass and McRae, 1983; Currie et al., 1984; Dzubay et al., 1984; Lewis et al., 1986; Valaoras et al., 1988). Lewis et al. (1986) and Dzubay et al. (1988) pre-treated their tracer data with mass balance techniques to remove nonautomotive contributions to lead and soil-derived contributions to other trace elements. Morandi et al. (1987) similarly employed preliminary regressions of multiple source tracers on unique source tracers to refine their regressors. Hopke et al. (1980), Thurston and Spengler (1985), and Pratsinis et al. (1988) inferred source types through factor analysis of elemental data and used the factor levels as source strengths in subsequent regression analyses.

Multiple regression analysis also has been tried as a method for apportionment of ambient concentrations of particulate sulfur among contributing sources. Some approaches, like those for total mass reviewed above, rely on chemical information. Regressors can be source strengths derived through mass-balance techniques (Rahn and Lowenthal, 1985) or raw tracer concentrations (Dzubay et al., 1988). Other approaches exploit a superior knowledge of temporal and spatial patterns in sulfur emissions. Regressors can be emission rates that fluctuate daily (White et al., 1978) or monthly (Sisler and Malm, 1990) or binary indicators that tell whether a source is operating ($=1$) or out of service ($=0$) (Murray et al., 1990). Alternatively, regressors can be the prior residence times of sampled air over specific geographical regions, estimated by counting end points of calculated back-trajectories (Iyer et al., 1987; Gebhart and Malm, 1990).

Aerosol chemical substances that have been apportioned among contributing sources by multiple regression include, in addition to sulfates, elemental and organic carbon (Daisey and Kneip, 1981; Shah et al., 1985; Lewis et al., 1988; Pratsinis et al., 1988) and total (gas plus particle phase) sulfur and nitrogen (White and Roberts, 1977; Malm et al., 1990). Regressions of all measured elements on a fixed set of marker elements have been used to complete the chemical profiles of sources for which only unique tracers were known (Currie et al., 1984; Lewis et al., 1986; Rheingrover and Gordon, 1988). Some regression analyses have taken light extinction (Khalil et al., 1983; Pitchford and Allison, 1984) or mutagenicity (Lewis et al., 1988) as the response variable, directly apportioning effects rather than the gravimetrically or chemically determined airborne particle mass. (Regression analysis is used more often to relate light extinction to the optically important airborne particle components, rather than to source tracers; that application was discussed in Chapter 4.)

Currie et al. (1984) tested the apportionment of primary particles among sources by multiple regression in a methods intercomparison study based on simulated data. Regression analyses generated independent estimates of the chemical composition of contributing sources that were reasonably consistent with the true chemical composition of those sources and provided estimates of source contributions to ambient particle mass concentrations comparable to those of CMB and other receptor-oriented methods. Among the sources accurately characterized was an "unknown source," the existence of which had not been disclosed to the participants in the methods comparison study. The simulated data did not incorporate certain aspects of real atmospheric problems, most nota-bly the variability of source effluent chemical composition from one observation to the next.

In the real atmosphere, regression-derived apportionments of predomi-nantly primary airborne particle fractions have survived various cross-checks against emissions data. Cass and McRae (1983) found that the empirically estimated tracer contents of highway, oil-ash, and crustal source contributions were consistent with careful prior estimates based on local emissions source tests. Kleinmann et al. (1980) were able to relate results for motor vehicle and oil combustion fractions over several years to annual variations in the lead and vanadium contents of gasoline and fuel oil. Lewis et al. (1986) related differences in the empirically estimated lead contents of motor vehicle exhaust particles in different studies to differences in the prevailing lead content of gasoline.

Perhaps the most convincing validation yet of the use of regression analysis as a source apportionment tool was produced by Lewis et al. (1988) in their work on mutagenicity. Those authors apportioned total carbon between wood smoke and motor vehicle exhaust, by regressing carbon on nonsoil potassium and lead. They then had several samples analyzed for ^{14}C content, which is a direct indicator of the fraction of the total carbon that is due to contemporary (wood smoke) rather than fossil-fuel (motor vehicle exhaust) sources. Figure C-1 shows that the independent, nonstatistical ^{14}C measurements nicely confirm the source apportionment results obtained by regression analysis. The closeness of the correlation ($r = 0.88$) is the more impressive when we note that the fluctuations in absolute carbon concentrations, which would contribute common variability to both determinations, have been factored out in this presentation.

The accumulation of published successes does not wholly dispose of what has been called "the file drawer problem" (Rosenthal, 1987). This is the problem that uncounted file drawers might be filled with regres-sion analyses that were never published because they failed available cross-checks. Regression analysis is of greatest interest in precisely those applications for which emissions data do not exist, and alternative approaches are infeasible. It is of little value to know that regression analysis sometimes gives valid results, if there is no way to identify

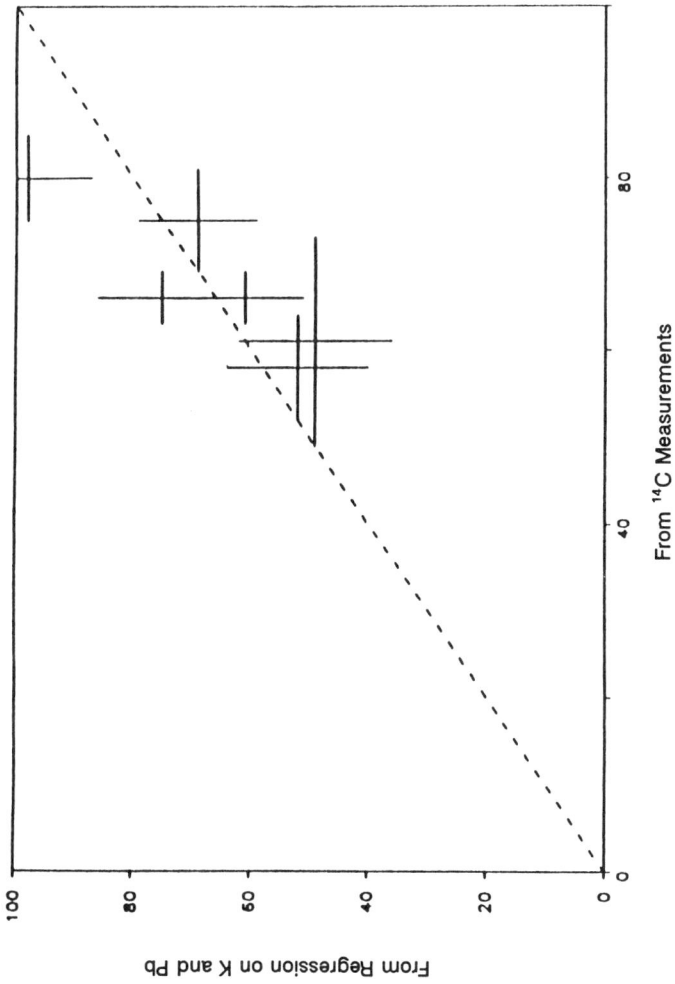

FIGURE C-1 Relative contribution of wood smoke to ambient fine-particle carbon in Albuquerque, New Mexico, during the winter. Sloping line indicates agreement between regression estimate and direct measurement. Data are from Lewis et al. (1988).

whether a particular analysis, which cannot be checked, is in fact producing the correct answer. Regression must be demonstrated to work nearly all the time in specified contexts, and that can be established only through formal trials that report all results, the failures and the successes.

The apportionment of predominantly secondary aerosol fractions by regression has yet to be rigorously tested. Apportionments of total mass have not addressed regression's performance on secondary material, because secondary substances such as sulfates and nitrates usually have been included as regressors in their own right (Kleinman et al., 1980; Cass and McRae, 1983; Dzubay et al., 1984; Lewis et al., 1986; Morandi et al., 1987; Dzubay et al., 1988). Sulfates and nitrates have thus been treated as explicit categories in the mass apportionments, with only the remaining, undifferentiated, predominantly primary material apportioned among sources types. Lewis et al.'s (1988) validation of their source apportionment study did not relate to secondary substances, because their data were from the winter, when photochemical conversion would have been minimal. No attempt has been made yet to replicate Currie et al.'s (1984) evaluation study with atmospheric conversion processes included in the simulated data base. By the nature of the secondary particle problem, cross-checks against source measurements are not helpful.

The support available for regression-derived apportionments of secondary particles is largely based on demonstrations of the internal consistency of the results obtained. Lowenthal and Rahn, 1988, for example, showed that the regression model of Rahn and Lowenthal (1985) yielded reproducible coefficients when fit to data from different years. Murray et al. (1990) noted that their regression coefficients yielded estimated source impacts that were least at their upwind monitor and greatest at the closer of two downwind monitors.

Statistical Assumptions and Consequences of Violation

Regression analysis determines coefficients f_j for the regression model described by Equation C-4 that optimize its fit to the data for ambient concentrations c_t and source strengths S_{jt}. Model fit can be characterized in various ways. Nearly all source apportionment analyses have employed ordinary least-squares (OLS) regression, which minimizes the mean squared difference between modeled and observed values of the response variable. The OLS approach models ambient concentrations as the sum of deterministic and random contributions:

$$c_t = \sum_j f_j S_{jt} + \epsilon, \qquad\qquad (C-5)$$

the deterministic component being the linear relationship of Equation C-4. The scatter in the observations is represented as a random error, ϵ in this relationship.

The scatter in Equation C-5 arises most fundamentally from the fact that the true coefficients f_{jt} generally fluctuate from observation to observation, as indicated in the original Equation C-1. Such fluctuations can be pronounced particularly in applications to visibility, where f_{jt} typically represents a ratio of secondary airborne particles to primary tracer. Such a ratio varies widely with the age of the emissions, atmospheric oxidant concentrations, atmospheric liquid water content, and other atmospheric factors. The approximation of the fluctuating relationship by a constant one can thus introduce significant errors in individual observations if the estimated value of f_j is later used to try to explain the relationship between individual observations of c_t and S_{jt}.

Additional scatter is introduced when source-strength estimates S_{jt} are unavailable for some categories of emissions. The standard practice is to treat untagged emissions collectively as "background" and represent the background in the regression equation as a constant term f_0, which represents the aggregate contribution of all untagged sources. In the geochemical community, the term "background" is usually reserved for contributions that are globally uniform or are predictable functions of latitude or altitude, for example. In regression analyses of air pollution, however, the background of untagged emissions typically varies just like any other component of the mix, and so is poorly predicted by any average value.

On the operational level, scatter arises from uncertainties in the determination of source tracer concentrations. Chemical concentrations are measured with only finite precision, and the precision can be rather poor because of the low primary pollutant concentrations measured in many regional haze analyses. Random errors in estimated source tracer concentrations clearly translate into random errors in predicted ambient concentrations. An analogous problem is that the ratio of tracer species emissions to total emissions from a source may fluctuate, reflecting inhomogeneities in fuels and feedstocks. Those fluctuations are similar to measurement errors in their effects on estimated source contributions to ambient samples.

Finally, the underlying relationship (Equation C-1), might involve a stochastic component, and be inherently imprecise itself. For example, the ambient mass or effect concentration c and the source strengths S_j might be measured in different air volumes. The comparison of path-averaged optical measurements with point measurements of aerosol chemical composition furnishes a common illustration of such a mismatch, as noted in Chapter 4.

The error in a regression relationship's reproduction of individual observations is of minor interest in most applications to source apportionment. Apportionments instead focus on the coefficients f_j, which represent empirical estimates for the mean values $m(f_{jt})$ of the unknown source characteristics f_{jt}. Those estimates are multiplied by the observed mean source strengths $S_j = m(S_{jt})$ to derive the estimated mean source contributions $f_j S_j = m(f_{jt} S_{jt})$. (That multiplication is grounded in the assumption that any fluctuations in the source characteristics are uncorrelated with variations in the source strengths, an assumption that is implicit in the representation of the source-ambient relationship as linear.)

Because of the empirical scatter in the relationship of ambient concentrations to source strengths, one cannot hope to determine mean source characteristics exactly with a finite number of observations. It seems reasonable to expect the estimation procedure to be unbiased, however, yielding results that are neither systematically high nor systematically low. It also seems reasonable to expect the estimates to be consistent, approaching the correct values as the number of observations increases. An advantage of the OLS approach over other forms of regression analysis is that the conditions required to establish the desirable properties are simply stated and do not involve the often unknown magnitudes of the relationship's random elements. For the OLS estimate—and for most other estimates that are often used—to be unbiased and consistent, it is sufficient that the error ϵ in Equation C-5 be statistically independent of the source strengths S_j (Goldberger, 1964).

The condition that ϵ be random, varying independently of the S_j, is a familiar assumption in statistical analysis, one seemingly legitimized by repeated invocation. However, a closer consideration of the physical relationship of ambient concentrations to estimated source strengths reveals several potential sources of correlation between ϵ and one or more of the S_j.

An absence of source-strength estimates for major contributors to the ambient mix is perhaps the most obvious source of probable bias in regression-derived apportionments. The potential magnitude of such bias can be illustrated by a simple calculation. Suppose that a particular source emits a tracer X from which its strength S_1 in the ambient atmosphere can be determined accurately. For simplicity, suppose also that the pollutant of interest is emitted in constant proportion to X and conserved in the atmosphere. The source-ambient relationship then can be written as $c_t = c_{bt} + f_X S_{1t}$, where c_{bt} is the fluctuating background contribution of untagged pollutant and $f_X = c_{Xt}/S_{1t}$ is the constant ratio of tagged pollutant to source strength. The sole source of scatter in the regression model $c_t = f_0 + f_1 S_{1t}$ is then the variation $c_{bt} - m(c_{bt})$ in background concentration. The regression coefficient f_1 is linear in $c = c_b + c_X$, and is therefore the sum $f_b + f_X$ of the regression coefficients of c_b and c_X on S_1. The bias in regression's attribution of pollutant to the tagged source is thus

$$\text{Estimate} - \text{True} = [f_b + f_X]m(S_1) - f_X m(S_1) = f_b m(S_1). \tag{C-6}$$

The regression coefficient f_b can be expressed in terms of more elementary statistics (e.g., Edwards, 1984) as

$$f_b = r(c_b, S_1)s(c_b)/s(S_1), \tag{C-7}$$

r and s being the usual correlation coefficient and standard deviations.

The quantities $s(c_b)$ and $r(c_b, S_1)$ appearing in Equation C-7 are unknown, but can be roughly estimated by empirical rules of thumb. One such guide, for ambient concentrations of pollutants with moderate atmospheric lifetimes, is that the standard deviation and mean are generally of comparable size (e.g., Hammerle and Pierson, 1975; Tuncel et al., 1985). (That regularity is related to the common observation that concentration distributions are approximately lognormal (Ott, 1990): the ratio s/m for a lognormal distribution is a relatively weak function of the geometric standard deviation s_g (Aitchison and Brown, 1957), with X at $s_g = 2.3$.) The substitution of $m(c_b)/m(S_1)$ for $s(c)/s(S_1)$ in Equation C-7 greatly simplifies the formula for bias:

$$\text{Estimate} - \text{True} = f_b m(S_1) =$$
$$r(c_b, S_1)[m(c_b)/m(S_1)]m(S_1) = r(c_b, S_1)m(c_b). \qquad \text{(C-8)}$$

The ambient correlation of unrelated emissions can be seen in the observed correlations of distinct source tracers or calculated source strengths where these are available. Such "spurious" correlations are clearly source- and site-specific, but are commonly substantial: Hammerle and Pierson (1975) found that $r = 0.79$ between lead (motor vehicles) and vanadium (fuel oil and soil dust) in the Los Angeles basin, for example; Lewis and Macias (1980) found that $r = 0.45$ between lead and selenium (coal) in West Virginia. Correlations with untagged emissions can be inferred by extrapolating from those observations, or by considering the common influence of meteorology on all emissions. As an example of the latter, Samson (1978, 1980) found that $r > 0.5$ between the eastern United States sulfate concentrations and the reciprocals of upstream wind speeds. Similarly, Patterson et al. (1981) found that $r = 0.4$ in the East between regional-average reciprocal visual range and regional-average air-mass residence time.

Our simple calculation is completed by setting $r(c_b, S_1) = 1/2$ in Equation C-8, as an approximate value. The bottom line is then that regression analysis can attribute incorrectly half of the untagged emissions to the tagged source: 2/3 of the ambient total can be attributed to a source that actually contributes 1/3, for example. That amounts to a sizable bias, unless the tagged emissions dominate all other contributions, in which case sophisticated data analyses are probably unnecessary in the first place.

Poor source-strength estimates are another straightforward source of bias in regression-derived apportionments. Errors in the measurement of a chemical signature, or fluctuations in its relationship to source strength, clearly degrade its performance as a predictor of ambient pollutant concentrations. This random empirical decoupling is manifested as a systematic depression of the corresponding regression coefficient.

The effect of imprecision on regression coefficients is easily understood in the case where only one source is considered. Suppose that estimates S'_1 of the source strength S_1 are accurate on average, but contain a random error δ_1. Consider the actual and ideal regressions $c = f'_0 + f'_1 S'_1$ and $c = f_0 + f_1 S_1$ of ambient concentration on estimated and true source strengths. The regression coefficient for S_1 is $f_1 = \text{cov}(c, S_1)/s^2(S_1)$, where $\text{cov}(x, y) = (n - 1)^{-1} \sum_t (x - m(x))(y - m(y))$ is the covariance (Edwards, 1984). Simple algebra shows the regression coefficient for S'_1 to be $f'_1 = \text{cov}(c, S_1 + \delta_1)/s^2(S_1 + \delta_1) = [\text{cov}(c, S_1) + \text{cov}(c, \delta_1)]/[s^2(S_1) + 2\,\text{cov}(S_1, \delta_1) + s^2(\delta_1)]$. Since the error δ_1 is random, we may assume its correlation, and hence covariance, with S_1 and c to be negligible. The formula for the degraded coefficient then simplifies to $f'_1 = f_1 F_1$, where

$$F_1 = s^2(S_1)/[s^2(S_1) + s^2(\delta_1)]. \qquad \text{(C-9)}$$

Equation C-9 shows that random errors in the estimation of a source strength depress the corresponding regression coefficient by a factor $F < 1$, sometimes called the reliability coefficient (Cochran, 1968), involving the ratio of error variance $s^2(\delta)$ to true variance $s^2(S)$. When the mean tracer concentration is near the detection threshold, the analytical precision $s(\delta_{anal})$ of the tracer measurement is by definition approximately half the mean. As noted earlier, the standard deviation $[s^2(S) + s^2(\delta)]^{1/2}$ of the measured tracer concentration is typically approximately equal to the mean. The reliability coefficient near the detection threshold is thus bounded above by 3/4 because of analytical error alone. Relative analytical error declines as concentrations increase, but other sources of imprecision need not. In particular, fluctuating signatures can yield imprecise estimates of source-strength at all concentrations. The imprecision of source-strength estimates is sometimes evident in the imperfect correlation of different tracers for the same source: a reliability coefficient of 3/4 corresponds to a correlation of $r = 0.87$.

The effect of imprecision is amplified in multiple regression analysis by the covariation of different sources' strengths. Consider, for example, the actual and ideal regression equations $c = f'_0 + f'_1 S'_1 + f'_2 S'_2$ and $c = f_0 + f_1 S_1 + f_2 S_2$. Suppose the estimates S'_1 and S'_2 to be precise and approximate, respectively ($F_1 \approx 1$ and $F_2 < 1$), and let $r = r(S_1, S_2)$ be the correlation among the true values. More simple algebra then shows the degraded coefficient of S'_2 (Cochran, 1968) to be given by

$$f'_2 = f_2 F_2 (1 - r^2)/(1 - r^2 F_2).$$ (C-10)

For $F_2 = 3/4$ and $r = 1/2$, as assumed earlier, Equation C-10 yields $f_2 = 0.7f_2$: the regression coefficient for the poorly characterized source is low by about 30%. The estimated contribution $f' m(S') = f'm(S)$ is low by the same amount. Of course, this estimate is an improvement over that obtained by ignoring the imperfect signature and treating the second source as an undifferentiated part of the background. We have already noted that regression analysis will underestimate the contributions of untagged sources by about 50% for the same assumed degree of covariance.

Just as regression analysis tends to attribute untagged emissions to tagged sources, it also tends to attribute poorly tagged emissions to well-tagged sources. The degraded regression coefficient corresponding to the well-characterized source in our two-source example (Cochran, 1968) is given by

$$\begin{aligned} f'_1 &= f_1 + f_2 r(s(S_2)/s(S_1))(1 - F_2)/(1 - r^2 F_2) \\ &\sim f_1 + f_2 r(m(S_2)/m(S_1))(1 - F_2)/(1 - r^2 F_2). \end{aligned}$$ (C-11)

Suppose the true contribution of the poorly characterized source to be twice that of the well-characterized source, so that $f_2 m(S_2) = 2f_1 m(S_1)$. The estimated contribution of the well-characterized source for $F_2 = 3/4$ and $r = 1/2$ is then $f'_1 m(S_1) = 1.3f_1 m(S_1)$, or about 30% high. Once again, that is an improvement over the 100% overestimate obtained earlier, under the same conditions, by completely ignoring the imperfect signature.

Other potential types of bias in regression-derived source apportionments are less predictable in their effects. Prominent among the types are fluctuations in the true coefficients f_{jt} that correlate with one or more source strengths. In regression analysis of secondary pollutants, f_{jt} and S_{jt} may be correlated because of their dependence on common environmental influences. Physically, f_{jt} and S_{jt} in that case are measures of conversion and (reciprocal) dispersion, respectively. Their statistical association depends on the empirical balance of competing influences. In dry air, for example, both conversion and dispersion of sulfur are promoted by strong insolation; this coupling would tend to generate negative correlations $r(f, S) < 0$. In other seasons and settings, sulfur conversion is promoted by fog and low stratus, which are associated with poor dispersion; this coupling would tend to generate positive correlations $r(f, S) > 0$. These examples are only two of many common influences that can be identified.

Spurious correlations can be detected sometimes by repeating the regression analysis while using a different response variable, one not expected to depend on the given source strengths. Newman and Benkovitz (1986), for example, used that technique to cast doubt on the physical relevance of a regression analysis presented by Oppenheimer et al. (1985). Oppenheimer et al. had shown that precipitation sulfate concentrations in the western United States were linearly related to SO_2 emissions from nonferrous metal smelters; Newman and Benkovitz pointed out that precipitation concentrations of elements that were not present in smelter emissions exhibited a similar relationship to SO_2.

Practical Guidelines

From the foregoing discussion, it is possible to identify some critical elements in measurement programs designed to support source apportionment studies that are based on regression analysis. The foremost objective of such a program must be to provide accurate estimates for the source strengths of all major contributors to the ambient mix. If a substantial fraction of the ambient total cannot be related to specific sources, then this undifferentiated background must be characterized directly.

Our discussion above shows that standard multiple regression analysis tends to overestimate the contributions of well-tagged sources to the ambient mix. Statistical procedures have been developed to compensate for imprecision in source-strength estimates (Fuller, 1987; White, 1989a,b), but it is clearly preferable to design measurement programs to provide the necessary precision in the first place. Posterior corrections have received little practical testing, and are sensitive to input statistics that themselves must be estimated. They cannot help, in any event, with sources whose strengths are not even roughly estimated because chemical or other signatures are altogether unavailable.

Some sources that lack endemic tags can be inoculated with artificial tracers to provide ambient source strengths that are measurable and that complete the data base. However, no artificial tracer, no matter how accurately and sensitively it can be measured, can substitute for balance in the experimental design. Tags are needed for all significant sources, not just the specific targets of regulatory or other interest (NRC, 1990).

When endemic or artificial tags cannot be found for a substantial fraction of the ambient pollutant concentration, our analyses show the necessity of characterizing this undifferentiated background by direct measurement, rather than estimating it as a by-product of the regression analysis. Of course, only total concentrations can be measured; the background concentrations can be measured only at times and places in which tagged emissions are absent. Such measurements are relevant, however, only if the background concentrations under these conditions are similar to those in the presence of tagged emissions. Since meteorological factors tend to impose a common temporal pattern on all ambient concentrations, as noted earlier, it is generally risky to estimate the background concentrations in one period from a measurement in another period. Simultaneous measurements made at different locations are often easier to defend.

When the tagged emissions form an identifiable plume, measurements made outside the plume provide unambiguous background data (e.g., White, 1977; Richards et al., 1981; White et al., 1983). If the measurements made to either side of the plume show similar concentrations, these background concentrations can be assumed representative within the plume as well. That interpretation must be invoked with care, of course, because valleys and other terrain features may channel untagged emissions along with the plume under certain conditions. Alternatively, the background can be measured upwind of a tagged source (e.g., Murray et al., 1990; NRC, 1990). For distant sources, this is a less desirable determination, because it is far from the receptors and lacks the consistency check that measurements to either side of the plume provide. Neither lateral nor upwind measurements can be consistently relied upon in extended stagnation episodes, when tagged emissions may diffuse throughout an entire airshed.

A final design consideration is the potential for reducing the risk of spurious correlations in regression analyses involving secondary pollutants by modifying the regression model to incorporate deterministic estimates of the atmospheric conversion of primary pollutants to form secondary reaction products. In this strategy, our mechanistic understanding of atmospheric chemical reaction processes is used to express the fluctuating ratio of secondary product concentration to source strength as a function $f_{jt} = g_j(\text{age}_t, \text{UV}_t, \text{RH}_t, \ldots; h_1, h_2, \ldots)$ of known external variables and unknown internal parameters. Regression analysis, possibly nonlinear, is then required to estimate only the constant parameters, as in applications to primary pollutants. Latimer et al. (1990) took the initial steps in this direction, but they employed a model that this committee judged to be unrealistic (NRC, 1990).

PLUME BLIGHT MODELS

Models for the visual appearance of plumes incorporate a model for computing pollutant concentrations in the plume and schemes for calculating radiative transfer processes that describe the visual aspects of the resulting plume. The pollutant reaction and transport codes that can be embedded in such a model range from simple Gaussian plume models to much more detailed numerical models that incorporate a mechanistic description of atmospheric chemical processes.

Plume blight models that are recommended for use by EPA have been built around the premise that plume dimensions and transport can be accurately represented by Gaussian plume equations. Such models contain modules for estimating the height of plume rise above its release point, a mathematical description of the expected plume transport and spread, estimation of the observer-plume orientation, and expected modulation of light intensity caused by the plume against the background.

The concentration χ of trace components at height z and lateral distance y from the axis of a plume can be estimated by using the Gaussian plume model for a continuous emission source whose effluent travels with constant wind speed u at plume elevation H (including plume rise) with lateral and vertical dispersion coefficients σ_y and σ_z, respectively, and is expressed as

$$\chi = \frac{Q}{2\pi\sigma_y\sigma_z u} \exp\left[-\frac{1}{2}\left(\frac{y}{\sigma_y}\right)^2\right] \times$$
$$\left\{\exp\left[-\frac{1}{2}\left(\frac{z+H}{\sigma_z}\right)^2\right] + \exp\left[-\frac{1}{2}\left(\frac{z-H}{\sigma_z}\right)^2\right]\right\}. \tag{C-12}$$

This equation is strictly for a conservative species, although the evaluation of chemically reactive species has been incorporated through temporal modification of Q.

The Gaussian plume model found within typical models for plume visual appearance can be replaced by a more advanced class of model known as a reactive plume model. Each of the models of this type cited below provide cross-wind resolution, at least in the horizontal. A grid system is defined within the air volume being modeled, and cross-wind diffusion proceeds according to Fick's law (flux proportional to the concentration gradient). An example of the strategy used for transport calculations is described in some detail by Stewart and Liu (1981).

The earliest of these models, by Eltgroth and Hobbs (1979), incorporates the conversion of SO_2 to sulfate particles by homogeneous photochemistry and by first-order catalysis that might occur on soot particles. The model represents the particle size distribution in terms of three dynamic lognormal modes that evolve in response to homogeneous nucleation, coagulation, condensation, and gravitational settling. Light-scattering calculations are based on this computed particle size distribution. Other reactive plume models have been developed by Hov and Isaksen (1981), Seigneur (1982), Seigneur et al. (1982), Hudischewskyj et al. (1987), and Joos et al. (1987).

The most recent reactive plume and aerosol model by Seigneur and his co-workers (Hudischewskyj and Seigneur, 1989) represents the present state of the art. It is assembled from free-standing modules for transport, chemistry, and aerosol dynamics, each the product of considerable evaluation and testing in its own right. (An example of the scrutiny to which individual modules have been subjected is the intercomparison of aerosol dynamics modules by Seigneur et al. (1986).) The gas-phase chemistry module is based largely on an update of the carbon bond mechanism (CBM-III) introduced by Whitten et al. (1980). Phase equilibrium, including the partitioning of water, is based on the model for an aerosol reacting system (MARS) introduced by Saxena et al. (1986). Aqueous-phase chemistry includes oxidation of SO_2 by H_2O_2 and O_2, the latter being catalyzed by MN^{2+} and Fe^{2+} (Saxena and Seigneur, 1987). The evolution of the particle size distribution, through homogeneous nucleation, coagulation, diffusion-limited condensation and evaporation, aqueous reaction, and sedimentation, is based on the sectional techniques introduced by Gelbard (1984).

MODELS FOR TRANSPORT ONLY AND FOR TRANSPORT WITH LINEAR CHEMISTRY

An analysis of wind flow during sampling periods is a necessary but insufficient test of source apportionment. There must be evidence of a reasonable probability of transport from the suspected source areas to the receptors during episodes of decreased visibility. Two options for assessment of the probability of transport are the application of models that: (1) assess only the transport of pollutants without regard to the chemical or physical processes affecting the pollutant, and (2) assess the transport but also include relatively simple linear chemical transformation processes.[2] Analyses can be conducted through investigation of wind flow either to a receptor region or from a source area. The following section discusses transport analyses using (1) backward trajectories, (2) wind field analyses, and (3) transport models that incorporate linear chemistry.

Back Trajectory Analysis

Backward trajectory analyses are a fundamental meteorological tool used to assess the spatial domain of source areas that could have contrib-

---

[2]A "linear" chemical transformation process is one in which the rate of chemical conversion is presumed to be linearly proportional to the amount of reactant. The process implies that as the amount of reactant (notably SO_2 for the issue of visibility degradation) is doubled, the rate of production of its end product (sulfates in the case of SO_2) is also doubled.

uted to a volume of sampled air. Trajectories can be stratified by concentration or direction to ascertain the consistency of the relationship between air movement from a source area and the resulting pollutant concentrations at a receptor site. Back trajectory analysis provides an estimate of the mean path followed by air en route to a sampling location. It is understood, but seldom articulated, that the estimate represents only the path with the highest probability that transport occurred along that line. The actual transport path is more accurately represented by a two or three-dimensional probability density function.

Trajectory calculations conducted over large regions of the United States are made by interpolation of available wind observations (or, in some cases, wind analyses from hydrodynamic models) in time and space from observations made every 12 hours at sites 400-500 km apart. In some urban areas, there are denser networks of wind stations that allow trajectories to be constructed from hourly observations taken at locations that are tens of kilometers apart. The most widely used trajectory models employ a linear interpolation in time and a $1/r^2$ spatial interpolation (where r is the distance from the trajectory to an observation point). A commonly used approach is the operational model of Heffter (1980), which employs available upper-air measurements assembled by the U.S. Air Force and the U.S. National Oceanic and Atmospheric Administration[3] to calculate trajectories for either pre-defined or "mixed-layer" thicknesses.

A variety of techniques have been developed that make use of ensembles of trajectories to estimate the most probable source areas that contribute to regionally transported pollutants. Ashbaugh (1983) and Ashbaugh et al. (1985) used trajectories calculated by the Heffter (1980) trajectory technique to calculate the residence times of air parcels over 3-day trajectories for relatively high-sulfate versus low-sulfate concentrations. Such techniques have the advantage of indicating both the direction and speed of wind over the course of an air parcel's transit from source to receptor. They also serve the purpose of identifying qualitatively which air masses are consistently related with high or low pollut-

---

[3]The input wind data are available from the National Climatic Center, Asheville, N.C., as NAMER tapes. They are prepared either in forward or backward modes.

ant concentrations. That information might be valuable in determining whether regions of consistently pristine air exist ("clean air corridors" as defined in the Clean Air Act Amendments of 1990).

The uncertainties inherent in individual trajectories can be reduced when considering an ensemble of trajectories. If it can be assumed that the errors in each trajectory calculation are stochastic and that the estimated paths are not biased, then the results of ensemble trajectory analysis (ETA) can be expected to identify source areas that consistently influence observed concentrations. ETA uses simple stratification of trajectories by concentration or through weighting of estimates of the probabilities that transport to a particular receptor site will occur from all of the possible surrounding source areas.

The estimation of transport relationships for a given sampling time should include representation of the spatial variability. The probability of a reactive, depositing pollutant (such as SO_2) arriving at two-dimensional point x in the horizontal plane of an airshed at time t, $A_r(x, t)$, can be expressed (Cass, 1978) as

$$A_r(x, t) = \int_{t-\tau}^{t} \int_{-\infty}^{+\infty} \int_{-\infty}^{+\infty} T(x, t|x', t')dx'dt', \qquad \text{(C-13)}$$

where $T(x, t|x', t')$, the potential mass transfer function in two dimensions, is defined by relationships of the general type

$$T(x, t|x', t') = Q(x, t|x', t') \, R(t|t') \times \qquad \text{(C-14)}$$
$$D(x, t|x', t') \, L(x, t|x', t'),$$

where $Q(x, t|x', t')$ is the transition probability density that an air parcel located at x' at time t' will arrive at receptor x at time t, $R(t|t')$ is the probability that the pollutant of interest in that air parcel will not react to form another species from time t' to time t, $D(x, t|x', t')$ is the probability that the pollutant will not be dry deposited between (x', t') and (x, t) and $L(x, t|x', t')$ is the probability that the pollutant will not be wet deposited during transport from (x', t') to (x, t). The integration is conducted over time period τ. The reaction and deposition terms (e.g., the last three probability functions on the right side of Equation C-14) have been quantified by many authors and can be modified to account for the interdependence of reaction, transport, and deposition processes and for the case of accumulating pollutants, like aerosol sulfate particles, that are formed by chemical reaction during transport in the atmosphere. Ignoring those terms allows the estimation of source-receptor relationships due to transport alone. Including those terms allows the estimation of source-receptor relationships incorporating linear transformation and removal.

The transition-probability density function, $Q(x, t|x', t')$, must be estimated with a transport model. For single or multiple-layer trajectory models, the axis of the computed trajectory can be assumed to represent the highest probability at any time that a particular upwind path is contributing to the trace substance composition at the monitor location. The spatial distribution of the transition-probability density function away from the axis of the trajectory can be adjusted to depend on meteorological conditions. As a first approximation, it was assumed by Samson (1980) that the "puff" of transition probability is normally distributed around each trajectory branch with a standard deviation that is increasing linearly in time upwind. Thus $Q(x, t|x', t')$ is assumed to be expressed as

$$Q(x, t|x', t') = \frac{1}{2\pi \sigma_x \sigma_y} \exp\left[-\frac{1}{2}\left(\frac{x''}{\sigma_x} + \frac{y''}{\sigma_y}\right)^2\right], \qquad \text{(C-15)}$$

where $x'' = X - x'(t')$ and $y'' = Y - y'(t')$; X and Y are the coordinates of the computational grid used to represent the airshed, and $x'(t')$ and $y'(t')$ are the coordinates of the centerline of the trajectory. It is assumed that σ_x and σ_y can be approximated by

$$\sigma_x(t') = \sigma_y(t') = at' \qquad \text{(C-16)}$$

with a dispersion speed, a, equal to a values believed to lie roughly between 1.8 and 5.4 km/hr (Draxler and Taylor, 1982).

Once the probability of transport from source to receptor has been calculated, it is possible to estimate the potential pollutant concentration distribution from a single source by using downwind (forward) trajectories or to estimate the potential for contribution to a specific sample by using upwind (backward) trajectories. A bias in transport can be calculated with techniques described by Poirot and Wishinski (1986) and Keeler and Samson (1989).

Quantitative transport bias analysis (QTBA) (Keeler and Samson, 1989) uses the estimated transport probability fields to identify and quantify the consistency of transport to a receptor. The ensemble of potential mass-transfer functions, calculated for each trajectory, are averaged over a sampling period to obtain an estimate of the mean potential mass transfer for that period. The spatial distribution of the field represents the "natural" potential for contribution to atmospheric pollutant concentrations if the source of that pollutant is spatially homogeneous.

The measured concentrations of trace substances are used to derive an implied transport bias. The potential mass transfer field for a given trajectory, $T(x, t|x', t')$, is integrated over the upwind time period of each trajectory to produce a two-dimensional probability of transport field. The resulting field, $T_k(x|x')$, for trajectory k is weighted by the corresponding pollutant concentration observed at the monitoring site at the time of arrival of the trajectory, $\chi_k(x)$, yielding a QTBA field, $T(x|x')$, calculated as

$$\bar{T}(x|x') = \sum_{k=1}^{K} T_k(x|x')\, \chi_k(x). \qquad \text{(C-17)}$$

From an individual receptor, the $T(x|x')$ field indicates the direction and preferred transport path associated with above-average concentrations, but it does not address the distance from the receptor to the contributing sources. It is conceivable, for example, that a particular wind-flow pattern could be conducive to local stagnation. The results of QTBA for a single site would suggest that the source was somewhere upwind along the corridor of the highest probability for delivering the above average concentrations but, in the case of stagnation would not further pinpoint the contributing source as possibly being quite close to the receptor air-monitoring site of interest. This shortcoming can be addressed through the use of concurrent measurements at multiple stations. By overlaying the QTBA fields for each receptor, one can identify systematic patterns of transport of higher concentrations from particular source areas to multiple receptors.

Transport-Only Analyses

The spatial distribution of pollutant concentrations downwind of emission sources may be estimated explicitly through the use of particle trajectory models. The transport of hundreds or thousands of particles released from the sources is simulated simultaneously with vertical mixing introduced at each time step. The degree of vertical mixing is dependent upon atmospheric conditions and the location of each particle relative to ground level or to elevated stable layers. Pollutant concentrations can be estimated through bookkeeping of the number of particles that fall within the air volume represented by each grid cell in the model. Over regional scales, several authors (e.g. McNider, 1981; Shi et al., 1990) have shown the importance of incorporating vertical mixing into particle trajectory modeling to explicitly include the potential for plume dispersion by wind velocity shear.

Transport analysis also can include direct Eulerian modeling. Brost et al. (1988) used the hydrodynamic model described by NCAR (1983; 1985; 1986) to simulate pollutant concentrations over regional scales downwind of specific sources. The resolution of such transport modeling is set by the selection of the computational grid cell size. The advection schemes used for Eulerian transport modeling often produce numerical diffusion that must be treated carefully to avoid unrealistic horizontal spreading of plumes.

Transport with Linear Chemistry

Assuming that conversion of SO_2 to sulfate particles occurs at a rate that is linearly proportional to the SO_2 concentration but that varies with such factors as time of day and season of year, equations like C-13 and C-14 can be used to describe pollutant transport and linear chemistry. The removal of a gaseous or particulate substances by wet and dry deposition is assumed to occur in linear proportion to the concentration of that species. The rate coefficient for wet, k_w, and dry, k_d, deposition is described by

$$k_w = \theta_i P/h, \qquad\qquad (C\text{-}19)$$

where θ is the washout ratio for substance i, and P is the precipitation rate (depth per unit time).

The use of linear chemistry models for estimating plume impact from individual sources or for evaluating contributions to observed concentrations presumes that conversion of SO_2 to sulfate is first order in SO_2 concentrations. This conversion is not unreasonable if gas-phase reactions are the mechanism for conversion, if sufficient hydroxyl radical concentration (OH) is available, and if actions are not taken that will change the OH concentrations. Likewise, for this approach to approximate the aqueous-phase conversion of SO_2 to form sulfates, there must be sufficient H_2O_2 or O_3 available in cloud water.

MECHANISTIC MODELS FOR TRANSPORT AND CHEMICAL REACTION

Mechanistic models attempt to incorporate mathematical descriptions of all the most important chemical and physical processes needed to properly investigate the atmospheric phenomena of interest. Observational data are used mainly to establish initial and boundary condition estimates and to evaluate model performance. The other classes of models discussed in this appendix are characterized by more extensive use of observations rather than fundamental process descriptions to establish source-receptor relationships.

The history of mechanistic modeling for air quality extends back approximately 20 years. The first models focused on photochemical oxidants (e.g., Reynolds et al., 1973; Demerjian, 1978) and, after many years of development, were applied for regulatory purposes (e.g., Reynolds et al., 1979). Such models focused exclusively on gas-phase chemical transformations, neglecting particles of central importance to visibility modeling. More recently, mechanistic modeling has been applied to problems involving secondary airborne particle formation and acid deposition (Chang et al., 1987; Russell et al., 1988a). In those cases, the formation of particles from acid gases by both gas-phase and in-cloud processes is considered, although developments needed to calculate particle size distributions are not yet complete. Extensive work on airborne particle processes, including realistic treatments of particle size distributions, has been carried out for zero-dimensional "boxlike" models-, those studies can provide guidelines for quantifying aerosol dynamics that could be used in regional scale three-dimensional models (Middleton and Brock, 1977; Gelbard et al., 1980). A summary of three-dimensional mechanistic models by Seigneur and Saxena (1990) is presented in Table C-1. The summary lists all the current models that are being (or could be) used as the basis for visibility modeling of the type discussed in this section.

Mechanistic visibility models are intended to calculate from first principles the impact of gases and particles on atmospheric optical properties. Such models are being developed, but many years might pass before they are available for routine regulatory purposes. In principle, these models use information on emissions, meteorology, and chemical transformations to calculate gaseous pollutant concentrations and particle concentrations or size distributions in a three-dimensional spatial domain. The results could be used to calculate the optical effects of the airborne particles.

Mechanistic modeling that incorporates comprehensive calculations of particle concentrations but not size distributions is the current state of the art. This modeling can be achieved by extending acid deposition models or certain regional photochemical smog models to calculate concentrations of primary particulate substances as well as products of gas-to-particle conversion (Russell et al., 1988a; Middleton and Burns, 1991). To use this approach, however, to determine optical characteristics requires assumptions about particle size distributions. Size distributions are required in visibility studies because the optical properties of particles strongly depend on particle sizes. To produce a comprehensive mechanistic model for visibility impairment, direct calculations of chemically resolved airborne particle size distributions are needed in conjunction with a theoretical treatment of scattering and absorption of light by particles to calculate the optical properties of aerosols. The current understanding of atmospheric aerosol processes requires considerable refinement before such models can be used with confidence. The construction of such models is under way, but many years will be required for model evaluation.

Traditionally, mechanistic models are classified as Lagrangian or Eulerian, the distinction being based on the reference frame used for the description of fluid motion. Lagrangian trajectory models quantify the

TABLE C-1 Overview of Three-Dimensional Air Quality Model

Model	Area of Application	References for Model Formulation	References for Model Performance Evaluation
RADM-II (Regional Acid Deposition and Oxidant Model)	Eastern North America	Chang et al., 1987	Middleton et al., 1988
ADOM (Regional Acid Deposition and Oxidant Model)	Eastern North America and northern Europe	Venkatram et al., 1988	Venkatram et al., 1988
STEM-II (Regional Acid Deposition and Oxidant Model)	Philadelphia, central Japan, Kentucky, and northeastern United States	Carmichael et al., 1986	Carmichael and Peters, 1987; Chang, 1987
ROM (Regional Oxidant Model)	Northeastern United States, and southeastern United States	Lamb, 1983	Schere, 1986
RTM-III (Regional Oxidant Model)	Northeastern United States, Minnesota, northern Europe, and San Joaquin Valley	Liu et al., 1984	Liu et al., 1984; Morris et al., 1987
UAPM (Urban Oxidant and Particulate Matter Model)	Los Angeles Basin	McRae et al., 1982; Pilinis and Seinfeld, 1988a,b	McRae and Seinfeld, 1983; Russell et al., 1988b

TABLE C-1 (continued)

Model	Area of Application	References for Model Formulation	References for Model Performance Evaluation
UAM/PARIS (Urban Oxidant and Particulate Matter Model)	More than 10 urban and nonurban areas in the United States and Europe	Reynolds et al., 1973, 1979; Seigneur et al., 1983	Roth et al., 1983; Seigneur et al., 1983
LIRAQ (Urban Oxidant Model)	San Francisco Bay Area, Monterey, St. Louis	MacCracken et al., 1978; Penner and Connell, 1987	Penner and Connell, 1987

Source: Seigneur and Saxena, 1990. Copyright ©1990. Electric Power Research Institute. EPRI EN-6649, Status of Subregional and Geoscale Models, Vol. 1. Air Quality Models. Reprinted with permission.

complex transport of trace chemicals by assuming that all the substances are uniformly mixed within a chemically isolated parcel of air that moves through the atmosphere following the mean motion of the air. In contrast, Eulerian models adopt a fixed two- or three-dimensional grid system, and continuity equations for chemical substances are solved at each grid point to calculate time-varying concentrations of several substances over a specified domain.

The Eulerian modeling approach provides the framework for most of the complex atmospheric photochemical models that represent the coupling and feedback among multiple physical and chemical phenomena. Within the Eulerian framework, it is possible to incorporate mathematical descriptions of numerous physical and chemical processes that are difficult to consider in the models with Lagrangian approaches, especially when the interaction of multiple sources with different spatial, temporal, and chemical characteristics must be considered and when model outputs are to represent concentration gradients over large geographical regions.

Processes that are included in mechanistic visibility models are outlined in Figure C-2. The modeled airshed is divided into grids of a size that depends on the terrain characteristics. The grids are initialized with a set of chemical concentrations. For each grid cell, hourly meteorological and gas and particle emissions data are specified for typical computational periods of 3-5 days. The models that use this information to calculate the time-dependent three-dimensional distributions of gases and particle size distributions.

Mechanistic models generally solve the following chemical conservation equation for each of the transported gas phase chemicals:

$$\frac{\partial C}{\partial t} = -\nabla \cdot VC + \nabla \cdot (K_e \nabla C) + P_{chm} - L_{chm} +$$
$$E + \left(\frac{\partial C}{\partial t}\right)_{clouds} + \left(\frac{\partial C}{\partial t}\right)_{dry}, \tag{C-20}$$

where C is the species volume mixing ratio, V is the three-dimensional velocity vector at each grid point in the model domain, K_e is the eddy diffusivity used to quantify the subgrid-scale fluxes due to subgrid-scale turbulence, P_{chm} and L_{chm} are gas-phase chemical production and loss terms, E is the emission rate, $(\partial C/\partial t)_{clouds}$ is the time rate of change due

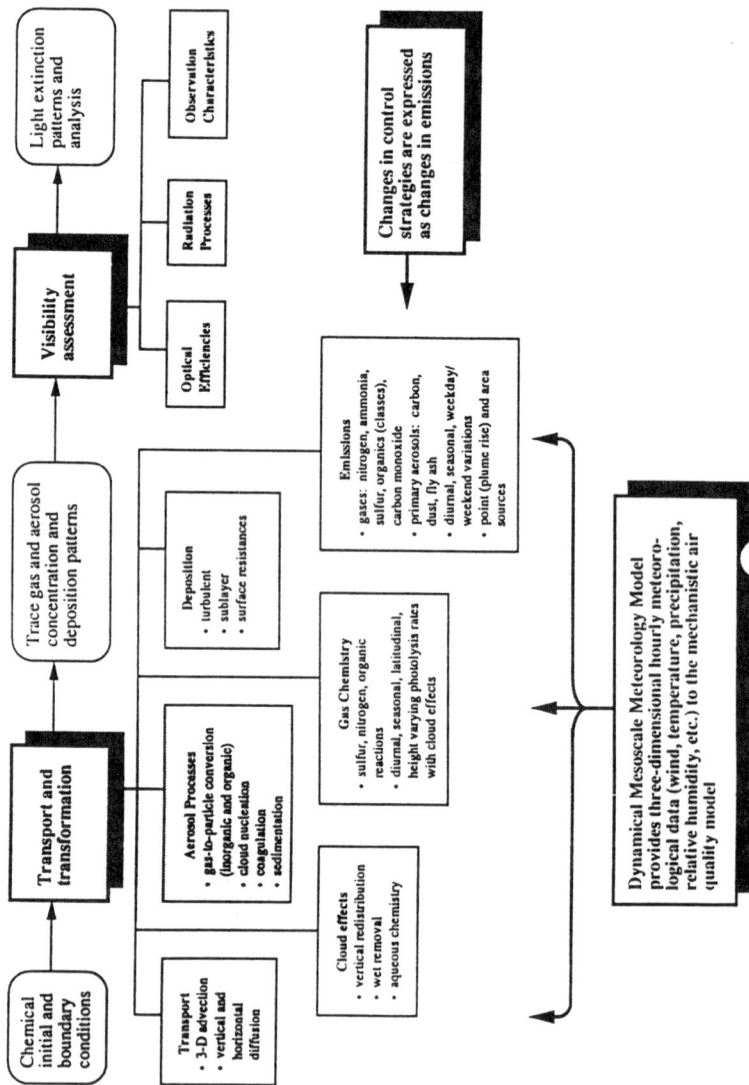

FIGURE C-2   Mechanistic visibility modeling system.

to cloud effects (including subgrid-scale vertical redistribution, aqueous chemical interactions, various nucleation and other droplet-related processes, and scavenging), and $(\partial C/\partial t)_{dry}$ represents the rate of change due to dry deposition.

A general equation for calculating Q_i, the concentration of the total aerosol mass in size category i, is given as:

$$\frac{\partial Q_i}{\partial t} = -\nabla \cdot VQ_i + \nabla \cdot (K_e\nabla Q_i) + \left(\frac{\partial Q_i}{\partial t}\right)_{grow} + E_i + \left(\frac{\partial Q_i}{\partial t}\right)_{coag} + \left(\frac{\partial Q_i}{\partial Q_i}\right)_{removal} , \quad (C-21)$$

The term on the left represents the change in the total concentration of the airborne particle mass in size category i over time. The first term on the right refers to advection and the second to turbulent diffusion of the airborne particles. The growth term represents the change in concentration due to changes in thermodynamic equilibrium and nucleation of new airborne particles and condensation of vapor onto existing particles due to production of new material via gas-phase chemical reactions. The coagulation term represents the change in concentration due to coagulation of particles. The removal term represents the concentration change due to sedimentation, particle scavenging, and wet and dry deposition as well as the effect of cloud processing on the size distributions. Finally, E_i represents the change in aerosol concentration due to direct particulate emissions.

For the proper treatment of visibility issues, it is essential that the models include appropriate descriptions of the aerosol processes. Mechanistic visibility models should involve two separate components. First, chemically resolved airborne particle size distributions need to be calculated at specified grid points. Information is required on the characteristics of primary particle emissions as well as on atmospheric aerosol processes that affect further evolution of particle size distributions. Processes that must be considered include advection, diffusion, coagulation, evaporative shrinkage or condensational growth of particles, gas-particle chemical reactions, cloud processing, and wet or dry deposition. There is a close coupling between the formation of secondary particles, which plays a major role in visibility impairment, and gas and in-cloud chemistry. Also, aerosol transport and removal by wet or dry deposition are dependent on local meteorology. Therefore, models that are used to determine airborne particle size distributions must be linked with meteorology and gas-phase chemistry models. Lorentz-Mie theory is used to calculate the optical characteristics of airborne particles by integrating over calculated particle size distributions. The particle-concentration prediction and the optical aspects of these computations involve mathematical approximations (introduced to speed up the calculations) and assumptions about the physical and chemical characteristics of atmospheric particles. These simplifications will affect the validity of calculated results. Therefore, the uncertainties associated with such model predictions need to be determined.

The phenomena that need to be considered in developing the aerosol models vary among the different chemical species. For example, although some sulfate particles are emitted directly by sources, most sulfate particles are formed in the atmosphere by the chemical transformation of SO_2 gas. Reactions that lead to sulfate particle formation can take place either in the gas phase or in liquid particles or cloud droplets. The size distribution and therefore, the optical properties of the secondary sulfate particles depend on the chemical transformation mechanism. Organic particles can be either primary (directly emitted as particles) or secondary (formed from gas-phase organic substances), although the relative contributions of both remain poorly understood. The particulate nitrate, ammonium, and water content are determined primarily by thermodynamic equilibrium between particles and gas-phase species. All particles are influenced by removal processes, although the extent of wet removal processes will depend on hygroscopicity and particle size. An important practical issue in developing mechanistic visibility models is the assessment of the detail and accuracy with which such processes need to be described to achieve satisfactory results.

A variety of approaches has been developed for calculating the evolution of atmospheric particle size distributions. Several have been compared by Seigneur et al. (1986). The approaches differ in accuracy, computational speed, and ability to describe the behavior of multicomponent aerosol systems and internally and externally mixed particles. Each of these factors needs to be weighed in selecting the optimal model for a given application.

Proper characterization of the relationship between aerosol concentrations, the properties of the visual environment, and the effect on the public of changes in visual air quality is central to correct prediction of the effects of visibility protection programs. The relationship between aerosol concentrations as determined by models and human perception generally is based on optical principles. Most work to date is based on the assumption that particles are spherical and that particles consist of homogeneous mixtures. Lorentz-Mie theory is then used to calculate the scattering and absorption of light by individual particles. The total extent of particle scattering and absorption is determined by integrating over calculated particle size distributions that are chemically resolved. Although the assumption is often reasonable for submicron particles, little experimental work has been done to examine its validity. The effect of nonspherical particles on light extinction needs to be considered in areas where coarse dust and fly ash are important contributors to visibility reduction. The scattering and absorption of light by air and NO_2, respectively, are straightforward calculations.

The optics calculations described above can be made for each of the grid points at which composition-dependent particle size distributions are calculated explicitly or implicitly by an air quality model. The calculations provide information on variations of atmospheric optical properties over the three-dimensional grid. Thus, for model calculations where particle size distributions are estimated, it is possible to use radiative transfer theory (Chandrasekhar, 1960) to calculate visibility indexes that depend on sight path, cloud cover and ground reflectance, color and texture of distant objects, and the angle between the observer and the sun, which requires data on the surrounding terrain and cloud cover. When composition-dependent size distributions are not calculated explicitly in the model, alternative approaches must be developed to link the aerosol composition information to optical factors. For example, in many field studies it has been observed that the atmospheric extinction coefficient per unit mass for certain airborne particle components (e.g., sulfate and elemental carbon particles) typically lies within a characteristic range of values. Those empirically determined extinction efficiency values could be multiplied by model predictions of aerosol species chemical concentrations to estimate the light-extinction coefficient that corresponds to a particular situation being modeled.

HYBRID MODELS

Hybrid models have been developed in the belief that no single mechanistic or receptor-oriented model can represent reality accurately under all circumstances and that each modeling approach has its own specific strengths and weaknesses. Hybrid (or composite) models offer the possibility of better resolution of source contributions by combining two or more receptor, trajectory, deterministic, or atmospheric chemistry models. For example, the multiscale source-receptor model suggested by Chow (1985) combines a regional scale trajectory model, a principal component analysis receptor model, a CMB receptor model, and an urban scale Gaussian dispersion model into a single composite modeling approach.

Linear-Chemistry—CMB Hybrid Models

Composite model research has been pursued because the current state of the art limits regulatory applications of conventional CMB receptor modeling to particulate matter that is directly emitted to the atmosphere. The remaining sulfate, nitrate, and organic compounds that are not attributed to primary emissions are classified as secondary substances and cannot be attributed to specific sources, thereby severely limiting conventional use of the CMB model in visibility studies.

If additional assumptions are made, hybrid CMB-atmospheric chemistry models can be used to extend conventional CMB modeling beyond the limits outlined in EPA's regulatory guidance. If one accepts the assumption that conversion of reactive gases (e.g., SO_2) to secondary particles (e.g., SO_4^{2-}) is complete and that the secondary substances have not been preferentially deposited en route to the receptor, secondary particles can be apportioned among contributing source types.

In real-world applications where conversion is not complete or the secondary particles are deposited during transport, CMB has been used only in research settings to quantitatively estimate secondary aerosol source contributions. Several investigators have proposed secondary aerosol hybrid receptor models that include SO_2-to-sulfate transformation and deposition terms, but none of the models has undergone thorough validation study (Stevens and Lewis, 1987; Dzubay et al., 1988).

Lewis and Stevens (1985) proposed a hybrid model that typifies the current state of hybrid models developed as a direct extension of the chemical element tracer approach that forms the basis of CMB receptor modeling methods. In that model, the secondary sulfate concentration from a specific source (M_{SO_4}) is estimated as

$$M_{SO_4} = M_P \times A \times T, \qquad (C\text{-}22)$$

where M_p is the mass concentration at the receptor of primary fine particles from the source, A is the ratio of the source mass emission rate for SO_2 and fine particles, and T describes both the transformation of SO_2 to sulfate in the atmosphere and its loss due to deposition at the earth's surface. A chemical reaction model is needed to specify the extent of conversion, T, of SO_2 to form sulfate.

M_p can be estimated by CMB receptor model applications or by use of multiple linear regression analysis. M_p also can be estimated using source tracers such as deuterated methane (CD_4) when the assumption is made that the tracer is associated uniquely with a specific source. Tracer applications are discussed below.

Lewis and Stevens theorize that these hybrid models are limited to distances of less than 200 km because of the following factors: (1) the concentration of source tracer elements used to estimate M_p must be above measurement detection limits; (2) particle fractionation effects during transport due to the differing size distributions of the chemical species must not occur; and (3) the estimate of plume age required to calculate T becomes less certain as the distance from the source increases.

Given those limitations, the principal assumptions inherent in the application of the hybrid model of Lewis and Stevens for secondary aerosol apportionment are as follows: (1) dispersion, deposition, and transformation processes are linear or pseudo-first-order processes in nature; (2) dispersion affects all three pollutants (SO_2, sulfate, and M_p) identically; (3) dry deposition is the only form of deposition that occurs (wet deposition, and oxidation by aqueous phase and heterogeneous processes are excluded); (4) deposition affects all the fine particles in the same way, but the rate of deposition of SO_2 might be different; (5) secondary sulfate is produced only by homogeneous gas-phase oxidation of SO_2; and (6) plume age can be estimated from available wind data.

If the path of the air parcel can be computed by trajectory analysis, then plume age can be estimated more exactly.

Many real physical situations of interest may occur outside the bounds of the above assumptions (e.g., heterogeneous SO_2 oxidation in clouds often is important).

A second type of composite model has been developed that employs CMB receptor modeling for attribution of primary airborne particles to their sources, accompanied by a separate deterministic model for sulfate formation and transport that is driven by atmospheric transport, reaction, and dilution calculations rather than by tracer concentration data (Harley et al., 1989). This approach employs the sulfate formation model of Cass (1981), which is based on gridded SO_2 and primary sulfate emissions, hourly wind speed, wind direction, mixing height, dry deposition rates, and measured or computed atmospheric pseudo-first-order rates for conversion of SO_2 to sulfates. The composite model has been applied to study the least-cost solution to the aerosol control problem in the Los Angeles basin (Harley et al., 1989).

Additional hybrid modeling can be envisioned in which tracer or CMB models are used for elements of the source attribution problem that are difficult to determine with a deterministic model (e.g., predictions of airborne soil-dust concentrations). More complete deterministic models for secondary airborne particle formation would then be used to compute sulfate, nitrate, and secondary organic particle concentrations along with those primary particle concentrations that are due to ducted emission sources.

Appendix D

Control Techniques

This appendix summarizes techniques for reducing emissions of the principal pollutants that impair visibility.[1] The sources of emissions are discussed in approximate order of their contribution to haze, as set forth in Table 6-2.

POWER PLANTS

Increasing environmental regulation in many industrialized countries has spurred the advancement of power-plant pollution control technologies over the past two decades. Research and development to render the technologies more efficient, more reliable, and less costly has accelerated in the United States in anticipation of new emission reduction requirements for existing power plants resulting from the 1990 Clean Air Act Amendments (Torrens, 1990). Power plants nationwide with 40,000-60,000 MW of electricity generation capacity probably will be retrofit with some form of control of SO_2 or NO_x before the year 2000 in response to new legislation (Dalton, 1991; EPRI, 1991a).

The Clean Air Act Amendments of 1990 were not aimed specifically at visibility. SO_2 reductions required by the legislation were aimed

---

[1]Unless specified otherwise, the basis for cost estimates of stationary source pollution controls provided in this appendix is the Electric Power Research Institute's Technical Assessment Guidelines (EPRI, 1989).

primarily at reducing emissions that cause acid deposition (i.e., acid rain). However, SO_2 reductions also should improve visibility in Class I areas in the United States, particularly in the East.

Table D-1 shows the results of an analysis funded by the U.S. Department of Energy (DOE) of power plants in relation to Class I "prevention of significant deterioration" (PSD) areas. The percentage of total coal-fired power-plant capacity affected by an SO_2 reduction requirement would be 31% within a 100-mile radius of a Class I area, increasing to 61% within 150 miles and 75% within 200 miles; 85% of total SO_2 emissions arise from boilers within 200 miles of Class I areas (Trexler, 1990).

TABLE D-1   Power Plants and Class I Areas

Power Plant	Boilers, No.	Capacity, 10^3 MW	SO_2 Emissions, 10^6 tons
Total coal-fired boilers	1,131	331.0	14.4
Within 100 miles of Class I area	348	116.6	5.1
Within 150 miles of Class I area	688	221.1	10.0
Within 200 miles of Class I area	849	263.6	12.2

Source: Trexler, 1990.

Sulfur Dioxide Control Technologies

Over 150 flue gas desulfurization (FGD) systems in power plants with approximately 72,000 MW of capacity are now operating in the United States to control SO_2 emissions (Dalton, 1990). About one-fifth of the total coal-fired capacity is covered by the FGD systems, of which about

92% are wet scrubbers and 8% are spray dryers. By the end of the decade, new FGD-equipped plants with an estimated 27,000 MW of generating capacity are expected to begin operating.

Wet Flue Gas Desulfurization

The predominant scrubbing technology used by U.S. utilities is lime or limestone wet FGD (scrubbing) and landfill disposal of by-products. The reagent is prepared (limestone is ground and lime is slaked) and mixed with water in a reagent preparation area. It is then conveyed as a slurry (approximately 10% solids in water) to an absorber (typically a spray tower) and sprayed into the flue gas stream. SO_2 present in the flue gas is absorbed in the slurry and collected in a reaction tank, where water is removed from the resulting calcium sulfite or sulfate crystals.

The U.S. utility industry's early experience with scrubbers was difficult. Inadequate understanding of the chemical reactions within the process led to frequent plugging and scaling of the scrubber components, corrosion and erosion of the construction materials, poor handling characteristics and large land requirements for sludge by-products, high capital, and high operating costs. Better understanding of system chemistry has led to increased reliability and improved performance. Additives permit SO_2 removal to exceed 90% at an added cost that is usually not prohibitive (compared with the average cost of removing the first 90%).

Future improvements in conventional lime or limestone FGD are illustrated by three examples:

- Process improvements such as the jet bubbling reactor used in the Chiyoda CT-121 process;
- Application of spray drying to high-sulfur coal;
- An advanced limestone FGD system producing gypsum with no reheat and no spare modules, for use at compact sites (i.e., space-constrained).

Each of the designs could achieve SO_2 control of 90-95% or better. The advanced system would offer 20-50% capital cost savings and 20-40% operating cost savings over conventional FGD systems available today. The jet bubbling reactor and the advanced limestone and gypsum designs could be used to make wallboard-grade gypsum, for sale or disposal.

Engineering and other improvements have reduced costs over the past decade, but wet FGD remains relatively expensive. Capital costs for conventional systems vary depending on fuel sulfur content, unit size, and other factors. Preliminary results from a recent Electric Power Research Institute (EPRI) cost estimation indicate that a state-of-the-art wet FGD system for medium-sulfur coal on a new plant could be built for less than $200/kW and annual operating costs would range from 5 to 10 mills/kW-hr (Torrens and Radcliffe, 1990).

Depending on existing plant conditions, especially available space and accessibility, retrofitting an FGD system can cost up to three times more than installing it in a new plant. For the different FGD systems now available and a moderately difficult installation (approximately 30% more costly than in a new plant), the range of capital requirements and total levelized costs over 30 years with no inflation are shown in Table D-2.

TABLE D-2   Capital Costs and Levelized Costs in 1990 of Retrofitting a FGD System[a]

FGD System	Capital Cost per kW	Cost per ton of SO_2 (constant dollars)
Wet FGD (range of system types)	$190-230	$440-500
Spray dryer	$175	$450

[a]The figures are in 1990 dollars and are subject to a ±20% uncertainty.
Source: EPRI, 1991b.

Spray Dry Flue Gas Desulfurization

Spray dry FGD is the other principal method of SO_2 control used today. Calcium oxide (quicklime) mixed with water produces a calcium hydroxide slurry, which is injected into a spray dryer, where it is dried by the hot flue gas and reacts with the gas to remove SO_2. The dry product is collected both at the bottom of the spray tower and in the downstream particulate removal device, where more SO_2 may be removed. Capital costs for dry FGD can be substantially less than for wet systems, especially for low-sulfur coal applications. Seventeen spray dry FGD systems were operating as of mid-1987, all on relatively low-sulfur coal (less than 2%).

High-sulfur spray dryer applications have not been demonstrated on a long-term commercial scale. Pilot testing has indicated that SO_2 removal of 80-90% is possible, and over 90% removal is possible under certain conditions. However, a fabric filter may be added to maintain particulate emission standards if SO_2 removal exceeds 90%, since the mass flow of solids to the electrostatic precipitator (ESP)—the principal particulate control technology in U.S. power plants—will at least double if a spray dryer is used.

Control of Plume Opacity

Stack-plume visibility, also known as plume opacity, has become a concern to the utility industry because some coal-fired stations with operating FGD systems have been cited for opacity in excess the New Source Performance Standards under the Clean Air Act, even though particulate mass emissions are within regulated limits. At other utilities, with scrubbed or unscrubbed stack emissions, the visible emissions have been higher than expected, and there is concern about visibility reduction in the local environment. Unacceptable opacity may be caused by scrubber-generated particulate matter, condensible particulate matter such as sulfuric acid in the flue gas, fine particles penetrating the particulate control device, or colored gases such as NO_2 in the flue gas. EPRI field tests have shown some contribution from all the above causes. However, the primary contributor to plume opacity at most units firing medium- (1-3%) and high- (greater than 3%) sulfur coal appears to be condensed sulfuric acid mist.

Sulfuric acid is formed in the furnace when sulfur from coal is converted to SO_2 and some of the SO_2 is oxidized to SO_3. The amount of SO_2 thus oxidized is generally about 0.5%, depending on excess oxygen and the presence of conversion catalysts in the fly ash. The process is not understood completely. As the flue gas cools, the SO_3 combines with water vapor in the gas to form H_2SO_4 but does not condense unless the temperature drops below the acid dew point (usually between 250 and 350°F depending on the H_2SO_4 concentration and the moisture content of the gas).

When the gas temperature drops below the acid dew point, condensation begins as molecular clusters of H_2SO_4 agglomerate and attach colliding water molecules to form submicron droplets. That can occur at the stack exit, where the gas cools because of mixing and heat transfer with the ambient air, or at the inlet to an FGD system, where the flue gas is reduced in temperature to saturation (typically 125-130°F) by water evaporation. Because of their small size, the H_2SO_4-based droplets are not removed by the FGD system.

There are three options for controlling acid mist plume formation: switching to lower-sulfur fuel, injecting an alkaline additive to react with the H_2SO_4, or adding a wet ESP. Switching to lower-sulfur fuel might require major changes to the boiler system and other plant operating systems. Because, coal contracts usually are of long term and lower-sulfur coal might cost much more than coal obtained locally.

The utility industry has applied additives to enhance ESP performance (without regard to H_2SO_4 removal) for many years. The additive injection systems are normally simple, requiring only a storage facility, transport system, and injection ports. A variety of reagents have been used to control acid mist, including magnesium oxide (MgO), organic amines, and ammonia (NH_3). MgO and NH_3 were evaluated during an EPRI field testing program. MgO appeared limited to approximately 50% removal efficiency with injection into both high- and low-temperature regions of the boiler flue gas system. NH_3, however, demonstrated consistently high removal efficiency. Care should be used when NH_3 is injected. Some precipitators have been disabled by the use of NH_3. It is not well understood why NH_3 injection works well at some sites and not at others. Also, ammonium compounds in the fly ash or scrubber waste can make sale or reuse more difficult. EPRI is also conducting research on other additives to control H_2SO_4.

The wet ESP has potential for acid mist control but has a high capital cost and has no proven application to technology for U.S. utilities.

As a result of installing FGD systems in response to the Clean Air Act Amendments, plants burning high-sulfur coal are likely to form sulfuric acid mist under certain operating conditions. That possibility should be anticipated in the design, so that appropriate control measures can be taken after start-up if necessary.

NO$_x$ Emissions Control

Combustion modification in the boiler can reduce NO$_x$ emissions by about 50%, depending on the type of boiler, at moderate cost ($10-$30/kW of capacity for capital cost, or $200-$400/ton of NO$_x$ removed for total annual cost). To achieve greater reductions requires some form of postcombustion NO$_x$ control (downstream of the boiler) at a cost of $100-$150/kW or $1,500-$4,000/ton of NO$_x$ removed (Kokkinos et al., 1991).

NO$_x$ Reduction by
Combustion Modification

Combustion modification (adjusting the fuel mixture so that there is enough oxygen to support combustion but not enough to combine with nitrogen) can reduce NO$_x$ emissions by 40-60% and has been applied widely to new coal-fired boilers. Twelve full-scale demonstrations for retrofit combustion modification also are in progress or planned, covering the range of boiler types in operation in the United States and sizes from 40 to 500 MW. Combustion modification is the most cost-effective method of meeting NO$_x$ emission reduction requirements for the majority of the U.S. boiler utilities.

Another combustion modification technique, and the only one suited to cyclone boilers, is reburning, which involves redirecting 10-20% of the total fuel to the upper furnace region of the boiler to create a zone for chemical reduction of NO$_x$ formed in the primary zone of combustion in the boiler. Full-scale demonstrations of reburning with either natural gas or pulverized coal are planned. Potential NO$_x$ reductions are in the range of 30-50%.

Capital costs for retrofit low NO$_x$ burners are estimated to be $10-$30/kW. Reburning costs depend on the choice of reburning fuel, and capital costs are $25-$45/kW.

Postcombustion NO_x Reduction

The main technology for postcombustion NO_x reduction, selective catalytic reduction (SCR), involves injection of NO_3 in the presence of a catalyst between the boiler and the air heater. NO_3 reacts with NO_x, reducing it to nitrogen and water. Europe has over 30,000 MW of coal-fired generating capacity equipped with SCR, and Japan has 6,000 MW. Retrofit installations in Europe and Japan typically achieve 60-80% NO_x removal, and residual NH_3 emissions of less than 5 ppm (usually 1-2 ppm). Capital costs in Europe average $125/kW. The costs are consistent with EPRI's capital cost estimates for hypothetical retrofit installations, which range from $100 to $150/kW. Levelized cost projections for both U.S. and Japanese plants are estimated at 4-9 mills/kW-hr (Cichanowicz et al., 1990).

The United States and Europe are paying increased attention to selective noncatalytic reduction (SNCR) technologies. The principle is similar to that of SCR, but no catalyst is involved. Results with urea injection show the potential for NO_x reductions of 30-50%, perhaps up to 75%, and NH_3 emissions of 5-10 ppm or less. Capital costs are estimated to be $5-$15/kW, and operating costs of less than 3 mills/kW-hr.

Clean Coal Technologies

Initial operational problems with FGD, combined with the high cost of the systems, stimulated development of less costly methods of SO_2 reduction at coal-fired power plants. That led to the introduction, during the mid-1980s, of clean coal technologies, which have evolved into a family of precombustion, combustion and conversion, and postcombustion technologies designed to improve technical pollution-control capabilities and flexibility at lower costs (Torrens, 1990).

Sorbent Injection

Sorbent injection removes SO_2 during combustion. The simplest technique is furnace sorbent injection (FSI) of lime or limestone. The pulverized coal and sorbent mixture is maintained suspended as a fluid in the boiler by a stream of air flowing upward. SO_2 removal of 35-55% has been measured in FSI demonstrations using lime. Injection of sorbent downstream of the furnace at approximately 1,000°F (540°C) shows similar removal efficiency but has not been demonstrated at full-scale. The U.S. Environmental Protection Agency (EPA) and the U.S. Department of Energy (DOE) have sponsored FSI programs at the 60-180 MW scale. A key issue may be the handling and disposal of highly alkaline waste material generated by FSI through the use of existing particulate control devices, usually an ESP.

The main advantage of sorbent injection is its low capital cost (estimated at $70-$120/kW). However, the overall cost per ton of sulfur removed is comparable to that of FGD at higher load factors: the economic benefit of sorbent injection would be greatest, therefore, for plants with lower capacity.

Fluidized Bed Combustion

Atmospheric fluidized bed combustion (AFBC) is now an established technology for industrial boilers (10-25 MW) and is being demonstrated for utility boilers (75-350 MW). The pulverized coal and sorbent mixture is maintained suspended as a fluid in the boiler by an upstream of air. AFBC boilers can meet the New Source Performance Standards (NSPS) for both SO_2 and NO_x,[2] without additional control equipment. Much greater reductions in SO_2 emissions from AFBC plants probably would require additional postcombustion technology. Like FSI, AFBC results in additional wastes that might prove difficult to handle in existing particulate control devices.

In pressurized fluidized bed combustion (PFBC), the boiler operates under a pressure of 10 atmospheres. The increased energy of the exit gases can drive both a gas turbine and a steam turbine (combined cycle),

---

[2]The NSPS for SO_2 require 90% removal for emissions of 0.6 = 1.2 lb/MBtu. Less than 90% removal is allowed if emissions are less than 0.6 lb/MBtu, as is generally the case for low-sulfur coal plants. The NSPS for NO_x are 0.6 lb/MBtu for bituminous coal and 0.5 lb/MBtu for subbituminous coal.

potentially boosting generating efficiency to over 40%. Four commercial demonstration units, each of 70-80 MW capacity, are being built at utility sites in Sweden, Spain, and the United States; each is repowering an existing plant. Hot gas cleanup is still under development and is needed if the combined cycle is to fill its potential for improving generating efficiency.

Combined NO_x and SO_x Reduction

Combined NO_x and SO_2 processes offer the potential to reduce SO_2 and NO_x emissions for less than the combined cost of SCR and conventional FGD. Most processes are under development and are not commercially available, although several are being demonstrated in the DOE Clean Coal Technology Program (DOE, 1991).

Integrated Gasification Combined Cycle

Several integrated gasification combined cycle (IGCC) processes are now in the developmental and demonstration stage. All result in a synthetic gas that is cooled by generating steam and desulfurized before combustion in a turbine. Heat is recovered as steam from the combustion turbine and is used to generate additional electricity. The advantages of IGCC include higher generating efficiency, the possibility of phased construction, and very low SO_2 and NO_x emissions (99% removal of SO_2 and one-tenth of new source requirements for NO_x emissions).

Integrated Gasification Fuel Cell

A further conversion technology under study is the integrated gasification fuel cell (IGFC) power plant. IGFC involves substituting a molten carbonate fuel cell for the combustion turbine, potentially increasing overall generating efficiency to above 45%. Optimal chemical and thermal integration of the fuel cell and a catalytic gasifier, with hot gas cleanup, could increase efficiencies even further, to 55-60%. To do so would require that both gasification and gas cleaning take place at temperatures near the 1,300°F operating temperature of the molten carbonate fuel cell.

Thirty-eight IGFC demonstrations are under way or planned with cost-sharing by government in the DOE Clean Coal Technology Demonstration Program (DOE, 1991).

Alternative Emission Reduction Methods

Switching to Low-Sulfur Coal

The 1990 Clean Air Act Amendments have raised the question as to whether the supply of low-sulfur coals is adequate to meet substantially higher demands from electric utilities and others seeking to reduce emissions. In the near term, there are concerns about the utilities' ability to develop mines and the transportation industry's capacity to ship the coal. At least until the strategies of utilities in response to the legislation become clearer, predictions of the course of low-sulfur coal prices are highly uncertain (EPRI, 1991a; NERC, 1991).

Even without fuel price increases, switching to low-sulfur coal could be costly, depending on boiler compatibility with the available low-sulfur coals and on the new infrastructure required to handle the switch. Also, reliance on low-sulfur coal alone is not sufficient to meet present NSPS.

Switching to Natural Gas

The capital costs of installing natural gas generation technologies are relatively low, and the ease of adding capacity in small increments gives gas-fired electricity an added advantage. Emissions reductions can be achieved quickly by using natural gas, especially since many emitters already have gas available on site. A substantial amount of new gas-fired generating capacity, mainly gas turbines for power to meet peak demands, is being planned by utilities and independent power producers.

Like oil markets, natural gas markets are uncertain. The natural gas market is now going through a major structural change, especially with regard to the rules governing transmission pipeline access. It is difficult to predict how gas supply, demand, and price will interact.

Another potential natural gas alternative is co-firing coal with natural gas in a coal-fired boiler. As the percentage of gas increases above 15-30%, the capital improvement costs could increase because of greater engineering changes required for co-firing capability.

Environmental Dispatching

For some utility systems, environmental dispatching (i.e., sending out electricity from the lowest-emitting plants first and the highest-emitting plants last) could reduce emissions. By allowing gas-fired units to be dispatched ahead of higher-emitting stations (especially during the summer when gas prices are seasonably low), a utility system could lower emissions without capital investment or modification of its coal-fired units. This shifting of load, which is a kind of fuel switching, generally would increase gas usage and fuel costs.

Energy Efficiency and Demand Management

Improvements in the way electricity is produced and used reduce pollutant emissions by preventing future growth in demand and by reducing present emissions. In recent years, many U.S. utilities have shifted toward demand-side management (DSM), which includes both energy efficiency and load shifting. Virtually all U.S. utilities are pursuing DSM to some degree. In 1990, nearly 15 million residential customers participated in DSM programs. Existing and planned DSM programs are expected to reduce summer peak demand in the year 2000 by approximately 43,000 MW—the equivalent output of 43 large power plants. That represent a reduction of about 6.5% in the demand forecast for 2000 (EPRI, 1990).

INDUSTRIAL BOILERS

Much of the technology discussed above for power plants can also be used to control emissions from industrial combustion sources. However, there are some important differences. The size, distribution, and fuel type of industrial boilers are quite diverse. Apart from the steel industry, industrial boilers generally are small emission sources compared with power-plant boilers, but they are often located in clusters. Nearly 9,000 combustors are in operation, mostly in the eastern United States. Coal is the economic fuel choice for the larger units, and natural gas and fuel oil are used widely in the smaller units. Less common fuels, such as agricultural waste (bagasse), sewage sludge, wood waste, residual and waste oil, and refuse (garbage), also are burned.

As in the utility industry, wet scrubbing technologies are available in industrial boilers for controlling SO_2 emissions from industrial combustion sources. Although most large utility boilers use lime or limestone scrubbing methods, about 90% of wet scrubbers installed on industrial boilers are sodium-based. These scrubbers use a solution of sodium hydroxide or carbonate to absorb the SO_2 from the flue gas. The absorbers are simple, easy to control, and require little maintenance. Over 95% removal efficiency is possible. Dual-alkali scrubbing is the next most common method for small industrial units. That process also uses a sodium solution for absorption but incorporates a precipitation step where lime is added to remove sulfate and regenerate sodium for recycling. The waste stream is gypsum. About 90% removal efficiency is typical. Capital costs for the sodium solution method are $40-$60/kW of heat input; dual-alkali investment costs are about twice as high.[3] Operating costs, including capital recovery, for sodium scrubbing are $400-$1,000/ton of SO_2 removed by sodium scrubbing, depending on the sulfur content of the fuel. With dual-alkali scrubbing, operating costs are similar because lower raw-material costs offset the higher capital recovery.

Lime spray-drying systems also are practical for use on industrial boilers and municipal waste burners. Removal efficiency of 80-90% is possible. Removal costs are $400-$800/ton of SO_2. Lime spray-drying is particularly useful in garbage combustion applications where the high-chloride content of the flue gas is corrosive to wet scrubbing systems; both SO_2 and HCl are removed. As an alternative to spray drying, the sorbent can be injected dry either directly into the combustor or

---

[3]Control cost estimates for industrial boilers were obtained from South et al., 1990 (SOS/T Report 25).

into the ductwork upstream of the dust collector. Injection is inexpensive compared with spray drying, but removal efficiency is low (only 30-40%).

Processes are available to recover SO_2 emissions in a useful form. MgO scrubbing appears to have good potential for industrial application because removal and recovery are separate operations. A centralized recovery plant can serve a number of small emission sources. Over 90% removal efficiency is possible. Costs are highly dependent on plant size, SO_2 concentration, proximity of the operation that regenerates MgO, and credit for recovered product (sulfuric acid or elemental sulfur). In many situations, MgO scrubbing would be competitive with sodium scrubbing.

Atmospheric fluidized bed combustion (AFBC) is an alternative to conventional coal burning that removes SO_2 during the combustion process as described earlier. AFBC is a well-developed technology for industrial use. In the United States, more than 80 units with capacities up to 65 MW are in operation on a wide variety of fuels. One method, reaching the commercial stage involves high-velocity flow to recirculate entrained solids to the combustor. Costs for SO_2 control by AFBC are comparable to those for wet scrubbing.

NO_x control methods for industrial sources are similar to those for large sources: combustion modification and flue gas treatment. Flue gas recirculation involves returning a side stream of furnace exhaust to the burner. Both temperature and oxygen concentration in the combustor are reduced, resulting in less favorable conditions for NO_x formation; reductions up to 50% are possible, depending on fuel type and practical limitations on recirculation rate. Improvements in burner design can reduce NO_x emissions by reducing flame temperature and adjusting air flow to the burner. Reductions of 30-70%, compared with conventional burners, have been achieved—the lower values with coal burners and the higher values with oil or gas.

Flue gas treatment by ammonia injection has been used in industrial boilers. The reaction rate is temperature dependent; reduction under optimal conditions (900-1100°C) is about 40%, but drops below 10% outside the favorable range. An effective temperature regime in the combustion systems is essential, and retrofit might not be feasible. When catalytic conversion is used in conjunction with ammonia injections, removal efficiencies up to 90% are possible. Industrial application has been mainly on oil and gas combustion sources.

New and developing technologies being demonstrated by the DOE

Clean Coal Technology Demonstration Program mentioned above will provide some additional options for industrial boilers (South et al., 1990).

NONFERROUS SMELTERS

Production of copper, lead, and zinc from natural ores historically has been a principal source of industrial SO_2 emissions. Metal sulfide in the ore is converted to SO_2 during the smelting process. If left uncontrolled, the concentration of SO_2 in the off-gas can be as high as 12-14%. However, controls have been developed to reduce emissions, and the smelting industry is no longer a major contributor to the national SO_2 problem. SO_2 emissions from the copper industry, which accounts for over 80% of smelter emissions, have been reduced from nearly 2 million tons in 1975 to less than 300,000 tons in 1988.

Use of sulfuric acid plants to recover SO_2 has become common. Smelter acid plants are distinguished from commercial plants that burn sulfur by the wide variability in inlet SO_2 concentration and the need for extensive gas cleaning. The off-gas from smelter equipment is collected in a variety of hood and duct arrangements and typically is passed through a cyclone collector for coarse dust removal, then cooled in a spray chamber before removal of small particles in an electrostatic precipitator. The gas is then scrubbed, cooled, and passed through a mist separator before it is dried by contact with recirculated acid. The clean, dry gas containing 4-8% SO_2 is then heated to reaction temperature in a series of heat exchangers ahead of the catalyst (vanadium pentoxide) beds, where the SO_2 is oxidized to SO_3. Heat liberated in the process is transferred to the inlet gas so that no additional energy is needed, provided that the SO_2 concentration is in the proper range. Strong gases, such as those from a fluidized bed roaster, must be diluted to within the operational limits of 4-8% SO_2.

The gas from the catalytic converter flows to an absorption tower where the SO_3 is absorbed in strong recirculating sulfuric acid to produce commercial 98% acid. The drying tower that treats the inlet gas produces 93% acid, also a commercial product. A single contact (one-phase absorption) acid plant is 97% efficient; a double contact unit (two towers in series) is over 99% efficient. Effective collection of acid mist is essential.

Use of new smelter technology also will reduce emissions from the smelting industry. For example, integration of steps into continuous processes reduces variation in SO_2 concentration in fugitive streams, and use of flash smelting technology, such as the Outokumpu flash furnace instead of the reverberatory furnace, reduces emissions and improves efficiency (ACSPCT, 1977; South et al., 1990; E. Trexler, pers. comm., MSCET, DOE, Washington, D.C., 1990).

PETROLEUM AND CHEMICAL INDUSTRIES

Crude oil contains from 0.1% to 5% or more sulfur by weight (Carrales and Martin, 1975). As crude oil is broken down into products by the refining process, much of that sulfur is converted to either elemental sulfur or sulfuric acid, thereby reducing the potential for SO_2 emissions to the atmosphere. However, sulfur recovery and sulfuric acid plants are not 100% efficient, and some emissions to the atmosphere do occur during sulfur recovery. In addition, certain refinery processes (such as fluid catalytic cracking), coking operations, flares, and heaters or other fuel-burning devices lead to SO_2 emissions (Bond, 1972; Danielson, 1973).

The processing, storage, and marketing of petroleum products lead to emission of volatile organic compounds (VOCs). At petroleum refineries and petrochemical plants, losses occur from process streams, refinery pipe flanges and valves, cooling towers, wastewater treatment units, and storage tanks (Danielson, 1973; EPA, 1985c). During marketing of petroleum products, organic vapors are displaced into the atmosphere whenever tank trucks, bulk terminal storage tanks, gasoline-station tanks, and eventually automobile and truck fuel tanks are filled.

Sulfur Oxide Control at
Refineries and Chemical Plants

Substantial emissions of sulfur oxides from refinery processes occur during fluid catalytic or thermal cracking processes. Available control techniques include (1) careful selection of cracking catalysts that will minimize the accumulation of sulfur-bearing coke on the catalyst, thereby reducing SO_2 emissions during catalyst regeneration, (2) desulfurization of the feedstock to the cracking units, and (3) installation of wet scrubbers on the exhaust of the cracking units (Hunter and Helgeson, 1976; SCAQMD, 1978). In the case of feedstock desulfurization or the use of scrubbers, control efficiencies of 80-90% or more are possible (see the earlier discussion of SO_x scrubbers). Catalyst replacement is less expensive but might remove less sulfur. Both catalyst and feedstock should be considered carefully.

Sulfur recovery systems at refineries gather gases bearing hydrogen sulfide (H_2S) and convert H_2S to elemental sulfur, usually via the Claus process (EPA, 1985c). H_2S is not completely converted to elemental sulfur in one-, two-, or three-stage Claus systems, which results in release of SO_2 in the effluent. A variety of Claus tail-gas treatment technologies have been developed. Approximately 99% control of SO_2 can be obtained with their use relative to uncontrolled Claus plants. The cost of such technologies is attractive compared with that of many other SO_2 reduction systems (Hunter and Helgeson, 1976; EPA, 1985c).

Elemental sulfur, H_2S, or spent acid from refinery systems can be used as feedstock for sulfuric acid production (Hunter and Helgeson, 1976; EPA, 1985c). A conventional uncontrolled single-stage contact sulfuric acid plant can emit about 2-5% of its sulfur input in the form of unabsorbed SO_2 or SO_3. By adding of a second set of catalytic converters or absorbers to the process stream, overall control can be raised above 99.7%. SO_2 scrubbers also can be used to meet similar control objectives.

Control of VOC Emissions from Petroleum Refining and Marketing

Noncondensable hydrocarbons generated by refinery processes generally are gathered at the refinery and burned to fuel process heaters. Fugitive emissions from leaking valves, flanges, compressor seals, wastewater treatment plants, and cooling towers, along with product spills lead to significant hydrocarbon release to the atmosphere. Through careful selection and maintenance of seals and packing materials, leaks from piping and compressors can be reduced. The degree of control varies greatly from site to site (Danielson, 1973). Approximately 90% reduction in uncontrolled VOC emissions from wastewater treatment plants is possible through use of vapor stripping systems and through covering of tanks (EPA, 1985c). Cooling-tower emissions are controlled by preventing hydrocarbon leaks from heat exchangers and condensers into cooling water lines.

Storage tanks at petroleum refineries and petrochemical plants can release hydrocarbon vapors in two ways (EPA, 1985c). First, as fixed-roof tanks are filled, hydrocarbon vapors that occupy the empty volume of the tanks are displaced into the atmosphere. Some losses also occur when tanks are emptied. Second, diurnal temperature changes cause the vapors in tanks to expand and contract, causing tanks to "breathe" even though their liquid levels have not changed. Hydrocarbon losses can be reduced if floating-roof tanks or similar devices are used to eliminate the vapor space above the liquid product in the tanks.

Hydrocarbon losses from filling fuel tanks during petroleum transport and marketing can be reduced by submerged filling of the tanks, which creates fewer vapors than does allowing the product to splash in the tank while filling. Further control is possible by forcing vapors displaced by the rising fuel to be condensed or returned to the tank. Submerged filling of service station gasoline tanks can reduce vapor losses by about one-third. Submerged filling combined with vapor return to the tank from which the gasoline is disposed can achieve up to 97% control relative to emissions that result from splash filling (EPA, 1985c).

Vehicle refueling at gasoline stations also generates hydrocarbon vapors as the rising fuel displaces vapors from an empty tank. Vapor recovery systems that return those vapors to the gasoline-station storage tanks offer control efficiencies of about 90% if used properly (EPA, 1985c).

DIESEL-FUELED MOTOR VEHICLES

An uncontrolled diesel engine emits 30-100 times more particulate matter than a comparable-sized gasoline engine. Because particles impair visibility and threaten human health, EPA has established limits on particle emissions from light- and heavy-duty diesel engines. Many of the limits were codified in altered form by the 1990 Clean Air Act Amendments.

Manufacturers are expected to have little difficulty complying with the

0.25 gram per brake horsepower hour (g/bhph) standard, which went into effect for 1991 model-year trucks. Compliance includes use of the following:

• Advanced, low-emitting, and fuel-efficient high-swirl direct injection engines.
• Higher fuel injection pressures and more precise control over the fuel injection process.
• Computerized electronic engine control systems, which will improve trade-offs between NO_x and particles by continuously adjusting the fuel injection timing. Reductions in particle emissions of up to 40% are possible with this approach.
• Turbocharging, which increases the amount of fuel that can be burned without excessive smoke, accompanied by intercooling, which reduces the adverse temperature effects of turbocharging and further increases maximal power potential. Turbocharging and intercooling can reduce NO_x and particle emissions and increase fuel economy and power output. Most heavy-duty diesel engines are now equipped with turbochargers, and most have intercoolers.
• Plans for reducing oil consumption, thereby reducing emissions of oil-derived particles.

Though sufficient to meet the 1991 standard, these improvements probably will not be adequate to achieve the higher standards scheduled to take effect later in the decade. Compliance with the 1993 urban bus standard of 0.1 g/bhph probably will require either alternative fuels or exhaust treatment systems, such as trap oxidizers. The 0.1 g/bhph standard will apply to all heavy-duty trucks in 1994; most manufacturers have indicated their intention to comply by using oxidation catalysts that require low-sulfur fuel. The measures are described below.

Trap Oxidizers

A trap oxidizer consists of a durable particulate filter in the engine exhaust stream and some means of cleaning the filter by burning off the collected particulate matter. The most challenging problem in developing trap oxidizers has been devising a system to burn accumulated particulate matter off the trap without damaging the exhaust system. There are two types of systems for doing so. So-called passive systems regenerate during normal vehicle operation. The most promising approaches require the use of a catalyst (either as a coating on the trap or as a fuel additive) to reduce the ignition temperature of the collected particulate matter. Regeneration temperatures as low as 420°C have been reported with catalytic coatings, and even lower temperatures are possible with fuel additives. Those temperatures, however, may not be low enough for regeneration to occur during normal operation of diesel trucks.

Active systems monitor the buildup of particulate matter in the trap and trigger actions to regenerate the trap when needed. Many approaches to triggering regeneration have been proposed—from diesel fuel burners and electric heaters to catalytic injection systems. Catalytic coatings also have several advantages in active systems and might make possible a simpler regeneration system.

Catalytic Converters

Due to reductions in the solid soot fraction of particulate emissions from diesel engines, the soluble organic fraction now accounts for 30-70% of the emitted particulate matter. A catalytic converter can be used to treat that emission. Particulate control efficiency of even 25-35% would be enough to bring many engines into compliance with the 0.1 g/bhph 1994 standard. The oxidation catalyst also greatly reduces emissions of VOCs, CO, odor, and gaseous and particle-bound toxic air contaminants, such as aldehydes, PNA, and nitro-PNA. Unlike the trap oxidizer, the catalytic converter is a relatively mature technology; millions of catalytic converters are used in gasoline vehicles, and diesel catalytic converters have been used in underground mining applications for more than 20 years.

The catalytic converter requires low-sulfur fuel; otherwise, the increase in sulfate emissions from the catalyst's conversion of SO_2 would more than counterbalance the decrease in the soluble organic fraction. Regulations mandating low-sulfur fuel (0.05% by weight) have been promulgated by EPA and are scheduled to take effect by October 1993. In addition to reducing direct emissions of SO_2 and sulfate particles, lowering the sulfur content of diesel fuel decreases formation of sulfate particles from SO_2 in the atmosphere.

GASOLINE-FUELED MOTOR VEHICLES

Control technology for gasoline-fueled vehicles continues to advance rapidly. Controls generally target NO_x, VOCs, and carbon monoxide (CO) emissions. VOCs and to a lesser extent NO_x contribute to visibility impairment.

Current federal emission standards allow no more than 0.41 g of VOCs per mile, 3.4 g of CO per mile, and 1.0 g of NO_x per mile. The standards usually are met through use of a three-way catalytic converter which oxidizes VOC and CO into water and carbon dioxide by using a platinum or palladium catalyst and reduces NO_x into elemental nitrogen and oxygen by using a rhodium catalyst.

The Clean Air Act Amendments of 1990 follow California's lead by tightening standards for new vehicles. For instance, beginning in model year 1994, a more stringent VOC standard (0.25 g of non-methane hydrocarbon per mile) will be phased in for light-duty vehicles. According to the California Air Resources Board, which has adopted the same standards with an earlier phase-in, the standards will not require any fundamental change in existing technology. Rather, manufacturers are expected to comply through reduced oil consumption, more precise electronic control and diagnostics, engine improvements, and improved catalytic coatings. Additional emission reductions are expected as a result of lower trace lead levels in unleaded gasoline and more advanced emissions control components, particularly more durable catalysts, better air-fuel management systems, and improved electronics.

The Clean Air Act Amendments also phase in a standard for NO_x of 0.4 g/mile already in force in California. In adopting this standard, the California Air Resources Board noted that it can be met by using three-way catalytic converters on engines if excess oxygen is controlled and the air to fuel ratio is kept at its ideal or stoichiometric ratio.

At present, manufacturers must show that their new vehicles will meet emissions standards for 5 years or 50,000 miles. The 1990 amendments follow California's lead by requiring manufacturers to demonstrate that their vehicles will meet slightly less stringent standards for 10 years and 100,000 miles. (Full warranty coverage and recall testing, however, will not initially apply for the whole of this period.) To meet the proposed 100,000-mile standards, manufacturers would have to ensure that deterioration of the emission control system in the second half of the vehicle's life is not greater than in the first half. California has concluded that low emission levels can be maintained for 100,000 miles by improving fuel quality and using advanced control systems, such as on-board diagnostics and increased precious metal loadings in catalysts.

The Clean Air Act Amendments of 1990 also establish a second tier of emissions standards for light-duty vehicles and trucks. Those so-called Tier II standards (non-methane hydrocarbons at 0.125 g/mile, NO_x at 0.2 g/mile, and CO at 1.7 g/mile) will take effect in model-year 2003 unless EPA makes specified findings. To some extent, the Tier II standards follow the lead of California, which decided in 1989 to adopt a second phase of tightened standards. California, however, will go further by requiring that some zero-emissions vehicles be produced. The 0.125 g/mile standard is based on possible emissions reductions from use of fuels with lower ozone-forming potential than is used today. Those fuels, along with electrically heated catalysts and other improvements, are expected to allow compliance with the 0.2 g/mile standard for NO_x.

In addition, the 1990 Clean Air Act Amendments establish new requirements for reformulated gasoline, which are expected to lower emissions of VOCs. Special requirements for fleet vehicles also might encourage development of alternative, less-polluting fuels, such as diesel-fuel substitutes discussed above.

The amendments supplement the requirements with other measures to decrease in-use emissions. Inspection and maintenance requirements have been strengthened; for instance, maximal expenditures for repair have been increased. EPA will now be able to require annual centralized inspections in areas with the worst ozone problems. EPA is also required to revise certification test procedures to determine whether 1994 and later model-year passenger cars and light-duty trucks are capable of passing state inspection emission tests. EPA also must review and revise certification procedures to ensure that motor vehicles are tested under conditions that reflect actual driving, including fuel condition, temperature, acceleration, and altitude. Recent evidence indicates that high acceleration testing might be critical. For example, the California Air Resources Board found one car that met the 0.41 g/mile standard for VOCs under the present test procedure but that emitted over 15 g/mile when accelerating.

The requirements of the Clean Air Act Amendments are expected to lower total mobile-source emissions at least for the rest of the century.

Much will depend, however, on how EPA and the states implement the amendments. For example, enhanced inspection and maintenance requirements could lower emissions more than tightening the tailpipe standards. States also could reduce emissions by adopting California standards for motor vehicles, which will continue to be stricter than the federal standards.

Fuel Modification

Substituting cleaner-burning alternative fuels for diesel fuel and gasoline has drawn increasing attention during the last decade. Alternative fuels now under consideration include natural gas, methanol made from natural gas, and, in limited applications, liquid petroleum gas. See NRC (1991b) for additional information.

Natural Gas

Clean burning, cheap, and abundant in many parts of the world, natural gas already fuels many vehicles in several countries. The major disadvantage of natural gas as a motor fuel is its gaseous form at normal temperatures and poor self-ignition qualities. That makes it less promising as a fuel for diesel engines. Natural-gas engines are likely to use up to 10% more energy than the diesels they replace.

Liquified Petroleum Gas

Liquified petroleum gas is already widely used as a vehicle fuel in the United States, Canada, The Netherlands, and elsewhere. As a fuel for spark ignition engines, it has many of the same advantages as natural gas, and an additional advantage of being easier to carry aboard the vehicle. Its major disadvantage is the limited supply.

Like natural gas, liquid petroleum gas in spark ignition engines is expected to produce essentially no particulate emissions (except for a small amount of lubricating oil), very little CO, and moderate VOC emissions. NO_x emissions are a function of the air-to-fuel ratio. Liquid petroleum gas does not burn as well under lean conditions as does natural gas, so the NO_x emission reductions achievable through lean-burn technology are expected to be somewhat lower.

Methanol

Methanol has many desirable combustion and emissions characteristics, including good lean-combustion characteristics, low flame temperature, and low photochemical reactivity. Methanol cannot be used in a diesel engine without some supplemental ignition source. Investigations to date have focused on the use of ignition-improving additives, spark ignition, glow plug ignition, or dual injection with diesel fuel. Converted heavy-duty diesel engines using each of these methods have been developed and demonstrated.

Methanol combustion does not produce soot, so particulate emissions from methanol engines are limited to a small amount of lubricating oil. Although flame temperature is lower for methanol than that for hydrocarbon fuels, it is not clear whether methanol use would lead to lower NO_x emissions (NRC, 1991b).

The potential for large increases in formaldehyde emissions with the widespread use of methanol fuel has raised considerable concern. Efforts to resolve that problem are focusing on developing special low-formaldehyde catalysts and on minimizing unburned methanol emissions. Those efforts, although promising, have not yet provided a solution to the problem in methanol diesel engines under all conditions.

PRESCRIBED FORESTRY AND AGRICULTURAL BURNING

Three factors contribute to the importance of prescribed burning as a source of visibility impairment in Class I areas: (1) the emissions are often from forests near Class I areas; (2) the emissions are often comparable in magnitude to those from stationary industrial sources; and (3) the emissions are effective in scattering light because of the fine size of smoke particles.

The use of fire as a land management tool has evolved only over the past 50 years (Walstad et al., 1990). Land managers use fire in reforestation programs to clear debris after timber harvest, improve rangeland characteristics, reduce fuel levels that might create a fire hazard, and improve wildlife habitats. Over 7 million acres of forest and range lands are burned annually in the United States, generating about 1.8 million metric tons (megagrams) of fine particles. Much of the burning occurs in the southeastern United States, far from Class I areas. However, prescribed fire emissions (about 150,000 metric tons of fine particles annually) are an important contributor to Class I area visibility impairment in Oregon and Washington (State of Washington Department of Ecology, 1983; State of Oregon Department of Environmental Quality, 1986). As a result, restrictions on forestry burning play a key role in both states' Visibility Protection State Implementation Programs (Core, 1989b; Pace, 1990).

Prescribed Forestry Burning

Emission Control Measures

Control measures to reduce prescribed burning emissions include (1) reduction of the number of acres burned; (2) reduction in fuel consumption; (3) burning under conditions of increased fuel moisture; (4) use of helitorch ignition methods; and (5) application of rapid mop-up techniques to expedite fire suppression during the smoldering phase.

Reduction in acres burned can be achieved through alternative treatment methods described below. Several land managers in Washington State have stopped using prescribed fire entirely because of concern about public sensitivity to smoke (State of Washington Department of Natural Resources, 1989).

Reduction in fuel consumption can be achieved through increased residue use. As forest conservation programs reduce timber harvest levels, the demand for chipped fuelwood for use in industrial boilers will increase. That demand could be met by increased use of residues, which have become an increasingly source of energy over the past 15 years. Nationally, the use of wood waste has increased from less than 1 quad (10^{15} British thermal units) in 1972 to 2.1 quads in 1982. Much of the residue is taken as firewood. In Oregon alone, it is estimated that as much as 300,000 tons per year of slash could be used as firewood rather than burned as slash. Hogged fuel boilers also are likely to use more wood residues as other energy sources become more costly.

Burning under conditions of increased fuel moisture has been shown to reduce emissions by about 30% (Sandberg, 1983). In most cases, the land manager's intent is to eliminate small residue to create an adequate number of seedling planting spots. There is little value in reducing the amount of large residue on the site. Successful burns eliminate small residue but do not burn large logs or the duff layer of twigs, needles, and leaves on the forest floor. Burning under increased fuel moisture reduces the amount of fuel consumed and therefore emissions.

Increased use of helitorch (or aerial) ignition can reduce emissions by up to 20% by achieving mass burn fire behavior (in which small-sized combustible material is burned quickly without sufficient radiative heat to combust large logs) (D.V. Sandberg, pers. comm., USDA Forest Service, Pacific Northwest, 1985).

Rapid mop-up of residual smoke following the active phase of a fire can also reduce emissions. Mop-up typically is conducted to minimize the risk of new ignitions of smoldering fuels that could result in an escaped fire; it also eliminates smoldering emissions that might be carried downslope into valley floors. Because emissions from the smoldering phase of a fire account for about 40% of the total emissions from a prescribed burn, mop-up can reduce emissions without compromising the land manager's objectives. Mop-up within 8 hours typically reduces overall emissions by about 10% (Freeburn, 1986).

Control Costs

The costs of alternative measures to reduce prescribed burning emissions were evaluated during the development of the Oregon Visibility Protection Program (Freeburn, 1986). That study showed that the cost of prescribed burning varies greatly as a function of harvest unit conditions, fuel characteristics, and land ownership but averages about $102 per acre on private land and $150 per acre on federal land (Marcus, 1981). The standard deviation of cost per unit is as high as 98% for western Oregon and Washington (Mills et al., 1985), but about one-half of the actual unit costs appear to lie within ±70% of the nominal burning cost (Freeburn, 1986).

The cost of mopping up a typical 20-acre unit with a crew of 15 persons working for 4 hours is in the range of $100-$200 per acre, depending on fuel characteristics. In comparison, the cost of alternative treatment, such as using herbicides, manual methods, or mechanical methods, is $250 per acre for private land and as much as $400 per acre for federal lands (Freeburn, 1986).

Alternative Treatment Methods

Several available alternatives to prescribed burning include the following:

- Manual methods (e.g., chain saws) to remove competing vegetation or create conditions favorable to a desired plant;
- Mechanical methods (e.g., tractors, cable systems, or crawlers equipped with circular blade devices);
- Biological methods (e.g., animals or insects) to control vegetation
- Explosives and herbicides.

The U.S. Forest Service has recently completed a thorough evaluation of each method and of the relative health risks to workers and the public (USDA, 1988).

Agricultural Burning

The burning of straw stubble following the harvesting of cereal grain, grass-seed fields, and other crops is different from forestry burning in that agricultural lands are more accessible, are typically level enough to be worked by farm machinery, and have much lighter fuel loading than forested lands.

Large reductions in particulate emissions are possible through burning grass-seed fields on alternate years (rather than annually), and through growing crops that do not require burning. The removal of stubble from the field before treatment by using tractor-mounted, propane-fueled torches also reduces emissions.

The use of straw to produce cattle feed, hardboard products, and paper has not proved to be economically feasible, nor has straw incineration for energy production. Mobile field sanitizers, designed to burn the stubble at high temperature as they pass over the field, have not proved to be either economically feasible or practical (Oregon State Department of Environmental Quality, 1988). As a result, the primary emphasis has been on improving smoke management programs to minimize effects of burning on the public and (in Oregon) on wilderness air quality and visibility.

Other Biomass Burning

Open burning of biomass is a common way to dispose of brush, stumps, and other residues following land clearing and highway right-of-way projects. In most of the United States, open burning outside urban areas is largely unregulated, requiring only a permit issued by a local fire district. Alternatives, such as open-pit incineration using air curtain destructors and grinding or chipping residues for use as mulch or boiler fuels, are seldom used.

Future Directions

There is a growing awareness that biomass burning can impair visibility in national parks and wilderness areas. Both the Oregon and Washington visibility protection programs include strategies to control emissions from biomass burning. In recognition of the importance of establishing "best available control measures" (BACM) for biomass burning, EPA's Office of Air Quality Planning and Standards has undertaken a program to describe BACM for biomass burning that affects serious PM_{10} nonattainment areas. Development of BACM documents is required by the 1990 Clean Air Act Amendments.

RESIDENTIAL WOOD COMBUSTION

Particulate emissions from residential space heating with wood, especially in urban areas, create a noticeable pall of smoke over many western communities during the winter months. Many ski resorts in the West (e.g., Colorado) and small towns located in forested regions near

Class I wilderness areas have a wood-smoke air-quality problem. Wood smoke is a significant contributor to PM_{10} nonattainment in many western communities. Regional emissions from wood smoke can greatly impair visibility in Class I areas during the fall, winter, and early spring.

A large, steady decline in emissions of residential wood combustion since the 1940s was reversed in 1973-74 when prices for oil, natural gas, and electricity increased sharply as a result of the Arab oil embargo. The strong resurgence in residential use of cordwood as a space-heating fuel has continued over the past decade, and about 1 million new wood stoves were sold each year during 1975-85. National fireplace and wood-stove PM_{10} emissions have been estimated to be at about 1 million tons per year (EPA, 1986b).

Wood-Stove Emissions

Pollutant emission rates from wood stoves are influenced by the following factors:

- *Wood-Stove Design.* The design of the appliance is very important (EPA, 1988b). Older, conventional wood-stove emissions are typically 21 g/kg (i.e., 21 g of pollutant per kilogram of wood burned) for PM_{10}, as compared with 4 g PM_{10}/kg for stoves with the newest "best existing stove technology" (BEST) (Crane, 1989).
- *Wood Moisture Content.* For wood stoves, cordwood that produces the lowest particulate emissions contains about 20-26% moisture. Wood moisture above or below this range results in higher emissions (EPA, 1988b).
- *Burn Rates.* Generally, the higher the burn rate, the lower the particulate and carbon monoxide emissions.
- *Heat Output Requirements.* The higher the home owner's heat output requirement, the greater the amount of fuel burned. Energy conservation through home weatherizing programs is the key to reducing fuel consumption, although the use of wood stoves with smaller fireboxes also reduces total emissions.

Wood-Burning Control Technologies

There are two approaches to reducing wood smoke from stoves and fireplaces: (1) improving the performance of wood heating systems through programs such as certification testing; and (2) burning less wood through wood-stove curtailment, home weatherization, and fuel-switching programs. Some of those strategies have multiple advantages. Woodstoves that have been certified, for example, reduce the amount of wood smoke per cord of wood burned while improving energy efficiency. Other examples are public information programs to teach proper wood-burning techniques and firewood-seasoning programs that result in better combustion (lower emissions) and increased energy efficiency.

To assist the states, EPA has issued a guidance document describing emission control measures for residential wood combustion (EPA, 1989b). That document describes four basic strategies: public information and awareness, improvements in wood-burning appliance performance, reduced dependence on wood, and wood-burning curtailment programs.

Public Education Programs

Local programs to educate the public about the wood smoke problem, to promote good burning practices, to urge reduced reliance on wood as a space-heating fuel, and to promote compliance with voluntary and mandatory wood-burning curtailment programs are considered essential to any control of residential wood combustion. EPA guidance allows a 50% or more emissions reduction credit for mandatory curtailment and up to a 50% credit for voluntary programs.

Wood-Stove Certification Programs

In 1983, Oregon became the first state to adopt a wood-stove certification program that required all new wood stoves sold in the state to be laboratory tested for emissions and efficiency to assure compliance with newly adopted emission standards (Kowalczyk and Tombleson, 1985). As a result, stoves sold after July 1986 were required to emit 50% less wood smoke than conventional stoves. After July 1988, new stoves were required to emit 70% less smoke.

After the Oregon program was adopted, EPA adopted a slightly more restrictive national certification program, which became effective in July 1990 (CFR Title 40). The national certification program will result in substantial emission reductions as old stoves are replaced with newer certified models.

Further emission reductions are possible by increasing the durability of the stoves to reduce sheet metal warpage that allows flue gases to bypass catalytic converters and by increasing the durability of the converters themselves (Crane, 1989).

Reduced Dependence on Wood

In some mountain communities, reduced dependence on wood through weatherization and fuel switching might be the only long-term strategy that will assure compliance with the PM_{10} NAAQS. Programs have been adopted that limit installation of new wood stoves, require phaseout of stoves, and prohibit stove use (EPA, 1989b).

Wood-Burning Curtailment

The most immediate short-term strategy to achieve the PM_{10} NAAQS is often adoption of voluntary or mandatory wood-burning curtailment programs. Mandatory curtailment programs are in operation in Boise, Idaho; Denver, Colorado; Juneau, Alaska; Missoula, Montana; Seattle and Yakima, Washington; Reno, Nevada; and Medford, Oregon; and voluntary programs operate in many other communities. Emission reductions of as much as 80% have been documented on winter days as a result of curtailment programs.

Future Conditions

As the federal wood-stove certification program and the state PM_{10} control strategies are implemented, emissions from wood stoves might decline if the energy market continues to offer home owners low prices on natural gas, fuel oil, and electricity. However, wood use depends on fuel prices, and reductions achieved through the above strategies may be offset by population growth and changing economic conditions.

FUGITIVE DUST

Fugitive particulate emissions originating from a variety of sources can severely impair visibility in Class I areas in arid parts of the western United States or near major agricultural areas. Global emissions of wind-blown dust are estimated to be of the order of 2 million tons per day, or about one-tenth of total global tropospheric particle emissions.

Fugitive emissions may be separated into process sources (those associated with industrial operations) and open dust sources. Process sources include emissions from storage and transfer of raw, intermediate, and waste aggregate materials; open dust sources include agricultural tilling, paved and unpaved road dust, wind erosion of soils in areas without ground cover, and construction activities.

Agricultural Operations

Agricultural tilling is often the largest anthropogenic source of fugitive dust. Emissions from tilling depend on the silt content of the soil (nominally 18%), wind speed, soil erodibility, soil moisture, and the portion of total particulate emissions that fall within the PM_{10} or fine particle ($PM_{2.5}$) fraction (EPA, 1988c).

Over the years, the U.S. Department of Agriculture Soil Conservation Service has taken a leading role in reducing topsoil loss by wind erosion through improved land management techniques, such as reducing the need for tilling and using wind barriers and strip-cropping farming methods. The Food Security Act contains two provisions to reduce dust emissions from agricultural tilling. The first required development of conservation plans by 1990 for all lands designated as "highly erodible" by wind. The second provision, the Conservation Reserve Program, has taken highly erodible cropland out of production and covered it with vegetation.

Paved and Unpaved Road Dust

In Class I areas downwind of major urban areas, fugitive emissions from paved and unpaved roads might impair visibility. Haul roads at open pit coal mines near Class I areas also might be of concern.

Fugitive dust is emitted whenever a vehicle travels over a paved or unpaved road surface. Soil dust loading, silt content, vehicle traffic volumes, and soil moisture are critical in determining emission rates of road dust.

Control strategies for reducing paved road dust include road sweeping and flushing, reducing soil tracked out onto the road network from construction sites or unpaved roads, reducing spills from haul trucks, and reducing wind erosion from lands adjacent to the roadway. Unpaved road dust strategies include roadway surface improvements such as watering, chemical stabilization, and paving.

Process Fugitives

Storage piles, raw material handling, and transfer operations also are sources of visibility-impairing dust emissions. Dust is emitted at several points in the storage cycle, including material-loading onto piles, disturbances by strong wind currents, and movement of trucks and loading equipment. Emission rates vary with the volume of the aggregate passing through the storage cycle, the age of the pile, moisture content, silt content, and friability of the material. Control strategies include improved material handling to reduce transfer needs, wind sheltering, moisture retention, chemical stabilization or the use of water sprays, and enclosure of the materials.

Although measures to reduce fugitive dust emissions are available and are being applied nationwide, it is impractical to control natural wind-entrained soils in much of the West. Where agricultural operations are an important (and controllable) source of dust, emissions are being reduced. The reductions can be calculated and verified. They are also cost effective in reducing wind erosion of topsoils.

The technology of fugitive dust emission control is relatively well known, especially as it applies to agricultural operations. It is difficult, however, to differentiate visibility impairment associated with anthropogenic soil dust emissions from that associated with natural causes.

FEEDLOTS AND OTHER
SOURCES OF AMMONIA

Of the 830,000 metric tons of ammonia (NH_3) emitted in the United States in 1980 by anthropogenic sources, emissions from livestock-waste management dominate at 540,000 metric tons/yr (Placet and Streets, 1987); animal excrement is estimated to be the major terrestrial source of ammonia worldwide. Other sources (in metric tons per year) are fertilizer production (110,000), agricultural application of anhydrous NH_3 (50,000), NH_3 synthesis (40,000), petroleum refineries (40,000), motor vehicles (30,000), stationary fossil fuel combustion (20,000), and coke manufacture (10,000). In polluted urban areas such as Los Angeles and Denver, where nitric acid concentrations are high (greater than 1 ppb), NH_3 emissions from feedlots can influence urban airborne particle composition and concentration. Studies made in 1978 in the Denver "brown cloud" showed that about 20% of the fine particle mass (i.e., particles less than 2.5 μm in diameter) was composed of ammonium nitrate (NH_4NO_3) (Countess et al., 1980); these particles accounted for about 17% of the visibility reduction (Wolff et al., 1981).

Söderlund and Svensson (1976) estimated that wild and domestic animals and humans produce globally about 27-50 million metric tons/yr of NH_3, but only a small faction of those emissions (2-7 million metric tons/yr) is attributable to wild animal wastes. Therefore, NH_3 emissions in national parks and wilderness areas should be very low.

It is possible, but probably unlikely, that visibility in the vicinity of national parks and wilderness areas could be impaired by NH_3 emissions from feedlots and other nearby anthropogenic sources. In such cases, control of the emissions might be desired but would be difficult because the openness of most feedlots prevents capture of NH_3 by the usual air treatment procedures employed in industrial operations (Bond, 1972). Chemical additives to the animal waste can lower NH_3 emissions; for example, added natural zeolites can lower NH_3 emissions by 50% (Miner, 1984).

Appendix E

Comparing Visibility Control Strategies

Regional haze arises from the combined emissions of many sources. As discussed in Appendix D, control technologies are available to reduce the emissions from individual sources. It often is not clear which combination of controls can improve visibility most effectively.

The task of control strategy synthesis can be considered at two levels. First, the design effort is directed at defining the range of technically feasible solutions to the problem at hand. Second, the most attractive of the feasible solutions is adopted in the form of an actual control program.

A technically feasible solution consists of a combination of control equipment or regulations that would achieve visibility at least as good as that required by the goals set for the control program. Because full restoration of natural visibility might not be feasible, a realistic goal must be chosen initially if the following analysis procedures are to be of any use at all. Because there are thousands of contributing sources, there usually are many combinations of controls that could achieve the same visibility improvements and some that could not achieve the desired visibility improvements. The purpose of identifying solutions as feasible or infeasible is to clarify for decision-makers the group of solutions that could achieve desired air quality improvements. Source apportionment models, such as those discussed in Chapter 5, provide the technical basis for testing candidate control programs to determine whether they would solve the problem at hand.

From the group of feasible solutions, a solution must be chosen. The chosen solution usually is viewed as more attractive than other potential solutions for some particular set of reasons: it is either the least expensive solution or the least burdensome solution from an administrative point of view. If the objective is to choose an economically attractive solution rather than an administratively convenient one, then the control-strategy design team has certain tools available to it that can be used to help identify cost-efficient control schemes. The tools are discussed below.

The search for feasible solutions to a regional visibility control problem begins by constructing a mathematical model for the regional visibility problem. This model is on a larger conceptual scale than has been discussed earlier. At the core of the regional model is one of the emissions-to-air-quality models described in Chapter 5 for computing source contributions to ambient pollutant levels. Adjoined to this emissions-to-air-quality model is a model for translating air pollutant levels into effects on visibility. Then these two models are subjected to model verification tests over a base-case historical period when meteorological conditions, emissions, ambient pollutant concentrations, and atmospheric optical properties are known. The purpose is to demonstrate that the chosen models can produce accurate results in the presence of well-defined inputs and to demonstrate that cause and effect relationships in that particular airshed are understood. Following confirmation of the models' technical performance, the available emissions controls that could be applied to the problem are used to compute the emission rates from the sources that would prevail in the presence of each of the controls. Then, the effect of those controls on air quality is tested by application through the completed air quality and visibility models.

Because the number of controls that must be tested for their effect on air quality and visibility is potentially quite large, care must be taken to structure an efficient search for feasible solutions. It usually is not practical to re-run an elaborate environmental model hundreds of times to learn about the properties of each control technique separately. Instead, for those models that are linear in emissions (e.g., most of the models with simplified chemistry), it is possible to perform the control evaluation without rerunning the models in their entirety. Transfer coefficients can be calculated for the linear models that state the pollutant concentration or light extinction *increment* at each receptor site per ton of emissions per day from each source. The transfer coefficients depend on meteorological conditions, the spatial location of the sources and receptors, and the age of the air parcels but they do not depend on emission rates. Therefore, they can be used outside the model to compute the future effect of each source in the presence of visibility controls. The transfer coefficients for each source, are multiplied by the new controlled emission rate from the source. Then, overall pollutant and light-extinction levels in the presence of a set of future controls can be synthesized quickly by adding the incremental effects of all controlled and uncontrolled sources on top of an estimate of background air quality and light extinction.

The search for feasible solutions (e.g., sufficient combinations of controls) can be pursued quickly by exhaustive enumeration or with the assistance of linear programming techniques.

For nonlinear environmental models (e.g., photochemical models for secondary particle formation), the search for feasible combinations of controls must be conducted by perturbing the model, either one source at a time, or by grouping controls with similar characteristics for evaluation (see Russell et al., 1988b). If the spatial distribution of the sources is thought to be less important than the overall level of emissions from the sources in a region, then the feasible control solutions can be approximated by progressively reducing all emissions of each chemical type by increments until a level of total emissions is found that is compatible with a solution to the visibility problem. (See Trijonis, 1972, and 1974 for examples of analysis with nonlinear models for ozone control).

The least expensive solution to a regional air quality problem in many cases can be found by examination of the group of feasible solutions. In the case of problems that are described by linear environmental models, cost-effectiveness indexes can be constructed from available data that describe the incremental cost per unit of air quality improvement that is attributed to each available control technique. The indexes (in micrograms per cubic meter of pollutant concentration improvement per million dollars per year spent on a particular type of control, or extinction coefficient increment per unit cost) can be constructed from the output of a linear environmental model by multiplying the air quality transfer coefficient (as defined above, e.g., in micrograms per cubic meter per ton per day emitted) times the cost of control at that source (tons per day abated per dollar spent). These cost-effectiveness indexes often can be rank-ordered to indicate which controls provide the least expensive way to improve air quality. Then, a number of those controls are selected to achieve a feasible solution.

Alternatively, a linear programming cost-minimization calculation can be done to find the least expensive set of controls. The problems can be formulated in several ways; for example,

$$
\begin{aligned}
\text{Minimize} \quad & C = cx, \\
\text{subject to} \quad & Bx \geq r, \\
\text{subject to} \quad & Ax \leq d, \\
\text{subject to} \quad & Dx \leq 1, \\
& 0 \leq x \leq 1,
\end{aligned}
$$

where C is the total cost of the control program; x is a vector of control method activity levels (if control measure i is adopted as part of the solution to the problem then $x_i = 1$; if control measure i is not used than $x_i = 0$); c is a vector whose elements state the annual cost of individual control measures if selected for application to the solution; r is a vector of pollutant concentration reduction requirements (or extinction coefficient reduction requirements) at each monitoring site; B is a matrix whose elements indicate the pollutant concentration reduction (or extinction coefficient reduction) at each monitoring site resulting from one unit of each control activity; A is a matrix of resource magnitude (e.g., natural gas) consumed when a control measure is selected; d is a vector of the limits on the physical resources available for the solution of the problem (e.g., limits, if any, on the total amount of natural gas that can be obtained in the region of interest); and D is a matrix of compatibility parameters that prevents the simultaneous application of two conflicting controls on the same source.

Usually, costs are stated as equivalent annualized costs of operation in which the capital costs and salvage value (if any) of the control equipment are spread over the life of the equipment by using discounted cash-flow calculations that reflect the capital costs. The objective is to select those controls (x) that minimize the total cost of the overall regional control strategy subject to attaining the required environmental improvement while not consuming resources that are unavailable and without prescribing combinations of controls that work against each other. A large number of feasibility studies have been conducted to show that this method of economic analysis can be performed by using data from actual cases of air pollution control; those studies are summarized by Cass and McRae (1981). A more recent study by Harley et al. (1989) might be particularly relevant to visibility control because it illustrates that a single least-cost control-strategy study can combine the results of receptor-oriented CMB models for source apportionment of primary particles with the results of linear chemical models for secondary sulfate formation.

Identification of the least expensive control strategy is more difficult in situations described by nonlinear environmental models than linear models. Nonlinear graphic solutions can be used to find the best strategy in some cases. (See Trijonis (1972, 1974) for a relevant example involving trade-offs between controls for hydrocarbons and for NO_x in the case of ozone abatement.) Further research into development of methods for optimizing selection of control programs for nonlinear air quality problems is warranted.

Certain economic incentive systems have been proposed as an alternative to the above engineering procedures for finding control programs that are economically attractive. Systems of transferrable licenses in which source owners trade rights among themselves to emit air pollutants can be used to limit total regional pollutant emissions.

In one form of a transferrable licenses system, an absolute limit is set on total regional pollutant emissions. The right to emit those pollutants is broken into small increments, and those rights are auctioned to the highest bidder. Those source owners with low control costs will choose to install control equipment rather than pay for expensive emissions licenses. Those with high control costs will bid up the price of emissions licenses rather than install controls. Payments for the licenses can be used then to offset part of the installation costs of those owners that chose control programs over emissions licenses. After initial bidding, licenses to emit can be bought and sold in an open market to facilitate adjustments between source owners.

The main advantage of such a system is that source owners can use their expertise in choosing control methods and minimizing total control costs.

It is often argued that such market-based systems require that government regulators need less technical information about air quality problems. The notion is probably false. To determine the magnitude of the number of emissions licenses and the geographical extent of trading that will not hinder the pollution control program, the regulator needs to know the science of emissions and air quality. An example of source-apportionment modeling used to establish the proper number of transferrable permit units and to test for anomalous effects of geographic distribution is given by Cass et al. (1982).

Although the procedures for selecting an optimal emission control strategy are widely available, it can be argued with some justification that they are seldom used effectively. There are a number of reasons for that. In many cases, legal mandates that carry short-time deadlines (a few months) for analysis of a complex problem simply preclude a careful technical analysis of the problem. Instead, regulatory boards are forced to guess about selecting a control program. In other cases, the human skills needed to analyze the problem are not available to the regulatory agencies. Given the large financial costs to society inherent in regional air pollution control, it is important to address the barriers to finding technically and economically sound solutions to regional visibility problems.